70th Year Anniversary of Carbon Nanotube Discovery—Focus on Real World Solutions

70th Year Anniversary of Carbon Nanotube Discovery—Focus on Real World Solutions

Editor

Muralidharan Paramsothy

Basel • Beijing • Wuhan • Barcelona • Belgrade • Novi Sad • Cluj • Manchester

Editor
Muralidharan Paramsothy
NanoWorld Innovations
(NWI)
Singapore
Singapore

Editorial Office
MDPI
St. Alban-Anlage 66
4052 Basel, Switzerland

This is a reprint of articles from the Special Issue published online in the open access journal *Nanomaterials* (ISSN 2079-4991) (available at: www.mdpi.com/journal/nanomaterials/special_issues/30th_Year_Carbon_Nanotube_Discovery).

For citation purposes, cite each article independently as indicated on the article page online and as indicated below:

Lastname, A.A.; Lastname, B.B. Article Title. *Journal Name* **Year**, *Volume Number*, Page Range.

ISBN 978-3-0365-9884-0 (Hbk)
ISBN 978-3-0365-9883-3 (PDF)
doi.org/10.3390/books978-3-0365-9883-3

Cover image courtesy of Muralidharan Paramsothy

© 2023 by the authors. Articles in this book are Open Access and distributed under the Creative Commons Attribution (CC BY) license. The book as a whole is distributed by MDPI under the terms and conditions of the Creative Commons Attribution-NonCommercial-NoDerivs (CC BY-NC-ND) license.

Contents

About the Editor vii

Preface ix

Muralidharan Paramsothy
70th Year Anniversary of Carbon Nanotube Discovery—Focus on Real-World Solutions
Reprinted from: *Nanomaterials* 2023, *13*, 3162, doi:10.3390/nano13243162 1

Dusan Vobornik, Maohui Chen, Shan Zou and Gregory P. Lopinski
Measuring the Diameter of Single-Wall Carbon Nanotubes Using AFM
Reprinted from: *Nanomaterials* 2023, *13*, 477, doi:10.3390/nano13030477 8

Boyang Zhang, Rui Li and Qing Peng
Controlling CNT-Based Nanorotors via Hydroxyl Groups
Reprinted from: *Nanomaterials* 2022, *12*, 3363, doi:10.3390/nano12193363 19

Yukang Zhu, Hongjie Yue, Muhammad Junaid Aslam, Yunxiang Bai, Zhenxing Zhu and Fei Wei
Controllable Preparation and Strengthening Strategies towards High-Strength Carbon Nanotube Fibers
Reprinted from: *Nanomaterials* 2022, *12*, 3478, doi:10.3390/nano12193478 25

Anna A. Vorfolomeeva, Svetlana G. Stolyarova, Igor P. Asanov, Elena V. Shlyakhova, Pavel E. Plyusnin, Evgeny A. Maksimovskiy, et al.
Single-Walled Carbon Nanotubes with Red Phosphorus in Lithium-Ion Batteries: Effect of Surface and Encapsulated Phosphorus
Reprinted from: *Nanomaterials* 2023, *13*, 153, doi:10.3390/nano13010153 49

Seenidurai Athithya, Valparai Surangani Manikandan, Santhana Krishnan Harish, Kuppusamy Silambarasan, Shanmugam Gopalakrishnan, Hiroya Ikeda, et al.
Plasmon Effect of Ag Nanoparticles on TiO_2/rGO Nanostructures for Enhanced Energy Harvesting and Environmental Remediation
Reprinted from: *Nanomaterials* 2023, *13*, 65, doi:10.3390/nano13010065 67

Lijo Francis and Nidal Hilal
Electrosprayed CNTs on Electrospun PVDF-Co-HFP Membrane for Robust Membrane Distillation
Reprinted from: *Nanomaterials* 2022, *12*, 4331, doi:10.3390/nano12234331 83

Gururaj M. Neelgund, Sanjuana Fabiola Aguilar, Mahaveer D. Kurkuri, Debora F. Rodrigues and Ram L. Ray
Elevated Adsorption of Lead and Arsenic over Silver Nanoparticles Deposited on Poly(amidoamine) Grafted Carbon Nanotubes
Reprinted from: *Nanomaterials* 2022, *12*, 3852, doi:10.3390/nano12213852 99

Lidia Orduna, Itziar Otaegi, Nora Aranburu and Gonzalo Guerrica-Echevarría
Ionic Liquids as Alternative Curing Agents for Conductive Epoxy/CNT Nanocomposites with Improved Adhesive Properties
Reprinted from: *Nanomaterials* 2023, *13*, 725, doi:10.3390/nano13040725 119

Joydip Sengupta and Chaudhery Mustansar Hussain
Decadal Journey of CNT-Based Analytical Biosensing Platforms in the Detection of Human Viruses
Reprinted from: *Nanomaterials* 2022, *12*, 4132, doi:10.3390/nano12234132 134

Yanhao Duan, Jian Wu, Shixue He, Benlong Su, Zhe Li and Youshan Wang
Bioinspired Spinosum Capacitive Pressure Sensor Based on CNT/PDMS Nanocomposites for Broad Range and High Sensitivity
Reprinted from: *Nanomaterials* **2022**, *12*, 3265, doi:10.3390/nano12193265 . . . . **153**

Shamil Galyaltdinov, Ivan Lounev, Timur Khamidullin, Seyyed Alireza Hashemi, Albert Nasibulin and Ayrat M. Dimiev
High Permittivity Polymer Composites on the Basis of Long Single-Walled Carbon Nanotubes: The Role of the Nanotube Length
Reprinted from: *Nanomaterials* **2022**, *12*, 3538, doi:10.3390/nano12193538 . . . . **165**

Ezgi Uçar, Mustafa Dogu, Elcin Demirhan and Beate Krause
PMMA/SWCNT Composites with Very Low Electrical Percolation Threshold by Direct Incorporation and Masterbatch Dilution and Characterization of Electrical and Thermoelectrical Properties
Reprinted from: *Nanomaterials* **2023**, *13*, 1431, doi:10.3390/nano13081431 . . . . **176**

Calin Constantin Lencar, Shashank Ramakrishnan and Uttandaraman Sundararaj
Carbon Nanotube Migration in Melt-Compounded PEO/PE Blends and Its Impact on Electrical and Rheological Properties
Reprinted from: *Nanomaterials* **2022**, *12*, 3772, doi:10.3390/nano12213772 . . . . **190**

N. M. Nurazzi, F. A. Sabaruddin, M. M. Harussani, S. H. Kamarudin, M. Rayung, M. R. M. Asyraf, et al.
Mechanical Performance and Applications of CNTs Reinforced Polymer Composites—A Review
Reprinted from: *Nanomaterials* **2021**, *11*, 2186, doi:10.3390/nano11092186 . . . . **208**

About the Editor

Muralidharan Paramsothy

Dr Muralidharan Paramsothy is an early-retired scientist based in Singapore. Before retiring by choice, he was active for almost 5 years in supporting condition monitoring as well as intelligent energy projects in the Agency for Science, Technology and Research (A*STAR). Prior to that, he was active for almost 10 years in contributing to nanomaterials and nano(science)technology endeavors in Singapore. Overall, he has extensively managed academic as well as industrial stakeholders. He has numerous internationally published peer-reviewed articles, the Nanoscale Electro Negative Interface Density or NENID theory, and intellectual property pertaining to optical fiber sensor utilization to his name. He has organized a fair number of academic symposia in the USA and Canada. He has also served more flexibly as a Guest Editor in themed topics in journals, as well as held less-flexible, traditional roles on journal editorial boards. He is driven by work that benefits everybody, not that which merely promotes the individual.

Preface

This Special Issue reprint highlights seventy years since the discovery of carbon nanotubes (CNTs) in 1952 by Russian scientists LV Radushkevich and VM Lukyanovich in what was then the Union of Soviet Socialist Republics or the USSR. It also reflects the popularization by the well-known Japanese scientist S Iijima, since 1991, of carbon nanostructures, including CNTs, towards nanotechnology as a forever impactful and inspiring field. For researchers, academics, and teachers of all levels, from novice to expert to guru, the articles contained within this Special Issue are themed around *sustainability from nanotechnology*, pertaining to diameter measurement, rotor system molecular simulation, ultra-high tensile strength, energy, filtration via membrane distillation, environmental remediation using adsorption, ionic liquids as curing agents, biosensors and bioinspired sensors, and electrical/mechanical properties of polymer nanocomposites. The CNT is a legacy 1D nanomaterial, only after which was 2D graphene able to emerge.

Muralidharan Paramsothy
Editor

Editorial

70th Year Anniversary of Carbon Nanotube Discovery—Focus on Real-World Solutions

Muralidharan Paramsothy

NanoWorld Innovations (NWI), 1 Jalan Mawar, Singapore 368931, Singapore; mpsothy@yahoo.co.uk

Seventy years ago in 1952, Russian scientists LV Radushkevich and VM Lukyanovich published clear images showing multiwalled carbon nanotubes (MWCNTs) with 50 nm diameters [1]. Their paper was written in Russian and published in *Zhurnal Fizicheskoi Khimii* (now the Russian Journal of Physical Chemistry A) at the height of the Cold War. It is noteworthy that S Iijima's work, published some 30 years ago in 1991, also generated unprecedented interest in carbon nanostructures, including CNTs, and has since fueled intense research in the area of nanotechnology [2]. *Radushkevich and Lukyanovich should be credited for the discovery that carbon filaments can be hollow and have a nanometer-size diameter, i.e., for the discovery of CNTs* [3].

CNTs are recognized for ultrahigh strength and deformability, high thermal conductivity, ballistic electrical conductivity, selected biocompatibility, unusual optical properties, and high surface area. Graphene and nanohorns are other well-known nanoscale forms of carbon, but CNTs remain distinguished by virtue of their one-dimensional or 1D structure, enabling directional tailoring of exceptionally favorable characteristics (enabling desired anisotropic properties) when required in application. Advancing the contemporary theme of sustainability, 14 representative articles have been published in this Special Issue. Of these, 3 articles highlight nanoscale fundamental aspects of CNTs, i.e., diameter measurement [4], rotor system molecular simulation [5] and ultra-high tensile strength [6]. Eleven articles are on applications utilizing CNTs pertaining to energy [7,8], filtration via membrane distillation [9], environmental remediation using adsorption [10], ionic liquids as curing agents [11], biosensors [12] and bioinspired sensors [13], and electrical/mechanical properties of polymer nanocomposites [14–17]. Geographically speaking, including in terms of mixed-country authorship, this Special Issue is of a truly global nature: 3 articles are from Canada, 1 article is from Saudi Arabia, 3 articles are from China, 2 articles are from Russia, 2 articles are from Spain, 3 articles are from India, 1 article is from Japan, 1 article is from United Arab Emirates (UAE), 2 articles are from USA, 1 article is from Germany, 1 article is from Turkiye, and 1 article is from Malaysia, with the Guest Editor being from Singapore, and who has published critical work on the CNT-polymer interface towards mechanical properties of the nanocomposite, up to 20 years ago [18,19].

Citation: Paramsothy, M. 70th Anniversary of Carbon Nanotube Discovery—Focus on Real-World Solutions. *Nanomaterials* **2023**, *13*, 3162. https://doi.org/10.3390/nano13243162

Received: 8 December 2023
Accepted: 14 December 2023
Published: 18 December 2023

Copyright: © 2023 by the author. Licensee MDPI, Basel, Switzerland. This article is an open access article distributed under the terms and conditions of the Creative Commons Attribution (CC BY) license (https:// creativecommons.org/licenses/by/ 4.0/).

Diameter Measurement

Making a valuable contribution to nanoscale research, Lopinski et al. [4] developed a method using atomic force microscopy (AFM) for reliable and accurate diameter measurements on 'real-world' supported samples, such as those used for device fabrication. To illustrate the utility of this diameter measurement method, it was applied to measure the diameter of commercially available, highly enriched semiconducting nanotubes. The measured diameter for the nanotubes in a poly(9,9-di-n-dodecylfluorene) single-wall CNT (PFDD/SWCNT) sample was shown to be larger than expected based on the diameters of the tubes present in the dispersion, suggesting the polymer remained on the SWCNTs upon deposition onto the substrate and rinsing with a solvent. The average thickness of the polymer on the nanotubes was estimated to be ~0.4 nm. In addition, the measured heights of tube–tube junctions were found to be smaller than the diameters of the two tubes that cross, indicating a compression of ~20% at junctions. *Finally, it was noted that the* measurement protocol developed should also be generally applicable to the measurement of vertical dimensions of other 1D or 2D nanomaterials.

Rotor Systems Molecular Simulation

Undertaking fundamental exploration, Qing Peng et al. [5] investigated the rotating and braking processes of a carbon nanotube transmission system using simulations of molecular dynamics. The effect of hydroxyl groups on the speed of response was examined. The energy dissipation during the whole process was discussed. The results showed that hydroxyl groups enhanced stability and reduced the response time of the system in both the acceleration and braking processes. The higher hydroxyl group ratio enabled the achievement of a better transmission system performance. The underlying mechanism was the presence of hydrogen bonds that form between hydroxyl groups, which resulted in higher interfacial interaction and a faster response. The analysis of the phonon density of the state showed that the vibration of O-H bonds in hydroxyl groups accelerated the energy dissipation of the system, which led to faster responses in the acceleration and braking processes. *The results showed that grafted hydroxyl groups resulted in stronger interaction, and therefore had the potential to enhance the response of the transmission system.*

Ultra-High Tensile Strength

In a critical review, Fei Wei et al. [6] analyzed the controllable preparation and tensile strength of CNTs at different scales. Operating based on recent years' research into the controllable preparation of CNT fibers (CNTFs), the significance of defect control and efficient treatment process to control tube–tube interactions was emphasized in research towards the development of ultra-high-strength CNTFs. The unique structure and large-scale preparation of CNTs, which constitute the basic unit of nanoscale assembly, was introduced. It was observed that the mechanical strength and toughness could be significantly improved by eliminating the tube–tube non-uniform interactions or other post-treatment interactions. Using the bottom-up method, the mass production of defect-free CNTs and their accurate assembly into the macro-scale products with fewer defects, fine alignment, and higher density was the priority, with the eventual aim of obtaining macro-scale assemblies with excellent tensile strength. Not only tensile strength, but also other mechanical properties such as fatigue, bending, and torsion, as well as electrical and thermal properties, all faced a similar problem of performance transfer across scales. *It was also observed that, with further analysis of the influencing factors and optimization strategies for the cross-scale transfer of other properties, including overcoming selected challenges, the excellent intrinsic properties of CNTs could be fully utilized, and CNTs could play a more significant role in many fields.*

Energy

Groundbreakingly, Bulusheva et al. [7] synthesized composite nanomaterials via the vaporization–condensation process using commercial red phosphorus and single-walled CNTs (SWCNTs). Under the synthesis conditions in play, phosphorus chains were formed inside open SWCNTs and the SWCNT surface was covered with red phosphorus, which was oxidized upon contact with air. The external phosphorus deposit was effectively removed using an aqueous solution of sodium hydroxide. The content of phosphorus in samples with only external phosphorus SWCNTs/P, only internal phosphorus P@SWCNTs and both types of phosphorus P@SWCNTs/P was 11%, 8% and 16%, respectively, according to XPS data. To reveal the effect of various combinations of SWCNTs and phosphorus in the composite on the electrochemical interaction with lithium ions, nanomaterials were tested as anodes in coin-cell batteries using lithium sheets as counter electrodes. The SWCNTs/P sample was found to perform better than the reference red phosphorus due to the presence of a conducting network of SWCNTs. Nanotubes practically did not contribute to the SWCNTs/P capacity because their surface was blocked by phosphorus. The creation of composite nanomaterials with internal phosphorus almost doubled the initial Coulombic efficiency as compared to SWCNTs/P and dramatically improved the specific capacity and cycling stability of the electrode. The presence of both external and internal phosphorus led to a gradual and slight degradation of the P@SWCNTs/P electrode over time, associated with slower diffusion processes in the internal volume of the material. *The excellent cycling stability of P@SWCNTs over 1000 cycles at a high current density of 5 A·g−1 was associated with the synergistic effect of highly capacitive phosphorus and conductive SWCNTs, the absence of inactive oxidized phosphorus deposits, and the presence of multiple channels for lithium-ion diffusion to encapsulated phosphorus.*

Uniquely, Archana et al. [8] successfully synthesized titanium dioxide/reduced graphene oxide/silver (TiO$_2$/rGO/Ag) hybrid nanostructures using a combination of solution processes and in situ growth. These hybrid nanostructures were then utilized as photoanodes for dye-sensitized solar cells (DSSCs) and as catalysts for photodegradation applications. Plasmon-enhanced DSSC devices demonstrated enhanced photovoltaic performance of 7.27% along with a higher short-circuit current of 16.05 mA/cm^2 and an incident photocurrent efficiency (IPCE) of 77.82% at 550 nm. The results suggested that the high photovoltaic performance of the plasmon-based TiO$_2$/rGO/Ag device could be attributed to (i) the large specific area of TiO$_2$/rGO/Ag, which led to high dye loading; (ii) TiO$_2$ mesospheres enhancing the light scattering effect of incoming light; and (iii) the incorporation of Ag nanoparticles (NPs) facilitating more induced photons and fast electron transport in the device. Upon natural sunlight irradiation, the prepared hybrid nanostructure showed a 93% improvement in the photocatalytic degradation of methylene blue (MB) dye within 160 min, and the effects of different scavengers on the obtained photocatalytic activity were systematically investigated. The effects of the optimum active surface area, the localized surface plasmon resonance (LSPR) properties of Ag NPs, and the enhanced electrical conductivity of the prepared ternary nanostructures combined to provide an enhanced visible-light-driven plasmonic DSSC device and photocatalyst for MB dye degradation. *The proposed plasmonic and hybrid-based nanostructures demonstrated an emerging strategy to establish large-scale applications of solar energy conversion technologies.*

Filtration via Membrane Distillation

Very relevantly, Hilal et al. [9] demonstrated that in membrane distillation (MD), the membrane characteristics could be tuned using an electrospray deposition technique to obtain desirable MD membrane properties, such as high hydrophobicity or high water contact angle (>120°), high liquid entry pressure (LEP), optimum pore size (~0.2 μm), narrow pore size distribution, etc., compared to the pristine electrospun membrane. CNT modification followed by heat pressing yielded mechanically robust nanocomposite membranes with improved membrane characteristics. A 3% increase in the water contact angle, 20% increase in the LEP, and 42.6% reduction in the mean flow pore size towards the optimum pore size were observed in the heat-pressed CNT-modified electrospun poly(vinylidene fluoride-co-hexafluoropropylene) (PVDF-Co-HFP) membranes. The tensile strength of the heat-pressed CNT-modified membrane was significantly improved by up to 120% compared to the electrospun PVDF-Co-HFP membrane. The presence of CNTs on the membrane surface before and after the MD process was observed. Water vapor flux enhancements of 15.7%, 20.6%, and 24.6% were observed at a ΔT of 20 °C, 30 °C, and 40 °C, respectively. Higher temperature polarization coefficient (TPC) values and percentage water vapor flux enhancements were observed at lower feed solution temperatures because of the higher heat loss at higher feed solution temperatures compared to the lower temperatures. Enhancements of 16% and 12% in the TPC values were observed at the feed solution temperatures of 35 °C and 55 °C, respectively. A > 99.8% inorganic salt rejection was observed through the use of quantitative analytical tools when conducting the direct contact membrane distillation (DCMD) process using a 3.5 wt. % simulated seawater feed solution. *Electrohydrodynamic atomization using appropriate nanomaterial dispersion can be recommended as an efficient tool for the surface modification of MD membranes.*

Environmental Remediation Using Adsorption

Futuristically, Neelgund et al. [10] successfully prepared an efficient adsorbent for the effective adsorption of Pb(II) and As(III) ions by grafting fourth-generation aromatic poly(amidoamine) (PAMAM) to CNTs and via the successive deposition of Ag nanoparticles. Thus, CNTs-PAMAM-Ag was able to adsorb 99% and 76% of Pb(II) and As(III) ions, respectively, within 15 min. The kinetics data obtained for the adsorption of Pb(II) and As(III) ions were well fitted with the pseudo-second-order model compared to the pseudo-first-order model. This revealed the occurrence of chemisorption due to sharing or exchanging electrons between Pb(II) or As(III) ions and CNTs-PAMAM-Ag. This could be the rate-controlling step in the process of adsorption. The multilinearity of the Weber–Morris plot demonstrated that intraparticle diffusion was not a rate-controlling step in the adsorption of Pb(II) and As(III) ions; instead, it was regulated by both intraparticle diffusion and the boundary layer effect. The proper fitting of kinetic data of Pb(II) and As(III) ion adsorption with the Langmuir isotherm model indicated the uniform distribution of active sites over the entire surface of CNTs-PAMAM-Ag and their homogeneity. In addition, it signified that the adsorption of Pb(II) and As(III) ions was dominated by monolayer binding on the homogeneous surface of CNTs-PAMAM-Ag. The adsorption ability of CNTs-PAMAM-Ag depended on the pH. *The CNTs-PAMAM-Ag was an ideal adsorbent for repeated use without losing its activity, and because of its significance in Pb(II) and As(III) ion adsorption, CNTs-PAMAM-Ag could be an efficient adsorbent with practically applicability for the adsorption of other heavy metals and other contaminants present in water.*

Ionic Liquids as Curing Agents

In a green approach, Guerrica-Echevarría et al. [11] showed the double role of ionic liquids (ILs) as effective curing and dispersing agents in the production of volatile amine-free epoxy/CNT nanocomposites with a better balance of mechanical, electrical, and adhesive properties. Three different ILs were tested, and all three generated good dispersion of the nanofiller, featuring individually dispersed CNTs as well as some small aggregates. Overall, with a percolation threshold of 0.001 wt.%, the IL trihexyltetradecylphosphonium dicyanamide (IL-P-DCA) system was the most effective. The addition of CNTs had no effect on the thermal or low-strain mechanical properties of the epoxy/IL systems. However, CNT addition improved the systems' adhesive properties. The lap shear strength of epoxy/IL-P-DCA system containing 0.025 wt.% CNTs was improved by 30% compared to that of epoxy/IL-P-DCA. This was the best improvement among the three cases. *This research proved that, using very small amounts of CNTs, it is possible to obtain electrically conductive, amine-free epoxy adhesives with similar mechanical properties but greater lap shear strength than the reference amine-cured epoxy system. Additionally, since ILs have a lower vapor pressure, and a significantly lower amount is needed to effectively cure epoxy resins, we therefore observed that replacing conventional epoxy resin curing agents (amines, anhydrides, etc.) with ILs was a major step forward in the development of more sustainable materials.*

Biosensors

Most relevantly and impactfully, Hussain et al. [12] summarized the advancements in CNT-based biosensors since the last decade in the detection of different human viruses, namely, SARS-CoV-2, dengue, influenza, human immunodeficiency virus (HIV), and hepatitis. It has been proven that viral infections pose a serious hazard to humans and also affect social health, including morbidity and mental suffering, as illustrated by the COVID-19 pandemic. The early detection and isolation of virally infected people are thus required to control the spread of viruses. *Due to the outstanding and unparalleled properties of nanomaterials, numerous biosensors have been developed for the early detection of viral diseases via sensitive, minimally invasive, and simple procedures. To that end, viral detection technologies based on CNTs have been developed as viable alternatives to existing diagnostic approaches, and the shortcomings and benefits of CNT-based biosensors for the detection of viruses have also been outlined and discussed.*

Bioinspired Sensors

Making practical and logical adoptions from natural biology, Wu et al. [13] proposed a sensitive capacitive pressure sensor with a broad detection range inspired by the skin epidermis. A simple and low-cost fabrication process was proposed for carbon nanotube/polydimethylsiloxane-based (CNT/PDMS-based) spinosum pressure sensors via the use of abrasive paper templates. The spinosum microstructure and doping content of CNTs effectively improved the performance of pressure sensors: high sensitivity (0.25 kPa^{-1}), wider pressure range (~500 kPa), fast response time (20 ms) and excellent stability over 10,000 cycles. Also, the effects of the mesh number of abrasive papers and CNT doping content on the sensing property were theoretically analysed via simulations and experiments. Further, a sensor array was manufactured for mapping the spatial distribution of pressure, which showed great potential for intelligent monitoring. *Finally, a new methodology was proposed in order to solve the tire–road contact pressure issue, as well as estimate related parameters by introducing a sensor array to tires. Practically and logically, a bioinspired, cost-effective, broad-range, high-sensitivity, and flexible sensor was fabricated, and a new patch to intelligently monitor the contact pressure of tires was also developed.*

Electrical/Mechanical Properties of Polymer Nanocomposites

In further groundbreaking research, Dimiev et al. (including A. Nasibulin) [14] controlled the permittivity of dielectric composites for numerous applications dealing with matter/electromagnetic radiation interactions. Polymer composites were prepared with a silicone elastomer matrix and Tuball™ SWCNTs using a simple preparation procedure. The as-prepared composites demonstrated record-high dielectric permittivity in both the low-frequency range (10^2–10^7 Hz) and in the X-band (8.2–12.4 GHz), significantly exceeding the literature data for such types of composite materials at similar levels of CNT content. Thus, with the 2 wt% filler loading, the permittivity values reached 360 at 10^6 Hz and >26 in the entire X-band. In similar literature, even the use of conductive polymer hosts and various highly conductive additives did not result in such high permittivity values. *The superior permittivity phenomenon was attributed to specific structural features of the SWCNTs, namely, length and the ability to constitute percolating networks in the polymer matrix in the form of neuron-shaped clusters. The low cost and large production volumes of Tuball™ SWCNTs, as well as the ease of the composite preparation procedure, opened the doors for the production of cost-efficient, low weight and flexible nanocomposites with superior high permittivity.*

Critically, Krause et al. [15] prepared poly(methyl methacrylate) (PMMA)/single-walled carbon nanotube (Tuball™ SWCNT) nanocomposites via melt mixing to achieve suitable SWCNT dispersion and distribution with low electrical resistivity, where the SWCNT direct incorporation method was compared with the results of masterbatch dilution. An electrical percolation threshold of 0.05–0.075 wt% was found, the lowest threshold value for melt-mixed PMMA/SWCNT nanocomposites reported so far. The influence of rotation speed and method of SWCNT incorporation into the PMMA matrix on the electrical properties and SWCNT macro dispersion was investigated. It was found that increasing rotation speed improved macro dispersion and electrical conductivity. The results showed that electrically conductive nanocomposites with a low percolation threshold could be prepared by direct incorporation using high rotation speed. The masterbatch approach led to higher resistivity values compared to the direct incorporation of SWCNTs. Also, the thermal behaviour and thermoelectric properties of PMMA/SWCNT nanocomposites were studied. The Seebeck coefficients varied from 35.8 µV/K to 53.4 µV/K for nanocomposites up to 5 wt% SWCNT. It was also observed that the addition of SWCNT increased the glass transition temperature (T_g) of PMMA by approximately 4 K. *Overall, the Tuball™ SWCNT material was observed to be an effective filler that could be used to obtain conductive nanocomposites with good dispersion and low electrical resistivity via melt-mixing. Also, the thermoelectric measurements indicated that PMMA/SWCNT nanocomposites could be used as a thermoelectric material.*

Also critically, Sundararaj et al. [16] investigated the effects of MWCNT concentration, mixing time, and compatibilizer addition on the migration of MWCNTs from the polyethylene (PE) phase to a polyethylene oxide (PEO) phase of a 60:40 PEO/PE blend, and the subsequent impact on electrical and rheological properties. Two-step mixing was used to pre-localize MWCNTs in the less thermodynamically favoured PE phase and observe migration into the thermodynamically favoured PEO phase. It was observed that MWCNTs migrated into the PEO phase as the mixing time increased at all concentrations of MWCNTs used. This migration was also supported by electromagnetic interference shielding effectiveness (EMI SE) and DC conductivity measurements, which showed significant reductions in electrical properties over time, suggesting a disruption of conductive networks as MWCNTs migrated into PEO. PEO/PE 40:60 samples containing 3 vol% MWCNTs showed a high conductivity of 22.1 S/m, which that suggested effective MWCNT networks were present at the onset of mixing. To arrest the migration of MWCNTs into PEO, a PE-graft-maleic anhydride (PEMA) compatibilizer was added to the PEO/PE blend. An improvement in the formation of MWCNT networks along the PEO/PE interface was observed at 5 min of mixing for the compatibilized polymer blend nanocomposite (PBN). Furthermore, major improvements in electrical conductivity (68.7 S/m) were observed. *Comparisons to the poly(vinylidene fluoride)/poly(ethylene) (PVDF/PE) system in previous research suggested that the viscosity of the destination phase, as well as the interfacial surface energies of the blend components, played significant roles in determining whether MWCNTs would successfully migrate across polymer/polymer interfaces or whether they would become trapped at the interface. Migration behaviour was shown to significantly influence the electrical and rheological properties of PBNs.*

Of interest in a rather over-arching sense as the first article in this SI, Nurazzi et al. (including R.A. Ilyas and A. Khalina) [17] reviewed the mechanical performance of CNT-reinforced polymer nanocomposites. It was observed that CNTs have excellent chemical and physical properties, making them ideal and promising for reinforcement in polymer nanocomposites. It was acknowledged that the mechanical properties of the CNT/polymer nanocomposites are influenced by the interactions between the nanofillers and the polymer matrices. It was noted that the main challenge is the tendency of the CNTs to agglomerate, resulting in poor dispersion properties, an issue that can cause the performance of the composite structures to deteriorate. Researchers have devised various methods for distributing and orienting CNTs. Further, it was also observed that dispersing a small amount of filler in the polymer matrix enhanced nanocomposite properties. Although numerous CNT nanocomposites have been investigated, consistent progress is still needed to obtain nanocomposites with the best performance. *Several dimensions, such as the amount of CNTs, size of CNTs, spatial distribution and orientation of CNTs, the suitability of surface modification of the CNT, and method of nanocomposite fabrication, all collectively affect the mechanical properties of the nanocomposite. The crucial nature of necessarily finding a collective balance among these multiple parameters for optimum performance of the CNT nanocomposite was duly reflected.*

Acknowledgments: The Guest Editor graciously thanks all the authors for submitting their work to the Special Issue and for its impactful completion with 29,400 views to date. A special thank you to all the academic editors and reviewers participating in the peer review process and enhancing quality and impact of the manuscripts deserves mention. The Guest Editor also makes an overwhelming and infinite gesture of gratitude to the editors and the editorial assistants who made the entire Special Issue creation possible and memorable. Thank you once again.

Conflicts of Interest: The author declares no conflict of interest.

References

1. Radushkevich, L.V.; Lukyanovich, V.M. O strukture ugleroda obrazujucegosja pri termiceskom razlozenii okisi ugleroda na zeleznom kontakte. *Zhurnal Fiz. Khimii* **1952**, *26*, 88–95. (Translated into Radushkevich, L.V.; Lukyanovich, V.M. About the structure of carbon formed by thermal decomposition of carbon monoxide on iron substrate. *Russ. J. Phys. Chem. A* **1952**, *26*, 88–95.).
2. Iijima, S. Helical Microtubules of Graphitic Carbon. *Nature* **1991**, *354*, 56. [CrossRef]

3. Monthioux, M.; Kuznetsov, V.L. Guest Editorial: Who Should Be Given the Credit for the Discovery of Carbon Nanotubes? *Carbon* **2006**, *44*, 1621–1623. [CrossRef]
4. Vobornik, D.; Chen, M.; Zou, S.; Lopinski, G.P. Measuring the Diameter of Single-Wall Carbon Nanotubes Using AFM. *Nanomaterials* **2023**, *13*, 477. [CrossRef] [PubMed]
5. Zhang, B.; Li, R.; Peng, Q. Controlling CNT-Based Nanorotors via Hydroxyl Groups. *Nanomaterials* **2022**, *12*, 3363. [CrossRef] [PubMed]
6. Zhu, Y.; Yue, H.; Aslam, M.J.; Bai, Y.; Zhu, Z.; Wei, F. Controllable Preparation and Strengthening Strategies towards High-Strength Carbon Nanotube Fibers. *Nanomaterials* **2022**, *12*, 3478. [CrossRef] [PubMed]
7. Vorfolomeeva, A.A.; Stolyarova, S.G.; Asanov, I.P.; Shlyakhova, E.V.; Plyusnin, P.E.; Maksimovskiy, E.A.; Gerasimov, E.Y.; Chuvilin, A.L.; Okotrub, A.V.; Bulusheva, L.G. Single-Walled Carbon Nanotubes with Red Phosphorus in Lithium-Ion Batteries: Effect of Surface and Encapsulated Phosphorus. *Nanomaterials* **2023**, *13*, 153. [CrossRef] [PubMed]
8. Athithya, S.; Manikandan, V.S.; Harish, S.K.; Silambarasan, K.; Gopalakrishnan, S.; Ikeda, H.; Navaneethan, M.; Archana, J. Plasmon Effect of Ag Nanoparticles on TiO_2/rGO Nanostructures for Enhanced Energy Harvesting and Environmental Remediation. *Nanomaterials* **2023**, *13*, 65. [CrossRef] [PubMed]
9. Francis, L.; Hilal, N. Electrosprayed CNTs on Electrospun PVDF-Co-HFP Membrane for Robust Membrane Distillation. *Nanomaterials* **2022**, *12*, 4331. [CrossRef] [PubMed]
10. Neelgund, G.M.; Aguilar, S.F.; Kurkuri, M.D.; Rodrigues, D.F.; Ray, R.L. Elevated Adsorption of Lead and Arsenic over Silver Nanoparticles Deposited on Poly(amidoamine) Grafted Carbon Nanotubes. *Nanomaterials* **2022**, *12*, 3852. [CrossRef] [PubMed]
11. Orduna, L.; Otaegi, I.; Aranburu, N.; Guerrica-Echevarría, G. Ionic Liquids as Alternative Curing Agents for Conductive Epoxy/CNT Nanocomposites with Improved Adhesive Properties. *Nanomaterials* **2023**, *13*, 725. [CrossRef] [PubMed]
12. Sengupta, J.; Hussain, C.M. Decadal Journey of CNT-Based Analytical Biosensing Platforms in the Detection of Human Viruses. *Nanomaterials* **2022**, *12*, 4132. [CrossRef] [PubMed]
13. Duan, Y.; Wu, J.; He, S.; Su, B.; Li, Z.; Wang, Y. Bioinspired Spinosum Capacitive Pressure Sensor Based on CNT/PDMS Nanocomposites for Broad Range and High Sensitivity. *Nanomaterials* **2022**, *12*, 3265. [CrossRef] [PubMed]
14. Galyaltdinov, S.; Lounev, I.; Khamidullin, T.; Hashemi, S.A.; Nasibulin, A.; Dimiev, A.M. High Permittivity Polymer Composites on the Basis of Long Single-Walled Carbon Nanotubes: The Role of the Nanotube Length. *Nanomaterials* **2022**, *12*, 3538. [CrossRef] [PubMed]
15. Uçar, E.; Dogu, M.; Demirhan, E.; Krause, B. PMMA/SWCNT Composites with Very Low Electrical Percolation Threshold by Direct Incorporation and Masterbatch Dilution and Characterization of Electrical and Thermoelectrical Properties. *Nanomaterials* **2023**, *13*, 1431. [CrossRef] [PubMed]
16. Lencar, C.C.; Ramakrishnan, S.; Sundararaj, U. Carbon Nanotube Migration in Melt-Compounded PEO/PE Blends and Its Impact on Electrical and Rheological Properties. *Nanomaterials* **2022**, *12*, 3772. [CrossRef] [PubMed]
17. Nurazzi, N.M.; Sabaruddin, F.A.; Harussani, M.M.; Kamarudin, S.H.; Rayung, M.; Asyraf, M.R.M.; Aisyah, H.A.; Norrrahim, M.N.F.; Ilyas, R.A.; Abdullah, N.; et al. Mechanical Performance and Applications of CNTs Reinforced Polymer Composites—A Review. *Nanomaterials* **2021**, *11*, 2186. [CrossRef] [PubMed]
18. Wong, M.; Paramsothy, M.; Xu, X.J.; Ren, Y.; Li, S.; Liao, K. Physical Interactions at Carbon Nanotube-Polymer Interface. *Polymer* **2003**, *44*, 7757. [CrossRef]
19. Paramsothy, M. Dispersion, Interface, and Alignment of Carbon Nanotubes in Thermomechanically Stretched Polystyrene Matrix. *J. Mater.* **2014**, *66*, 960. [CrossRef]

Disclaimer/Publisher's Note: The statements, opinions and data contained in all publications are solely those of the individual author(s) and contributor(s) and not of MDPI and/or the editor(s). MDPI and/or the editor(s) disclaim responsibility for any injury to people or property resulting from any ideas, methods, instructions or products referred to in the content.

Article

Measuring the Diameter of Single-Wall Carbon Nanotubes Using AFM

Dusan Vobornik, Maohui Chen, Shan Zou and Gregory P. Lopinski *

Metrology Research Center, National Research Council, Ottawa, ON K1A 0R6, Canada
* Correspondence: gregory.lopinski@nrc-cnrc.gc.ca

Abstract: In this work, we identify two issues that can significantly affect the accuracy of AFM measurements of the diameter of single-wall carbon nanotubes (SWCNTs) and propose a protocol that reduces errors associated with these issues. Measurements of the nanotube height under different applied forces demonstrate that even moderate forces significantly compress several different types of SWCNTs, leading to errors in measured diameters that must be minimized and/or corrected. Substrate and nanotube roughness also make major contributions to the uncertainty associated with the extraction of diameters from measured images. An analysis method has been developed that reduces the uncertainties associated with this extraction to <0.1 nm. This method is then applied to measure the diameter distribution of individual highly semiconducting enriched nanotubes in networks prepared from polyfluorene/SWCNT dispersions. Good agreement is obtained between diameter distributions for the same sample measured with two different commercial AFM instruments, indicating the reproducibility of the method. The reduced uncertainty in diameter measurements based on this method facilitates: (1) determination of the thickness of the polymer layer wrapping the nanotubes and (2) measurement of nanotube compression at tube–tube junctions within the network.

Keywords: atomic force microscopy; carbon nanotubes; diameter; nanometrology

---

Citation: Vobornik, D.; Chen, M.; Zou, S.; Lopinski, G.P. Measuring the Diameter of Single-Wall Carbon Nanotubes Using AFM. Nanomaterials 2023, 13, 477. https://doi.org/10.3390/nano13030477

Academic Editors: Muralidharan Paramsothy and Arthur P. Baddorf

Received: 21 December 2022
Revised: 17 January 2023
Accepted: 21 January 2023
Published: 24 January 2023

Copyright: © 2023 by the authors. Licensee MDPI, Basel, Switzerland. This article is an open access article distributed under the terms and conditions of the Creative Commons Attribution (CC BY) license (https://creativecommons.org/licenses/by/4.0/).

1. Introduction

Advances in methods for the scalable manufacture, purification, and dispersion of single-wall carbon nanotubes (SWCNTs) have dramatically accelerated the development of applications based on this material. In particular, various approaches for separating semiconducting and metallic tubes have enabled demonstration of a variety of SWCNT-based electronic devices [1–7]. The increased availability of highly purified SWCNT samples containing a limited number of chiralities has highlighted the need for improved methods for assessing these materials. With fewer chiralities present, these samples are expected to have narrower diameter distributions, motivating the need for accurate measurements of this key dimensional parameter. While transmission electron microscopy (TEM) is a powerful method to measure the diameter of individual tubes [8–10], this approach is costly and requires special substrates. Raman spectroscopy is commonly used to determine SWCNT diameters through the frequency of the radial breathing mode, which is related to diameter [11,12]. However, recent studies comparing diameter distributions measured by Raman and TEM methods reveal discrepancies, highlighting the limitations of the Raman approach [13,14].

Atomic force microscopy (AFM) is a widely used technique capable of measuring the vertical dimensions of nano-objects with subnanometer accuracy [15]. While AFM has been commonly used to image SWCNTs [16–18], fewer studies have focused on using this method for measuring the tube diameter [19–21]. In contrast to TEM, AFM can be performed on a wide variety of substrates, including those relevant for device applications. One limitation is that AFM-measured heights often differ from their actual height due to interactions of the AFM tip with the sample. These issues have been studied in detail for soft biological samples such as oligonucleotides, where it was shown that the AFM measurements with common settings can underestimate the actual height by over 50% [22–24]. In the case of SWCNTs, the effects of both substrate–tube and tip–tube interactions during AFM imaging were examined with molecular dynamics calculations, concluding that for tube diameters less than 2 nm, applied forces are not sufficient to significantly alter the measured height [19]. In this work, we show that even at the moderate forces commonly used for AFM imaging, compression of the SWCNTs does occur, which can result in a significant underestimation of the diameter extracted from the images. A straightforward experimental procedure has been developed to minimize and/or correct for this effect and extract accurate values of the SWCNT diameter. Another issue that contributes to the uncertainty of diameter measurements, the roughness of both the nanotubes and the substrate, has also been considered. A simple analysis protocol has been developed to reduce these uncertainties to <0.1 nm.

To demonstrate the utility of our protocol for AFM-based diameter measurements, we applied it to measure the diameter distribution of individual nanotubes in networks of highly semiconducting enriched polyfluorene/SWCNT dispersions on silicon oxide substrates. Good agreement was obtained for diameter distributions measured on the same sample with two different commercial AFM instruments and analyzed by two different analysts. The reduced uncertainty associated with diameter measurements using this method facilitates determination of the thickness of the polymer layer wrapping the nanotubes, which has not been reported previously. In addition, analysis of height measurements at tube–tube crossings within the network demonstrate compression at the junctions. Quantifying this compression is expected to contribute to further understanding of electron transport at these junctions that limit the performance of devices based on SWCNT networks [25]. The method for extraction of vertical dimensions of nano-objects from AFM images outlined here is not specific to carbon nanotubes but can also be applied to a variety of 1D and 2D nanomaterials to ensure that AFM measurements result in accurate diameter and/or thickness measurements.

2. Materials and Methods

2.1. Sample Preparation

Four different SWCNT dispersions were utilized to prepare nanotube networks in this study. Ultrahigh purity SWCNTs (>99.9% semiconducting) poly(9,9-di-n-dodecylfluorene) (PFDD)-wrapped nanotubes (IsoSol S-100) were purchased from NanoIntegris (Boisbriand, QC, Canada) in two forms, a PFDD/SWCNT dispersion (10 mg/L) in toluene and solvent-free bucky paper which was redispersed in toluene to the same nanotube concentration. The diameter range of the SWCNTs in both these samples are specified as 1.2–1.4 nm. The polymer-to-nanotube ratio was measured by UV–Vis–IR absorption to be 4:1 and 1:1, respectively. Semiconducting SWNCTs in an aqueous surfactant solution (IsoNanotubes-S 99%, diameters of 1.2–1.7 nm) were also obtained from NanoIntegris. The final tube type measured was SWCNTs dispersed in dimethyl sulfoxide (DMSO) at 100 mg/L (SEER ink, Linde North America, Bridgewater, NJ, USA), for which the diameter range was given as 1.4–2 nm.

The PFDD/SWCNTs in toluene and aqueous SWCNT dispersions were deposited onto piranha-cleaned ~1 cm^2 silicon substrates (Si(100) with 100 nm thermal oxide, Silicon Quest International, San Jose, CA, USA). PFDD/SWCNT dispersions were deposited directly onto clean SiO$_2$ for 5 to 15 min, followed by rinsing with toluene for 20 to 60 s and drying with nitrogen. Aqueous surfactant SWCNT dispersions were deposited on Poly-L-Lysine (PLL)-coated SiO$_2$ for 15 min, rinsed with water, and dried with nitrogen. The Linde SWCNT dispersion in DMSO was handled in a nitrogen-purged glovebox, where 100 µL of 2 mg/L DMSO-nanotube dispersion was deposited on a freshly cleaved 1 cm^2 square of highly oriented pyrolytic graphite (HOPG). The sample was then heated to 100 °C for approximately 15 min, causing the DMSO to evaporate. For all samples, AFM measurements were performed within a day of preparation, but several samples were measured multiple times at different dates during several months following their preparation with no significant changes in overall morphology or measured diameters.

2.2. AFM Measurements

Most of the AFM data were obtained with a MultiMode AFM with a NanoScope V controller (Bruker Nano Surfaces Division, Santa Barbara, CA, USA) in Peak Force QNM mode. In this mode, the AFM tip oscillates at a given (2 kHz) frequency, with the highest (peak) repulsive interaction force used as the feedback signal. Silicon nitride ScanAsyst-Air AFM probes (Bruker AFM Probes, Camarillo, CA, USA) were used in all peak force feedback measurements. The AFM tip diameter and cantilever spring constants (as specified by the supplier) were 2 nm and 0.4 N/m, respectively.

To demonstrate the wide applicability of the measurement methods presented here, a Nanowizard II BioAFM (JPK Instruments, Berlin, Germany) in intermittent contact mode was also used to measure SWCNT diameters. Silicon cantilevers (HQ:XSC11/AL BS, Cantilever D, typical radius of 8 nm, resonance frequency of 350 kHz, spring constant of 42 N/m, MikroMasch, Watsonville, CA, USA) were used for imaging in air at RT (20–23 °C). Force was carefully minimized by first gradually reducing the free amplitude to the highest level where stable imaging could be performed (evaluated by comparing trace–retrace lines while scanning the sample) and adjusting the feedback set point to a value as close as possible to the free amplitude. The typical set point obtained in this way was 85–90% of the free amplitude, which was set to 600 mV.

Images used to measure diameters were typically 1×1 µm^2 in size, acquired with 512×512-pixel resolution and at a scan rate of 1 Hz. In this way, the lateral pixel size was approximately 2×2 nm^2. The measured diameters of single nanotubes were in the range from 1.2 to 2 nm, and based on our tests, the radii of AFM tips for peak-force AFM were in the 2 to 15 nm range (tips were replaced when their radii exceeded 15 nm, even if they were still scanning properly). Based on the combination of the tip and nanotube radii, the pixel size of 2×2 nm^2 was deemed sufficient to have reliable nanotube height measurements. The pixel size should not be larger than the smallest tip size since that could lead to two adjacent pixels across the nanotube being both off the highest point of the nanotube, leading to underestimation of the tube height.

Traceable calibration grids (STS3-180P, STS3-440P, STS3-1000P, and STS3-1800, VLSI Standards Inc., Milpitas, CA, USA) were used on a regular basis (approximately every 6 months) for calibration verification and adjustments of our AFM instruments. On top of that, we regularly imaged atomic steps on HOPG. These steps have a known height of 0.34 nm. In our tests, we always measured the height of the smallest steps to be 0.34 nm +/− 0.01 nm, consistent with the expected value, further confirming the precision of our calibration (see Figure S1). The step height is much closer to the value of the SWCNT diameters than that of the calibration grids, thus ensuring that the AFM performs accurately at the relevant height scales.

2.3. Analysis of AFM Data

All image processing and analyses were performed using Gwyddion (v2.45, Czech Metrology Institute), a free, widely available open-source software [26]. For the intermittent contact mode AFM, images were additionally flattened using the first-level flattening with the JPK Data Processing Software (v5.1.8, JPK instruments, Berlin, Germany) prior to using Gwyddion software for further image processing. Image processing consisted of flattening the background to correct for drifts and zeroing the z scale. Images were flattened using the polynomial background removal function in Gwyddion, where special care was taken to only use the pixels corresponding to the substrate for the flattening and not those corresponding to the nanotubes or contaminants. This was ensured by first setting a mask based on a height threshold highlighting any features higher than the substrate, and then specifying to use only the unmasked pixels for the polynomial fitting and flattening.

3. Results and Discussion
3.1. Applied Force in AFM Imaging Leads to Nanotube Compression

To investigate the effect of applied force on measurements of nanotube diameter, the force was gradually increased while imaging the same network area. Figure 1 illustrates one such test on a network prepared from a 4:1 PFDD/SWCNT dispersion deposited on a SiO$_2$ substrate. The lowest force that consistently resulted in stable imaging, regardless of the AFM tip or of the particular nanotube sample, was 0.2 nN. Using this as the initial force, sequential images were obtained at increasing peak-force feedback values followed by a final image back at 0.2 nN, as shown in Figure 1a. It is apparent from the figure that even at the highest AFM applied force used here (2 nN), the imaging remained non-destructive as the nanotubes remain intact and do not appear to have been moved during imaging. Upon reducing the peak force back to 0.2 nN, the image appears essentially identical to the initial one, indicating that the tip was kept unchanged while acquiring the entire set of images. The maximum of the attractive van der Waals force between the tip and the sample was also monitored, remaining at −50 pN. Since the attractive van der Waals force is directly proportional to the tip diameter, this indicates that the tip was unchanged [27].

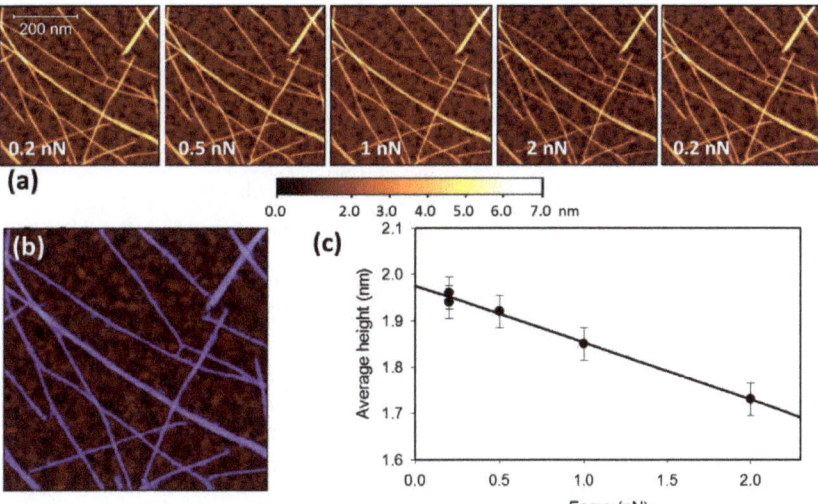

Figure 1. (a) Consecutive images of the same area of a PFDD/SWCNT network on a SiO$_2$ substrate, obtained at different peak-force feedback values. (b) The blue mask shows all the pixels whose height was included in obtaining the average network height. (c) Average network height as a function of applied force along with a linear fit to the data.

For each image in Figure 1a, the average network height was determined as described previously [27]. This was done rapidly by averaging the height of all the pixels above a certain height threshold as indicated by the colored mask (Figure 1b), which identifies the pixels corresponding to nanotubes. Once the network–substrate separation height threshold is selected, the Gwyddion software allows a one-click extraction of the average height of all colored pixels (nanotube network height), as well as the average height of all non-colored pixels (substrate average height). The final average network height is obtained by subtracting the average substrate height from the network height. Figure 1c illustrates that increasing the force decreases height in a linear manner, facilitating extrapolation of the data to extract the height at zero applied force. Control experiments involving imaging terraces on HOPG show no change in step height for a similar range of applied forces (see Supplementary Materials Figure S1).

Figure 2 shows average height versus applied force data for four different commercially available SWCNT samples (representative images are shown in Figure S2). Despite the differences in nanotube types and diameters, all graphs show a force-dependent behavior. Linear fits to this data, shown in Figure 2, allow extraction of the corresponding zero-force height and the "compression" slope. The differences in compression slopes can be qualitatively interpreted as a result of the differences between the four samples (polymer or surfactant on surface of the tubes, different diameter distributions, etc.), but more data are required to draw firm conclusions. Here, we only note that independent of the details of a particular carbon nanotube sample, the forces typically applied in AFM imaging (e.g., 0.2–2 nN) can significantly compress the tubes, leading to potential errors in measured diameters.

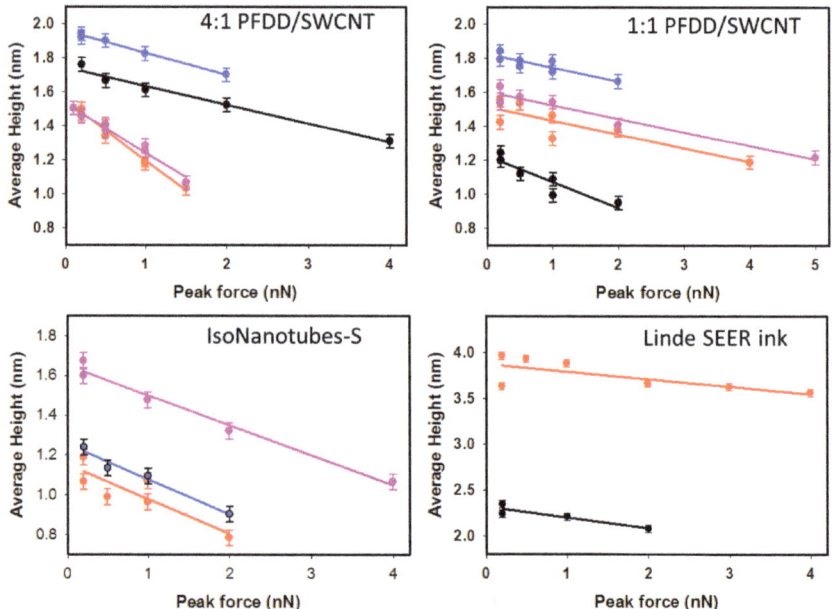

Figure 2. Average carbon nanotube network height as a function of AFM peak force measured on several different areas of four different SWCNT samples. In each graph, data points of the same color are obtained from consecutive AFM images of the same area.

The radial deformability of CNTs in AFM imaging has been demonstrated previously within a similar force range [28–30], but this compression effect has often been neglected in studies where AFM was used to measure diameters of carbon nanotubes [19,21]. Our results clearly indicate that imaging of SWCNTs results in compression of the tubes, translating to an underestimation of the nanotube diameter, even at the moderate forces commonly employed in AFM measurements. This compression effect can lead to errors easily reaching 30% or more of the actual nanotube height. However, measuring the height of the tubes as a function of applied force allows extraction of the average height of the tubes at zero force. The data shown in Figures 1 and 2 also indicate that for our set-up, using a force of 0.2 nN or less minimized this compression error.

3.2. Substrate and Nanotube Roughness Contribute to Diameter Uncertainty

While the method used above is useful for rapidly demonstrating the average diameter of an ensemble of nanotubes, it does not permit measurement of the diameter of an individual SWCNT (except in the case where there is only a single tube in the image). The conventional method to determine the heights of individual features from an AFM topography image involves extracting a height profile perpendicular to the object being measured. Here we show how this approach can lead to significant variations in the measured diameter due to the roughness of both the substrate and the nanotube.

Figure 3 shows a typical 1 μm² AFM image of a PFDD/SWCNT network on SiO₂ along with parallel pairs of height profiles extracted from the measured data. For each pair, one profile is along the top of the nanotube and the other is on the substrate adjacent to the tube. The profiles on the bare SiO₂ substrate (the red points in the figure) are observed to be more uneven, with height variations sometimes exceeding 1 nm. The standard deviation of the substrate height profile data ranges from 0.31 to 0.18 nm for the four profiles shown in the figure. The nanotube profiles (blue points) also appear uneven, but with considerably smaller height variations that seldom exceed 0.5 nm and with standard deviations ranging from 0.13 to 0.10 nm. While this is less roughness than that of the substrate, it is still significant and appears to be uncorrelated to that of the nearby substrate. This is consistent with studies quantifying the persistence length of carbon nanotubes, showing they behave like rigid rods within a length range that exceeds one hundred nanometers [31] and thus do not conform to shorter range fluctuations in the substrate height shown here. The height variations on the nanotubes are also not periodic, as would be expected if the polymer was wrapping the tube in a helical manner [32]. The height variations on the tubes exceed the random noise limits of our AFM setup, as height measurements with standard deviations below 0.1 nm are typically observed on atomically flat surfaces such as HOPG or mica at the same imaging parameters. Moreover, we observed similar tube roughness for all four different SWCNT samples studied. While the source of this roughness cannot be definitively assigned, for PFDD-wrapped nanotubes, it probably arises from variations in the conformation of the polymer coating the tube. In the case of the IsoNanotubes-S or the Linde SEER tubes, it is likely due to the surfactant or salt, respectively, used to disperse the tubes in the solution. Atmospheric contaminants binding to the tubes may also contribute to the observed roughness.

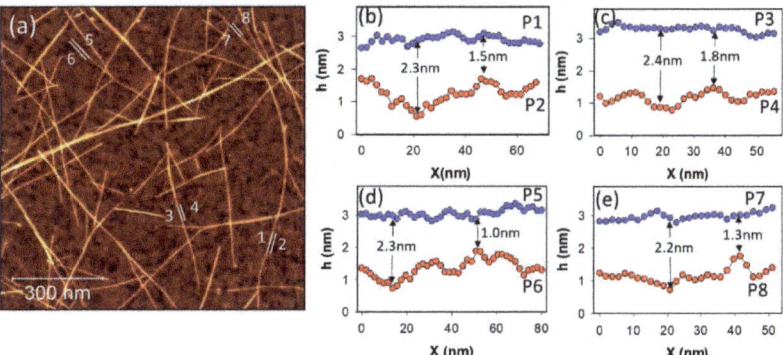

Figure 3. Roughness of profiles on nanotubes and on the SiO₂ substrate. (**a**) A typical 1 × 1 μm² AFM image of a network made of PFDD/SWCNTs with numbered white lines, along which the profiles P1–P8 shown in (**b**–**e**) were extracted. The data points corresponding to profiles along the top of the nanotubes are shown in blue, while adjacent substrate profiles are shown in red. Differences in the height of the substrate and the nanotube at selected points are indicated to illustrate the variability.

The profiles in Figure 3b–e show that the combined roughness of the substrate and nanotubes can lead to significant inaccuracies if the conventional AFM diameter measurement method is used. This method involves extracting cross-sections perpendicular to the nanotubes and measuring the height of the substrate from that of the highest point on the nanotube. Examining the profiles in Figure 3 shows that the difference between the tube apex and the substrate can vary substantially at different points along the tube, with some examples of these differences indicated in the figure. Although the difference between the average heights of the tube and substrate for the four profiles shown range from 1.7 to 2.2 nm, the individual height differences measured at each x-value range from 1 to 2.5 nm. This illustrates how large errors could result from an overly simplified approach to diameter measurements.

3.3. Analysis Method to Rapidly Extract SWCNT Diameters

One way to decrease the diameter measurement errors that result from roughness is to use an averaged diameter value based on multiple cross-sections. This can be accomplished easily using the Gwyddion analysis software, which offers an automated way to average up to 128 contiguous, single-pixel-wide cross-sections in one click. We find that using even 10-pixel-wide cross-sections reduces the uncertainty significantly, with larger widths offering even greater reduction. To further minimize this uncertainty, the averaging of the substrate baseline height can be improved by utilizing all the substrate pixels over the entire image. This can be done by choosing a height threshold to select the nanotube features (define a mask), as described above. Then, the average height value of all the pixels that are not masked is used as the average height of the substrate, which is then subtracted from the maximum height of the averaged nanotube profile to obtain the diameter.

The decrease in uncertainty realized by using this averaging analysis method was evaluated by randomly selecting an AFM image of a SWCNT network on SiO_2, and then repeating the analysis on it several times, each time extracting the diameters of the same tubes. In each of the images, diameters of a number of nanotubes (a total of 20) were measured 5–7 times using the above-described methods. Finally, the diameter standard deviation was calculated for each of the nanotubes using all of the extracted diameter values for that nanotube. The same diameter values for each of the nanotubes could be reliably obtained, with the standard deviation reduced to below 0.1 nm (see Figure S3 for an example of this evaluation). This analysis procedure is effective in that it allows for a greater degree of confidence in the measured diameter values with minimal additional analysis.

3.4. Diameter Measurements of PFDD/SWCNTs

Upon designing an effective method to increase the accuracy of the AFM measurements of SWCNT diameters by optimizing both our experimental (minimized forces) and analysis (averaging) approach, we set out to apply it to measure the diameter of PFDD/SWCNTs, providing insight into the interaction of the polyfluorene polymer with the nanotubes. PFDD/SWCNT networks were formed on SiO_2 substrates as described in the experimental section, and AFM images were obtained using a minimized peak-force value of 0.2 nN. Averaged cross-sections of 120 single nanotubes were then extracted using 30-pixel-width cross-sections. For each nanotube cross-section, the averaged substrate level was subtracted from the maximum height to obtain the diameter. To verify the universality of this approach, the same sample was also imaged with a different AFM instrument (see the Methods section for details), using the conventional intermittent contact mode with amplitude-based feedback. For these measurements, the AFM applied force was minimized by minimizing the free amplitude, and then using the smallest amplitude set point that facilitated stable imaging. The same analysis method (averaged cross-sections and average substrate baseline extraction) was used to measure diameters of an additional 126 nanotubes. Figure 4 shows the distribution of diameters that were obtained on the two instruments (with two different analysts carrying out the analysis). The blue-colored histogram corresponds to the peak-force AFM measured diameters, and the red histogram corresponds to the intermittent contact-mode AFM measurements. The two histograms are remarkably similar, with the average diameter value of 1.75 ± 0.23 nm from the peak force data and 1.67 ± 0.22 nm obtained using the intermittent contact mode AFM. With both methods, the measured diameter values ranged from 1.1 nm up to 2.2 nm.

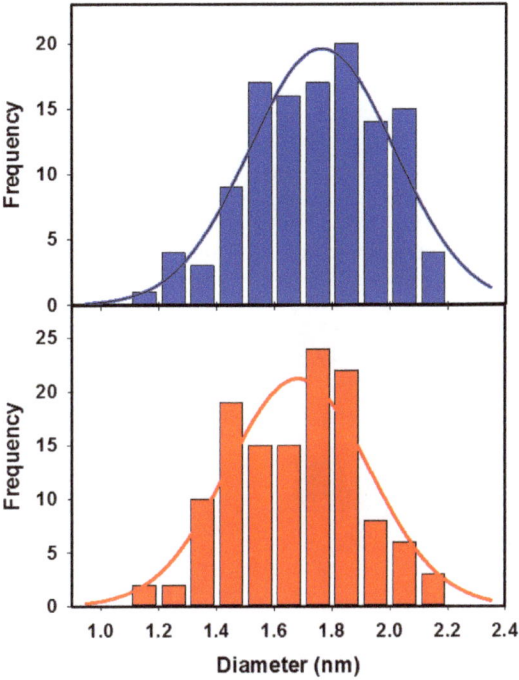

Figure 4. Histograms of the distribution of diameters measured on PFDD/SWCNTs using peak-force (Bruker) AFM (**blue**) on 120 nanotubes and intermittent contact-mode (JPK) AFM (**red**) on an additional 126 nanotubes. The red and blue solid lines represent Gaussian fits to the respective histogram data.

The nanotubes in the PFDD/SWCNT dispersion have been previously measured to have a narrow diameter distribution ranging from 1.2 to 1.4 nm [33]. Raman observations of the radial breathing mode for samples prepared in this work confirmed that the nanotube diameters were in this range (data shown in Figure S4). The larger average diameter obtained by the AFM measurements indicates that the polymer remained on the majority of the tubes even after all the preparation and rinsing procedures. While there was a range of diameters observed indicating that some (very few) nanotubes may be bare, while others carried more polymer, we can estimate the average added polymer thickness by subtracting 1.3 nm (the median of the diameter range for the unmodified nanotubes) from the average of the measured diameter values. Based on these measurements, the estimated average polymer thickness was in the range of 0.45–0.37 nm, suggesting a continuous coating of polymer along the nanotubes.

3.5. Measuring Compression at SWCNT Junctions

The protocol for accurate measurements of carbon nanotube heights can also be used to investigate junctions between tubes in these random networks. These junctions are of interest since electron transport across these junctions is the factor limiting the conductivity of random nanotube networks. Compression at SWCNT junctions has been modelled theoretically [34] and observed experimentally [35] in early studies on unmodified tubes.

To measure compression at a nanotube junction, the maximum height at the crossing point is compared with the sum of the two diameters of the tubes that cross. Figure 5 shows an AFM image depicting several junctions. Cross-sectional height profiles extracted from this image to measure the height of the junction (black) are also shown, along with the height of the two individual tubes (red and blue) making up the junction. The height of the junction is seen to be considerably smaller than the sum of the individual tubes. A histogram showing the comparison of measured and expected heights for 20 such junctions is shown in Figure 5c. All the junctions, apart from one case (junction #2 in the histogram), show a measured junction height that is smaller than expected, with the height reduction ranging from 8–34%. Averaging over all twenty junctions, the amount of compression is determined to be 19 ± 8%, similar to the previous observation [35]. Of the junctions analyzed, most are between single tubes based on the expected height being less than ~4 nm, although some with larger heights (such as junctions #3, #9, #10, and #17) likely involve small tube bundles. In the case of junction #2, which showed no significant compression, the tubes are seen to cross at a small angle. Additional data are required to determine the dependence of compression on the crossing angle.

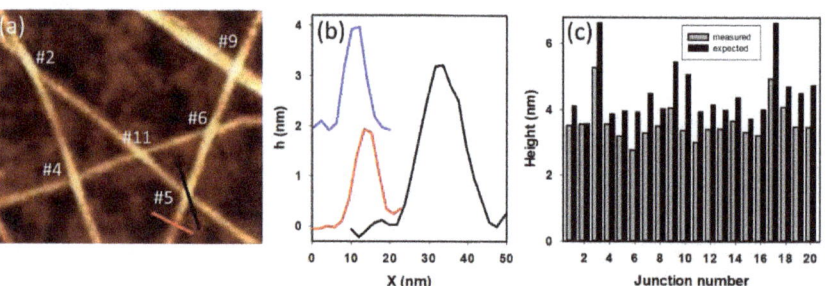

Figure 5. (**a**) AFM image of PFDD/SWCNTs showing several nanotube crossings. (**b**) Height profiles across the individual tubes (blue and red) and the crossing point (black) for junction #5. (**c**) Histogram showing comparison of measured junction heights and expected values based on the sum of the individual tube heights.

4. Conclusions

Two issues that can significantly affect the accuracy of AFM height measurements of single-wall carbon nanotubes have been identified. Measurements on several different SWCNT samples indicate that under forces commonly used for AFM imaging, nanotubes can be readily compressed, leading to a significant underestimation of the diameter. A simple procedure, consisting of imaging the same area several times at different forces and plotting height vs. force graphs, can verify the extent of the compression effect and recover the actual nanotube height via extrapolation to zero-force height. In addition, we showed that the common AFM height analysis method based on drawing a cross-section across each of the nanotubes to extract the diameter of nanotubes can lead to significant errors arising from the roughness of both the nanotubes and the substrate. An analysis protocol based on averaging a larger number of pixels was proposed and shown to be easily implemented using a free and open-source SPM analysis software. This enables decreasing the variation of measured heights to less than 0.1 nm. It is important to note that these results were obtained for moderately dense semiconducting SWCNT networks on SiO_2 that are functional as a channel material in transistors. While lower uncertainties may be achieved for isolated, pristine (no polymer or surfactant used for purification and dispersion) nanotubes deposited on ideally flat substrates such as mica or HOPG, the goal here was to develop a method for diameter measurements on "real-world" samples such as those used for device fabrication.

To illustrate the utility of this diameter measurement method, it was applied to measure the diameter of commercially available, highly enriched semiconducting nanotubes. The measured diameter for the nanotubes in a PFDD/SWCNT sample was shown to be larger than expected based on the diameters of the tubes present in the dispersion, suggesting the polymer remained on the SWCNTs upon deposition on the substrate and rinsing with a solvent. The average thickness of the polymer on the nanotubes was estimated to be ~0.4 nm. In addition, measured heights of tube–tube junctions were found to be less than the diameter of the two tubes that are crossing, indicating compression of ~20% at the junctions. Finally, we note that the measurement protocol developed here should also be generally applicable to measurement of vertical dimensions of other 1D or 2D nanomaterials.

Supplementary Materials: The following supporting information can be downloaded at: https://www.mdpi.com/article/10.3390/nano13030477/s1, Figure S1: HOPG atomic steps calibration and F vs height, Figure S2: Representative images of networks made of different nanotubes, Figure S3: Evaluating the repeatability of the analysis method. Figure S4: Raman spectra of the radial breathing mode of a PFDD/SWCNT sample.

Author Contributions: Conceptualization and methodology, D.V., S.Z. and G.P.L.; data acquisition, D.V. and M.C.; data analysis, D.V., M.C. and G.P.L.; writing—original draft preparation, D.V. and G.P.L.; writing—review and editing, all authors; supervision, S.Z and G.P.L. All authors have read and agreed to the published version of the manuscript.

Funding: This research received no external funding.

Data Availability Statement: Data available upon request.

Conflicts of Interest: The authors declare no conflict of interest.

References

1. Gomulya, W.; Costanzo, G.D.; De Carvalho, E.J.F.; Bisri, S.Z.; Derenskyi, V.; Fritsch, M.; Fröhlich, N.; Allard, S.; Gordiichuk, P.; Herrmann, A.; et al. Semiconducting Single-Walled Carbon Nanotubes on Demand by Polymer Wrapping. *Adv. Mater.* **2013**, *25*, 2948–2956. [CrossRef] [PubMed]
2. Kim, B.; Geier, M.; Hersam, M.C.; Dodabalapur, A. Inkjet Printed Circuits on Flexible and Rigid Substrates Based on Ambipolar Carbon Nanotubes with High Operational Stability. *ACS Appl. Mater. Interfaces* **2015**, *7*, 27654–27660. [CrossRef] [PubMed]
3. Brady, G.J.; Way, A.J.; Safron, N.S.; Evensen, H.T.; Gopalan, P.; Arnold, M.S. Quasi-Ballistic Carbon Nanotube Array Transistors with Current Density Exceeding Si and GaAs. *Sci. Adv.* **2016**, *2*, e1601240. [CrossRef]
4. Lefebvre, J.; Ding, J.; Li, Z.; Finnie, P.; Lopinski, G.; Malenfant, P.R.L. High-Purity Semiconducting Single-Walled Carbon Nanotubes: A Key Enabling Material in Emerging Electronics. *Acc. Chem. Res.* **2017**, *50*, 2479–2486. [CrossRef] [PubMed]
5. Zaumseil, J. Semiconducting Single-Walled Carbon Nanotubes or Very Rigid Conjugated Polymers: A Comparison. *Adv. Electron. Mater.* **2019**, *5*, 1800514. [CrossRef]
6. Bishop, M.D.; Hills, G.; Srimani, T.; Lau, C.; Murphy, D.; Fuller, S.; Humes, J.; Ratkovich, A.; Nelson, M.; Shulaker, M.M. Fabrication of Carbon Nanotube Field Effect Transistors in Commercial Silicon Manufacturing Facilities. *Nat. Electron.* **2020**, *3*, 492–501. [CrossRef]
7. Wang, J.; Lei, T. Enrichment of High-Purity Large-Diameter Semiconducting Single-Walled Carbon Nanotubes. *Nanoscale* **2022**, *14*, 1096–1106. [CrossRef]
8. Qin, C.; Peng, L.-M. Measurement Accuracy of the Diameter of a Carbon Nanotube from TEM Images. *Phys. Rev. B* **2002**, *65*, 155431. [CrossRef]
9. Mustonen, K.; Laiho, P.; Kaskela, A.; Zhu, Z.; Reynaud, O.; Houbenov, N.; Tian, Y.; Susi, T.; Jiang, H.; Nasibulin, A.G.; et al. Gas Phase Synthesis of Non-Bundled, Small Diameter Single-Walled Carbon Nanotubes with near-Armchair Chiralities. *Appl. Phys. Lett.* **2015**, *107*, 013106. [CrossRef]
10. Zhao, X.; Sun, S.; Yang, F.; Li, Y. Atomic-Scale Evidence of Catalyst Evolution for the Structure-Controlled Growth of Single-Walled Carbon Nanotubes. *Acc. Chem. Res.* **2022**, *55*, 3334–3344. [CrossRef] [PubMed]
11. Kuzmany, H.; Plank, W.; Hulman, M.; Kramberger, C.; Gruneis, A.; Pichler, T.; Peterlink, H.; Kataura, H.; Achiba, Y. Determination of SWCNT Diameters from the Raman Response of the Radial Breathing Mode. *Eur. Phys. J. B* **2001**, *22*, 307–320. [CrossRef]
12. Jorio, A.; Saito, R. Raman Spectroscopy for Carbon Nanotube Applications. *J. Appl. Phys.* **2021**, *129*, 021102. [CrossRef]
13. Tian, Y.; Jiang, H.; Laiho, P.; Kauppinen, E.I. Validity of Measuring Metallic and Semiconducting Single-Walled Carbon Nanotube Fractions by Quantitative Raman Spectroscopy. *Anal. Chem.* **2018**, *90*, 2517–2525. [CrossRef]
14. Castan, A.; Forel, S.; Fossard, F.; Defillet, J.; Ghedjatti, A.; Levshov, D.; Wenseleers, W.; Cambre, S.; Loiseau, A. Assessing the Reliability of the Raman Peak Counting Method for the Characterization of SWCNT Diameter Distributions: A Cross Characterization with TEM. *Carbon* **2021**, *171*, 968–979. [CrossRef]
15. Bellotti, R.; Bartolo, G.; Luigi, P. AFM Measurements and Tip Characterization of Nanoparticles with Different Shapes. *Nanomanuf. Metrol.* **2022**, *5*, 127–138. [CrossRef]
16. Lee, C.W.; Pillai, S.K.R.; Luan, X.; Wang, Y.; Li, C.M.; Chan-Park, M.B. High-Performance Inkjet Printed Carbon Nanotube Thin Film Transistors with High-k HfO_2 Dielectric on Plastic Substrate. *Small* **2012**, *8*, 2941–2947. [CrossRef]
17. Hennrich, F.; Li, W.; Fischer, R.; Lebedkin, S.; Krupke, R.; Kappes, M.M. Length-Sorted, Large-Diameter, Polyfluorene-Wrapped Semiconducting Single-Walled Carbon Nanotubes for High-Density, Short-Channel Transistors. *ACS Nano* **2016**, *10*, 1888–1895. [CrossRef]

18. Rother, M.; Schießl, S.P.; Zakharko, Y.; Gannott, F.; Zaumseil, J. Understanding Charge Transport in Mixed Networks of Semiconducting Carbon Nanotubes. *ACS Appl. Mater. Interfaces* **2016**, *8*, 5571–5579. [CrossRef]
19. Alizadegan, R.; Liao, A.D.; Xiong, F.; Pop, E.; Hsia, K.J. Effects of Tip – Nanotube Interactions on Atomic Force Microscopy Imaging of Carbon Nanotubes. *Nano Res.* **2012**, *5*, 235–247. [CrossRef]
20. Chen, J.; Xu, X.; Zhang, L.; Huang, S. Controlling the Diameter of Single-Walled Carbon Nanotubes by Improving the Dispersion of the Uniform Catalyst Nanoparticles on Substrate. *Nano-Micro Lett.* **2015**, *7*, 353–359. [CrossRef]
21. Timmermans, M.Y.; Estrada, D.; Nasibulin, A.G.; Wood, J.D.; Behnam, A. Effect of Carbon Nanotube Network Morphology on Thin Film Transistor Performance. *Nano Res.* **2012**, *5*, 307–319. [CrossRef]
22. Santos, S.; Barcons, V.; Christenson, H.K.; Font, J.; Thomson, N.H. The Intrinsic Resolution Limit in the Atomic Force Microscope: Implications for Heights of Nano-Scale Features. *PLoS ONE* **2011**, *6*, 23821. [CrossRef] [PubMed]
23. Cerreta, A.; Vobornik, D.; Di Santo, G.; Tobenas, S.; Alonso-Sarduy, L.; Adamcik, J.; Dietler, G. FM-AFM Constant Height Imaging and Force Curves: High Resolution Study of DNA-Tip Interactions. *J. Mol. Recognit.* **2012**, *25*, 486–493. [CrossRef]
24. Cerreta, A.; Vobornik, D.; Dietler, G. Fine DNA Structure Revealed by Constant Height Frequency Modulation AFM Imaging. *Eur. Polym. J.* **2013**, *49*, 1916–1922. [CrossRef]
25. Zorn, N.F.; Zaumseil, J. Charge Transport in Semiconducting Carbon Nanotube Networks Charge Transport in Semiconducting Carbon Nanotube Networks. *Appl. Phys. Rev.* **2021**, *8*, 041318. [CrossRef]
26. Nečas, D.; Klapetek, P. Gwyddion: An Open-Source Software for SPM Data Analysis. *Open Phys.* **2012**, *10*, 181–188. [CrossRef]
27. Vobornik, D.; Zou, S.; Lopinski, G.P. Analysis Method for Quantifying the Morphology of Nanotube Networks. *Langmuir* **2016**, *32*, 8735–8742. [CrossRef] [PubMed]
28. Yu, M.F.; Kowalewski, T.; Ruoff, R.S. Investigation of the Radial Deformability of Individual Carbon Nanotubes under Controlled Indentation Force. *Phys. Rev. Lett.* **2000**, *85*, 1456–1459. [CrossRef] [PubMed]
29. Minary-Jolandan, M.; Yu, M.F. Reversible Radial Deformation up to the Complete Flattening of Carbon Nanotubes in Nanoindentation. *J. Appl. Phys.* **2008**, *103*, 073516. [CrossRef]
30. Deborde, T.; Joiner, J.C.; Leyden, M.R.; Minot, E.D. Identifying Individual Single-Walled and Double-Walled Carbon Nanotubes by Atomic Force Microscopy. *Nano Lett.* **2008**, *8*, 3568–3571. [CrossRef]
31. Fakhri, N.; Tsyboulski, D.A.; Cognet, L.; Weisman, R.B.; Pasquali, M. Diameter-Dependent Bending Dynamics of Single-Walled Carbon Nanotubes in Liquids. *Proc. Natl. Acad. Sci. USA* **2009**, *106*, 14219–14223. [CrossRef] [PubMed]
32. Gao, J.; Loi, M.A.; de Carvalho, E.J.F.; dos Santos, M.C. Selective Wrapping and Supramolecular Structures of Polyfluorene - Carbon Nanotube Hybrids. *ACS Nano* **2011**, *5*, 3993–3999. [CrossRef]
33. Ding, J.; Li, Z.; Lefebvre, J.; Cheng, F.; Dubey, G.; Zou, S.; Finnie, P.; Hrdina, A.; Scoles, L.; Lopinski, G.P.; et al. Enrichment of Large-Diameter Semiconducting SWCNTs by Polyfluorene Extraction for High Network Density Thin Film Transistors. *Nanoscale* **2014**, *6*, 2328–2339. [CrossRef] [PubMed]
34. Yoon, Y.; Mazzoni, M.S.C.; Choi, H.J.; Ihm, J.; Louie, S.G. Structural Deformation and Intertube Conductance of Crossed Carbon Nanotube Junctions. *Phys. Rev. Lett.* **2001**, *86*, 688–691. [CrossRef]
35. Janssen, J.W.; Lemay, S.G.; Kouwenhoven, L.P.; Dekker, C. Scanning Tunneling Spectroscopy on Crossed Carbon Nanotubes. *Phys. Rev. B* **2002**, *65*, 115423. [CrossRef]

Disclaimer/Publisher's Note: The statements, opinions and data contained in all publications are solely those of the individual author(s) and contributor(s) and not of MDPI and/or the editor(s). MDPI and/or the editor(s) disclaim responsibility for any injury to people or property resulting from any ideas, methods, instructions or products referred to in the content.

Article

Controlling CNT-Based Nanorotors via Hydroxyl Groups

Boyang Zhang [1], Rui Li [1,*] and Qing Peng [2,3,4,*]

1. School of Mechanical Engineering, University of Science and Technology Beijing, Beijing 100083, China
2. Physics Department, King Fahd University of Petroleum and Minerals, Dhahran 31261, Saudi Arabia
3. K.A.CARE Energy Research and Innovation Center at Dhahran, Dhahran 31261, Saudi Arabia
4. Interdisciplinary Research Center for Hydrogen and Energy Storage, King Fahd University of Petroleum and Minerals, Dhahran 31261, Saudi Arabia
* Correspondence: lirui@ustb.edu.cn (R.L.); pengqing@imech.ac.cn (Q.P.)

Abstract: Nanomotor systems have attracted extensive attention due to their applications in nanorobots and nanodevices. The control of their response is crucial but presents a great challenge. In this work, the rotating and braking processes of a carbon nanotube (CNT)-based rotor system have been studied using molecular dynamics simulation. The speed of response can be tuned by controlling the ratio of hydroxyl groups on the edges. The ratio of hydroxyl groups is positively correlated with the speed of response. The mechanism involved is that the strong hydrogen bonds formed between interfaces increase the interface interaction. Incremental increase in the hydroxyl group concentration causes more hydrogen bonds and thus strengthens the interconnection, resulting in the enhancement of the speed of response. The phonon density of states analysis reveals that the vibration of hydroxyl groups plays the key role in energy dissipation. Our results suggest a novel routine to remotely control the nanomotors by modulating the chemical environment, including tuning the hydroxyl groups concentration and pH chemistry.

Keywords: transmission system; carbon nanotube (CNT); hydroxyl groups; response speed; energy dissipation

1. Introduction

With the development of nanotechnology, nano machines and nano devices have attracted extensive attention. Many nano machines such as molecular car motors [1], elevators [2] and shuttles [3] have been designed. Carbon nanotubes are one of the most important candidates in developing micro-electromechanical and nano-mechanical systems owing to their excellent mechanical characters, unique structures, high flexibility, and super-lubrication between multi-walled carbon nanotubes. Multi-walled carbon tubes have been applied to design nano tweezers [4,5], nano gears [6], gigahertz oscillators [7–9], nano bearings [10–13], nano motors [14–17], and nano bump [18].

Nano transmission systems that transfer motion and energies based on carbon nanotubes have also attracted extensive attention. There are two main kinds of transmission systems. One is to investigate the relative movement of inner and outer tubes in the axial direction. Barreiro et al. [19] studied the relative movement of the short outer tube relative to the long inner tube of multi-walled carbon nanotubes under axial thermal gradient. Santamaría-Holek et al. [20] proposed a model combining the actions of friction, van der Waals, and thermal forces and the effects of noise to explain the motion of a carbon nanotube along the other coaxial carbon nanotube. Another approach has been to design a nano-rotation transmission system using the interface interaction. Cai et al. [21,22] combined the carbon nanotube motor with a multi-walled carbon nanotube bearing to form a transmission system. Based on this design, Qiu et al. [23] developed a multi-level transmission system. Yin et al. [24,25], Gao et al. [26], Zhang et al. [7], Song et al. [27], Shi et al. [28] also studied transmission systems following a similar design. The above research all focused on using the interaction between hydrogen groups to achieve transmission. The hydrogen bond formed between hydroxyl groups could increase the interface interaction [29,30]. Our previous work [31] showed that hydroxyl groups could enhance the transmission efficiency owing to the strengthening of the interaction. We speculate that hydroxyl groups might also enhance the response of the transmission system. Therefore, in this work, we have investigated the acceleration, braking process and energy dissipation of the transmission system based on double-walled carbon nanotubes grafted with hydroxyl groups via molecular dynamics simulation. The effect of hydroxyl groups on the response of the system is evaluated.

2. Model and Method

The model of the transmission system is shown in Figure 1. The system includes two identical double-walled carbon nanotubes, which are the motor on the left and the rotor on the right, respectively. The double-walled carbon nanotubes (DWCNT) are applied, which include SWCNT (5, 5) and SWCNT (10, 10). Their diameters are 0.69 and 1.38 nm, respectively. Both ends of the outer tube SWCNT (10, 10) are fixed to avoid movement. Hydroxyl groups are grafted on the end of the inner tube between interfaces. The number of hydroxyl groups to the number of C atoms on the corresponding ends is defined as the hydroxyl group ratio. The length of inner and outer tube is 5.90 and 4.91 nm, respectively. Zhu et al. [12] pointed out that the energy dissipation between tubes in DWCNT was approximately proportional to the contact area. Changing the length of carbon nanotubes does not influence the energy dissipation rate. Therefore, the length of carbon nanotubes is kept the same during simulations.

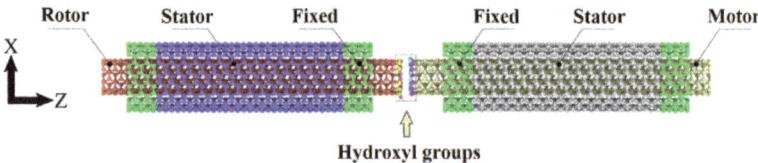

Figure 1. Illustration of the transmission system consisting of two DWCNTs.

The interaction among the C atoms of carbon nanotubes is described by AIREBO [32]. An OPLS_AA force field [33,34] is applied to describe C-O-H on the end of the inner carbon nanotube. Van der Waals force between interfaces is described by the 12-6 Lennard-Jones potential [35]. A DREIDING field [36] is applied to calculate the hydrogen bond between motor and rotor. The MD time step is 0.001 ps. This value is carefully selected as a compromise between numerical stability and computing resources. This value is also commonly adopted and successfully applied in similar transmission systems based on carbon nanotubes [23]. The Nose-Hoover method is applied to keep the temperature at 300 K.

The simulation process has three stages. At first, the whole system is relaxed for 200 ps. At the second stage, the four layers of atoms on the left end of the motor rotate at a constant frequency. The rotor on the right also begins to rotate because of the interaction between interfaces. At the last stage, the motion of the motor is removed to simulate the deceleration process. Consequently, the rotor gradually slows down to stop.

3. Results and Discussion

3.1. Transmission

Owing to the interaction between interfaces, the rotor begins to rotate when a constant speed is applied to the left-end of the motor. Figure 2a shows the rotation frequency of the rotor in five cases as pristine DWCNTs, the interface grafted with 40%, 60%, 80% and 100% hydroxyl groups. The results show that the rotor reaches a stable state with rotation frequency 200 GHz in about 30 ps in all cases, which is consistent with the rotation frequency of the motor. The maximum amplitude of vibration occurs in the case with original

DWCNTs, a phenomenon that can be attributed to the interaction between interfaces [31]. When pristine carbon nanotubes are applied, the interaction between interfaces is only the van der Waals force. However, hydrogen bonds form between interfaces in other cases, as shown in Figure 2b. The higher the hydroxyl group ratio is, the larger the number of hydrogen bonds that form.

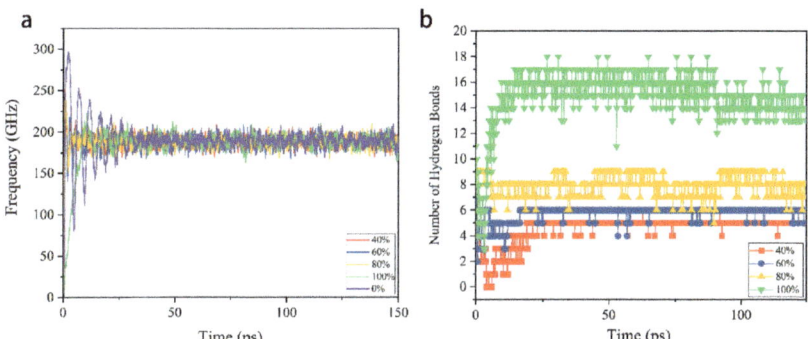

Figure 2. (a) The rotation frequency of rotor when pristine CNTs, CNTs with 40%, 60%, 80%, 100% grafted hydroxyl groups are applied. (b) The hydrogen bonds in five cases.

3.2. Braking Process

The braking process starts from 200 ps when the motion is removed from the motor. Owing to the interaction between interfaces, the rotor gradually decelerates. The rotation frequency of the rotor during the deceleration process is shown in Figure 3a, which includes the cases with pristine carbon nanotubes, with 40%, 60%, 80%, and 100% grafted hydroxyl groups on the interface. For the case with pristine DWCNTs, the rotor stops rotation at about 1000 ps. The rotor with higher hydroxyl groups stops earlier due to the stronger interaction between interfaces. More hydrogen bonds form when higher ratios of hydroxyl groups are grafted, as shown in Figure 3b. The stability of transmission systems during the braking process are also examined. The results show that the vibration of the centroid of the rotor in the x direction is below 0.04 nm, which implies that the rotor stabilizes in all cases.

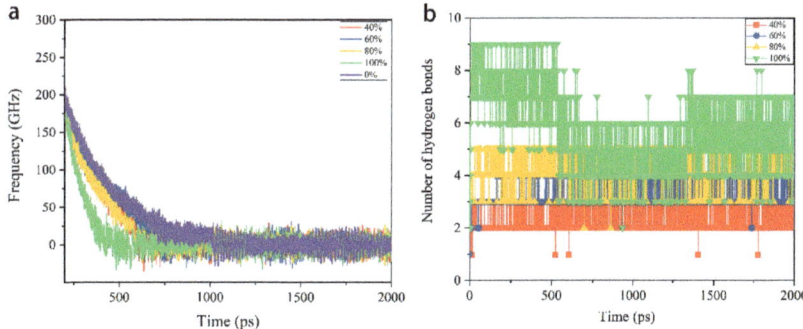

Figure 3. (a) The rotation frequency of rotor during braking process when pristine CNTs, CNTs with 40%, 60%, 80%, 100% grafted hydroxyl groups are applied. (b) The number of hydrogen bonds when 40%, 60%, 80%, 100% hydroxyl groups are grafted, respectively.

3.3. Energy Dissipation

To further explore the energy dissipation behavior of the transmission system during the whole process, the phonon density of state (DOS) of the rotor are calculated in three cases namely pristine carbon nanotubes, and those grafted with 40% and 100% hydroxyl groups, respectively, as shown in Figure 4. The subfigures (a–c) are in relaxation stage, 200–250 ps of rotation stage and 400~450 ps of deceleration stage, respectively. Konstantin [37] pointed out that that multi-walled carbon nanotubes generally have radial breathing mode (RBM), D band, and G band, where D and G bands represent the defects and in-plane stretching vibration of carbon nanotubes. The peaks of D band and G band in Raman spectra are at 150 cm^{-1} (inner tube), 300 cm^{-1} (outer tube), 1350 cm^{-1} and 1582 cm^{-1}, respectively. The peaks of hydroxyl groups in Raman spectra mainly include the vibration of C-O bond, O-H bond and out-of-plane bending vibration of O-H bonds, which are at 3200 cm^{-1}, 1200 cm^{-1} and 660 cm^{-1}, respectively. According to the equation k = f/c, where k, f and c are the wave number, the frequency, and the speed of light, respectively. Therefore, the frequencies of RBM, D band and G band of the multi-walled carbon nanotube are 4.50 (inner tube) and 9.00 (outer tube), 40.50 and 47.46 THz. The three peaks of the hydroxyl group are at 19.80, 36.00 and 86.00 THz, respectively.

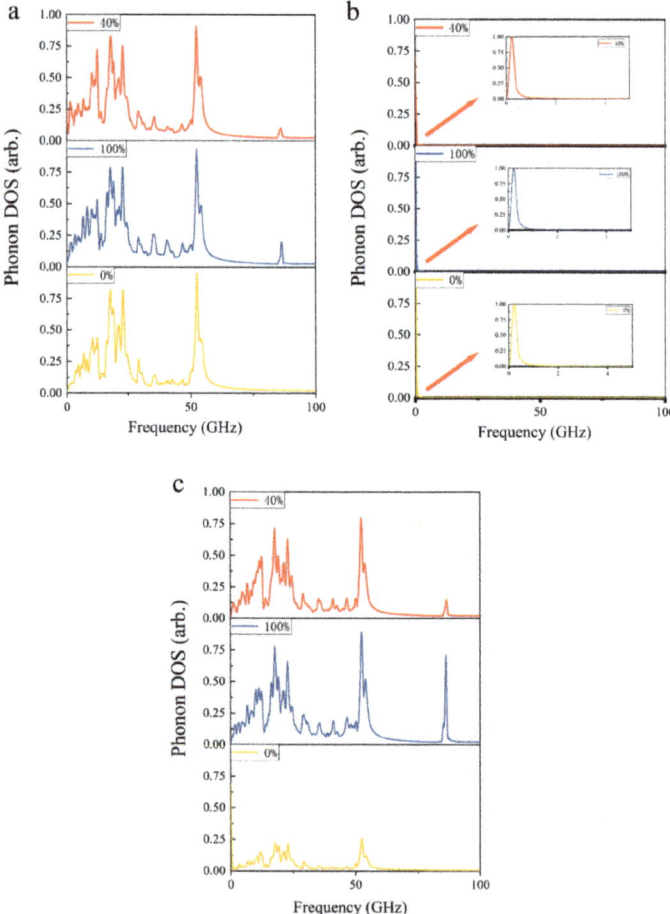

Figure 4. The phonon DOS of the transmission system in three stages when pristine carbon nanotubes, carbon nanotubes grafted with 40% and 100% hydroxyl groups are applied, respectively: (**a**) relaxation, (**b**) acceleration, and (**c**) deceleration.

In the relaxation stage, as shown in Figure 4a, the peaks of Phonon DOS of the system are similar in all three cases. Compared to pristine carbon nanotube, the most distinct difference in the cases with hydroxyl groups is the peak position at 86 THz. This clearly implies that the O-H bond vibrates in the out-of-plane direction. The other peaks represent RBMs, 2RBMs and G band with a slight frequency shift. Owing to the ideal carbon nanotubes applied in simulation, the D band is not obvious in the case with original carbon nanotubes, whereas it can be observed in the case grafted with 100% hydroxyl group. In the rotation stage, there is no peak in phonon DOS of the system because the carbon nanotube is in high constant speed rotation. During the deceleration stage at 400–450 ps, the phonon DOS is similar to the relaxation stage. The vibration of carbon nanotubes is weakened, although the out-of-plane bending vibration of O-H bonds is more obvious. Our results show that the vibration of O-H bonds accelerates the energy dissipation of the system and leads to a faster braking process.

4. Conclusions

The rotating and braking processes of a carbon nanotube transmission system have been investigated via molecular dynamics simulations. The effect of hydroxyl groups on the speed of response is examined. The energy dissipation during the whole process is discussed.

The results show that hydroxyl groups enhance the stability and reduce the response time of the system in both the acceleration and braking process. The higher the hydroxyl group ratio is used, the better the performance of the transmission system achieved. The underlying mechanism is the presence of hydrogen bonds that form between hydroxyl groups. These hydrogen bonds result in higher interface interaction and a faster response.

The analysis of the phonon density of state shows that the vibration of O-H bonds in hydroxyl groups accelerates energy dissipation of the system, which leads to faster response in the acceleration and braking process. Our results show that grafted hydroxyl groups result in stronger interaction, and therefore have potential in enhancing the response of the transmission system.

Author Contributions: B.Z. and R.L. conducted simulations, analyzed the results and wrote the original manuscript. Q.P. conducted phonon DOS calculation. R.L. and Q.P. edited the manuscript. All authors have read and agreed to the published version of the manuscript.

Funding: The work is supported by Fundamental Research funds for the Central Universities (FRF-IDRY-20-008). Q.P. We would like to acknowledge the support provided by the Deanship of Scientific Research (DSR) at King Fahd University of Petroleum and Minerals (KFUPM) for funding this work through project No. DF201020.

Institutional Review Board Statement: Not applicable.

Informed Consent Statement: Not applicable.

Data Availability Statement: Data available on request.

Conflicts of Interest: The authors declare no conflict of interest.

References

1. Shirai, Y.; Morin, J.-F.; Sasaki, T.; Guerrero, J.M.; Tour, J.M. Recent progress on nanovehicles. *Chem. Soc. Rev.* **2006**, *35*, 1043–1055. [CrossRef] [PubMed]
2. Badjić, J.D.; Balzani, V.; Credi, A.; Silvi, S.; Stoddart, J.F. A molecular elevator. *Science* **2004**, *303*, 1845. [CrossRef] [PubMed]
3. Brouwer, A.M.; Frochot, C.; Gatti, F.G.; Leigh, D.A.; Mottier, L.C.; Paolucci, F.; Roffia, S.; Wurpel, G.W.H. Photoinduction of fast, reversible translational motion in a hydrogen-bonded molecular shuttle. *Science* **2001**, *291*, 2124. [CrossRef] [PubMed]
4. Kim, P.; Lieber, C.M. Nanotube nanotweezers. *Science* **1999**, *286*, 2148–2150. [CrossRef]
5. Akita, S.; Nakayama, Y.; Mizooka, S.; Takano, Y.; Okawa, T.; Miyatake, Y.; Yamanaka, S.; Tsuji, M.; Nosaka, T. Nanotweezers consisting of carbon nanotubes operating in an atomic force microscope. *Appl. Phys. Lett.* **2001**, *79*, 1691–1693. [CrossRef]
6. Han, J.; Globus, A.; Jaffe, R.; Deardorff, G. Molecular dynamics simulations of carbon nanotube-based gears. *Nanotechnology* **1997**, *8*, 95. [CrossRef]
7. Zheng, Q.; Jiang, Q. Multiwalled carbon nanotubes as gigahertz oscillators. *Phys. Rev. Lett.* **2002**, *88*, 045503. [CrossRef]

8. Guo, W.; Guo, Y.; Gao, H.; Zheng, Q.; Zhong, W. Energy Dissipation in gigahertz oscillators from multiwalled carbon nanotubes. *Phys. Rev. Lett.* **2003**, *91*, 125501. [CrossRef]
9. Legoas, S.; Coluci, V.; Braga, S.; Coura, P.; Dantas, S.; Galvao, D.S.J. Molecular dynamics simulations of carbon nanotubes as gigahertz oscillators. *Phys. Rev. Lett.* **2003**, *90*, 055504. [CrossRef]
10. Cumings, J.; Zettl, A. Low-friction nanoscale linear bearing realized from multiwall carbon nanotubes. *Science* **2000**, *289*, 602–604. [CrossRef]
11. Bourlon, B.; Glattli, D.C.; Miko, C.; Forró, L.; Bachtold, A. Carbon nanotube based bearing for rotational motions. *Nano Lett.* **2003**, *4*, 709–712. [CrossRef]
12. Zhu, C.; Guo, W.; Yu, T. Energy dissipation of high-speed nanobearings from double-walled carbon nanotubes. *Nanotechnology* **2008**, *19*, 465703. [CrossRef]
13. Cook, E.H.; Buehler, M.J.; Spakovszky, Z.S.; Solids, P.O. Mechanism of friction in rotating carbon nanotube bearings. *J. Mech. Phys. Solids* **2013**, *61*, 652–673. [CrossRef]
14. Cai, K.; Yu, J.; Shi, J.; Qin, Q.-H. Robust rotation of rotor in a thermally driven nanomotor. *Sci. Rep.* **2017**, *7*, 46159. [CrossRef] [PubMed]
15. Cai, K.; Wan, J.; Qin, Q.H.; Shi, J. Quantitative control of a rotary carbon nanotube motor under temperature stimulus. *Nanotechnology* **2016**, *27*, 055706. [CrossRef]
16. Fennimore, A.; Yuzvinsky, T.; Han, W.-Q.; Fuhrer, M.; Cumings, J.; Zettl, A. Rotational actuators based on carbon nanotubes. *Nature* **2003**, *424*, 408–410. [CrossRef]
17. Bailey, S.; Amanatidis, I.; Lambert, C. Carbon nanotube electron windmills: A novel design for nanomotors. *Phys. Rev. Lett.* **2008**, *100*, 256802. [CrossRef]
18. Joseph, S.; Aluru, N.R. Pumping of confined water in carbon nanotubes by rotation-translation coupling. *Phys. Rev. Lett.* **2008**, *101*, 064502. [CrossRef]
19. Barreiro, A.; Rurali, R.; Hernández, E.R.; Moser, J.; Pichler, T.; Forró, L.; Bachtold, A. Subnanometer motion of cargoes driven by thermal gradients along carbon nanotubes. *Science* **2008**, *320*, 775. [CrossRef]
20. Santamaría-Holek, I.; Reguera, D.; Rubi, J.M. Carbon-nanotube-based motor driven by a thermal gradient. *J. Phys. Chem. C* **2013**, *117*, 3109–3113. [CrossRef]
21. Cai, K.; Yin, H.; Wei, N.; Chen, Z.; Shi, J. A stable high-speed rotational transmission system based on nanotubes. *Appl. Phys. Lett.* **2015**, *106*, 021909. [CrossRef]
22. Zhang, X.-N.; Cai, K.; Shi, J.; Qin, Q.-H. Friction effect of stator in a multi-walled CNT-based rotation transmission system. *Nanotechnology* **2017**, *29*, 045706. [CrossRef] [PubMed]
23. Wei, Q.; Jiao, S.; Zheng, C.; Jicheng, Z.; Ning, W. A two-class rotation transmission nanobearing driven by gigahertz rotary nanomotor. *Comput. Mater. Sci.* **2018**, *154*, 97–105.
24. Yin, H.; Cai, K.; Wei, N.; Qin, Q.H.; Shi, J. Study on the dynamics responses of a transmission system made from carbon nanotubes. *J. Appl. Phys.* **2015**, *117*, 234305. [CrossRef]
25. Yin, H.; Cai, K.; Wan, J.; Gao, Z.L.; Chen, Z. Dynamic response of a carbon nanotube-based rotary nano device with different carbon-hydrogen bonding layout. *Appl. Surf. Sci.* **2016**, *365*, 352–356. [CrossRef]
26. Gao, Z.L.; Cai, H.F.; Shi, J.; Liu, L.N.; Chen, Z.; Wang, Y. Effect of hydrogenation and curvature of rotor on the rotation transmission of a curved nanobearing. *Comput. Mater. Sci.* **2017**, *127*, 295–300. [CrossRef]
27. Song, B.; Cai, K.; Shi, J.; Xie, Y.M.; Qin, Q.H. Coupling effect of van der Waals, centrifugal, and frictional forces on a GHz rotation-translation nano-convertor. *Phys. Chem. Chem. Phys.* **2019**, *21*, 359–368. [CrossRef]
28. Shi, J.; Cao, Z.; Wang, J.B.; Shen, J.H.; Cai, K. Stable rotation transmission of a CNT-based nanogear drive system with intersecting axes at low temperature. *Surf. Sci.* **2020**, *693*, 121548. [CrossRef]
29. Li, R.; Wang, S.; Peng, Q. Tuning the slide-roll motion mode of carbon nanotubes via hydroxyl groups. *Nanoscale Res. Lett.* **2018**, *13*, 138. [CrossRef]
30. Chen, Y.; Wang, S.; Xie, L.; Zhu, P.; Li, R.; Peng, Q. Grain size and hydroxyl-coverage dependent tribology of polycrystalline graphene. *Nanotechnology* **2019**, *30*, 385701. [CrossRef]
31. Li, R.; Liu, J.; Zheng, X.; Peng, Q. Achieve 100% transmission via grafting hydroxyl groups on CNT nanomotors. *Curr. Appl. Phys.* **2021**, *29*, 59–65. [CrossRef]
32. Stuart, S.J.; Tutein, A.B.; Harrison, J.A. A reactive potential for hydrocarbons with intermolecular interactions. *J. Chem. Phys.* **2000**, *112*, 6472–6486. [CrossRef]
33. Hughes, Z.E.; Shearer, C.J.; Shapter, J.; Gale, J.D. Simulation of water transport through functionalized single-walled carbon nanotubes (SWCNTs). *J. Phys. Chem. C* **2012**, *116*, 24943. [CrossRef]
34. Damm, W.; Frontera, A.; Tirado–Rives, J.; Jorgensen, W.L. OPLS all-atom force field for carbohydrates. *J. Comput. Chem.* **2015**, *18*, 1955–1970. [CrossRef]
35. Ruoff, R.; Hickman, A. Van der Waals binding to fullerenes to a graphite plane. *J. Phys. Chem.* **1993**, *97*, 2494–2496. [CrossRef]
36. Mayo, S.L.; Olafson, B.D.; Goddard, W.A. DREIDING: A generic force field for molecular simulations. *J. Phys. Chem.* **1990**, *94*, 8897–8909. [CrossRef]
37. Iakoubovskii, K.; Minami, N.; Ueno, T.; Kazaoui, S.; Kataura, H. Optical characterization of double-wall carbon nanotubes: Evidence for inner tube shielding. *J. Phys. Chem. C* **2008**, *112*, 11194–11198. [CrossRef]

Review

Controllable Preparation and Strengthening Strategies towards High-Strength Carbon Nanotube Fibers

Yukang Zhu [1], Hongjie Yue [1], Muhammad Junaid Aslam [1], Yunxiang Bai [2,*], Zhenxing Zhu [1,*] and Fei Wei [1,*]

1. Beijing Key Laboratory of Green Chemical Reaction Engineering and Technology, Department of Chemical Engineering, Tsinghua University, Beijing 100084, China
2. CAS Key Laboratory of Nanosystem and Hierarchical Fabrication, CAS Center for Excellence in Nanoscience, National Center for Nanoscience and Technology, Beijing 100190, China
* Correspondence: baiyunxiang0101@163.com (Y.B.); zxing@mail.tsinghua.edu.cn (Z.Z.); wf-dce@mail.tsinghua.edu.cn (F.W.)

Abstract: Carbon nanotubes (CNTs) with superior mechanical properties are expected to play a role in the next generation of critical engineering mechanical materials. Crucial advances have been made in CNTs, as it has been reported that the tensile strength of defect-free CNTs and carbon nanotube bundles can approach the theoretical limit. However, the tensile strength of macro carbon nanotube fibers (CNTFs) is far lower than the theoretical level. Although some reviews have summarized the development of such fiber materials, few of them have focused on the controllable preparation and performance optimization of high-strength CNTFs at different scales. Therefore, in this review, we will analyze the characteristics and latest challenges of multiscale CNTFs in preparation and strength optimization. First, the structure and preparation of CNTs are introduced. Then, the preparation methods and tensile strength characteristics of CNTFs at different scales are discussed. Based on the analysis of tensile fracture, we summarize some typical strategies for optimizing tensile performance around defect and tube–tube interaction control. Finally, we introduce some emerging applications for CNTFs in mechanics. This review aims to provide insights and prospects for the controllable preparation of CNTFs with ultra-high tensile strength for emerging cutting-edge applications.

Keywords: carbon nanotubes; carbon nanotube fibers; tensile strength; defect control; controlled preparation

1. Introduction

Materials are the basis of the evolution of human civilization. The pursuit of the ultimate properties of materials, such as super strength and super toughness, has strongly promoted the development of human culture. In 1895, Konstantin Tsiokovsky, a Soviet scientist, put forward building a "sky castle" at the top of a giant tower, which later evolved into the concept of "space elevator". By connecting the earth and the space station with a cable, people can achieve space sightseeing and transport items to the space station [1]. However, the biggest challenge of this concept is finding light and strong cable that can overcome its gravity. A variety of nanostructures can be composed of single carbon elements, such as fullerenes (0D), carbon nanotubes (1D), and graphene (2D). Carbon nanotubes (CNTs) are cylinders rolled from single or multi-layer graphene sheets. Single-walled carbon nanotubes (SWCNTs) are cylinders rolled from a single-layer graphene sheet, while double-walled carbon nanotubes (DWCNTs) and multi-walled carbon nanotubes (MWCNTs) are composed of two and multiple layers of rolled graphene sheets, respectively. As one of the strongest chemical bonds in nature [2,3], the in-plane σ covalent bond of graphene formed by sp^2 hybridization endows CNTs with extremely high axial Young's modulus (~1.1 TPa) and tensile strength (~120 GPa). Theoretical calculations have shown that CNTs are the most probable material to help mechanical materials achieve a breakthrough and even realize the "space elevator" dream [1]. However, CNTs with

extremely excellent mechanical properties are nanoscale solids, and practical applications require macro-scale materials. It is the prerequisite for CNTs to play a significant role in practical applications that they can maintain excellent mechanical properties after being assembled from a single nanoscale unit to a macroscopic aggregate.

In recent years, rapid progress has been made in the preparation and mechanical properties optimization of carbon nanotube fibers (CNTFs). Different spinning methods of CNTFs have been put forward and improved upon, and CNTFs with a tensile strength comparable to carbon fibers (CFs) have been prepared [4–7]. However, their mechanical properties are still far lower than single CNTs [7], which also shows an unsatisfactory phenomenon of property transfer across scales. Theoretical calculations and experimental results show that the tensile strength of CNTs with a nanoscale diameter can exceed 100 GPa [8,9]. CNT bundles with a diameter of 10–100 nm can have a tensile strength of up to 80 GPa [10]. CNTFs with a diameter of more than 1 μm, as a representative of macro assemblies of CNTs, have a maximum tensile strength of only 9.6 GPa [4], which is far lower than the intrinsic mechanical strength of CNTs. The reasons for such cross-scale tensile strength transfer are mainly due to defect accumulation and the lack of ideal tube–tube interactions during CNT assembly. Defects can have a fatal effect on the strength of CNTs [11,12]. With the increase in fiber size, defects also accumulate across scales. As shown in Figure 1b, the improvement of strength for CFs and CNTFs is closely related to the reduction in defect size [13,14]. For CFs, the tensile strength was increased from about 1 GPa to 10 GPa when the defect size was reduced from the micron to the nano scale. For CNTs, due to the fewer defects in structure compared with CFs, less attention was paid to their precise structural control, especially defects, resulting in their tensile strength having long been at a lower level. Until the 2000s, a series of achievements were made in the prepration of defect-free CNTs, which have shown extraodinary tensile strength performance both at the single-tube and bundle levels [9,15]. Therefore, the preparation of ideal solids such as defect-free or defectless CNTs is the basis for preparing CNTFs with high tensile strength. At the same time, many studies have shown that the mechanical properties of CNTFs can be significantly influenced by the tube–tube interactions involving orientation, length, and density. It is of great significance to regulate the tube–tube interactions and precisely control the atomic defects for the improvement of CNTs' mechanical tensile strength [4,16–18].

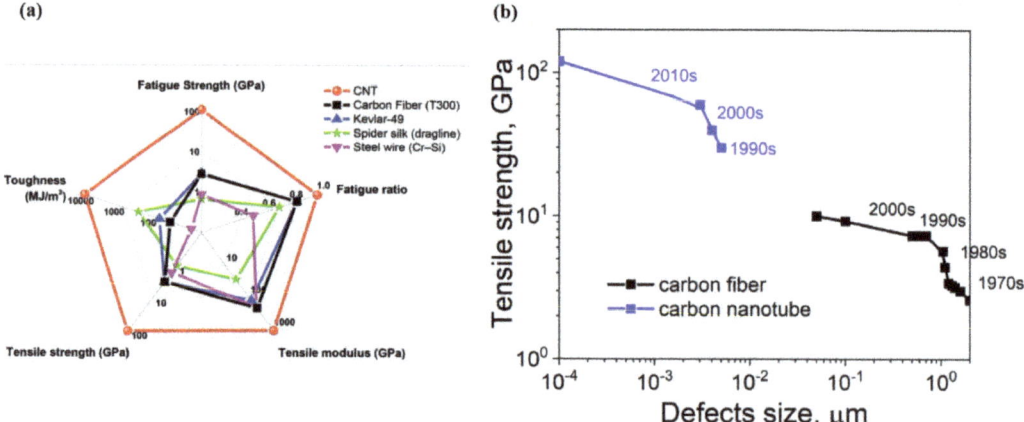

Figure 1. (a) Comparison of mechanical performance between CNTs and some high−performance materials. Reproduced with permission from [15]. Copyright 2020, American Association for the Advancement of Science. (b) Evolution of the tensile strength of CNTs and CFs at different defect sizes. The data are collected in [4,7,10,15].

In the past few decades, CNTFs have attracted extensive attention from academia and industry, and there are many reviews on the development and application of CNTFs [19–23]. However, few reviews have focused on the controllable preparation and strength optimization of CNTFs at different scales, which will be specifically highlighted in this review. First, the intrinsic mechanism of the excellent mechanical properties of CNTs and the preparation of CNTs are introduced. Then, the development and characteristics of techniques for fabricating CNTFs at different scales are discussed. Furthermore, we introduce the tensile strength of CNTFs from the nanoscale to the microscale, showing the recent development in the transfer of tensile properties across scales. Then, we analyze strategies for the strength optimization of CNTFs at different scales, particularly the aspects of defect and tube–tube interaction control. Finally, we provide an outlook for the practical applications of CNTFs with high mechanical strength. As a result, we aim to provide insights and prospects for the controllable preparation and performance optimization of macro CNTFs with high tensile strength in the future.

2. The Structure and Preparation of Carbon Nanotubes

2.1. The Structure of Carbon Nanotubes

CNTs can be seen as a graphene sheet curled into a cylinder with a nanoscale diameter [24,25]. Figure 2a shows the structure of single-walled carbon nanotubes (SWCNTs). Carbon atoms in CNTs are linked by sp^2 hybrid covalent bonds, which is one of the strongest chemical bonds in nature [2,3], providing graphite materials with extremely high in-plane Young's modulus and tensile strength. The special tubular full-atomic-surface (FAS) structure composed of carbon hexagons avoids in-plane hanging bonds, folds, and concentrated local stresses in the tube wall. As a result, CNTs can exhibit excellent mechanical properties far beyond other materials (tensile strength ~120 GPa, Young's modulus ~1.1 TPa, elongation at break ~16%, toughness ~8 GJ/m^3) [8,26–29]. Similar to other engineering mechanics materials, the existence of defects will destroy the structural perfection of CNTs and affect the mechanical properties. The negative effect is even more pronounced for CNTs. For instance, a single vacancy defect could reduce the tensile strength of CNTs by 26% [30], and a single topological defect could lower that by 50% [12]. Despite that, the formation of topological defects is often accompanied by a high energy barrier, which can effectively protect the sp^2 structure of CNTs. Ding et al. found that the formation energy of the five-membered and seven-membered ring pairs of topological defects was as high as 4.4 eV [31], which could effectively protect the sp^2 structure of CNTs. Such topological protection is an important reason why CNTs are less prone to defects than other materials, such as steel or concrete.

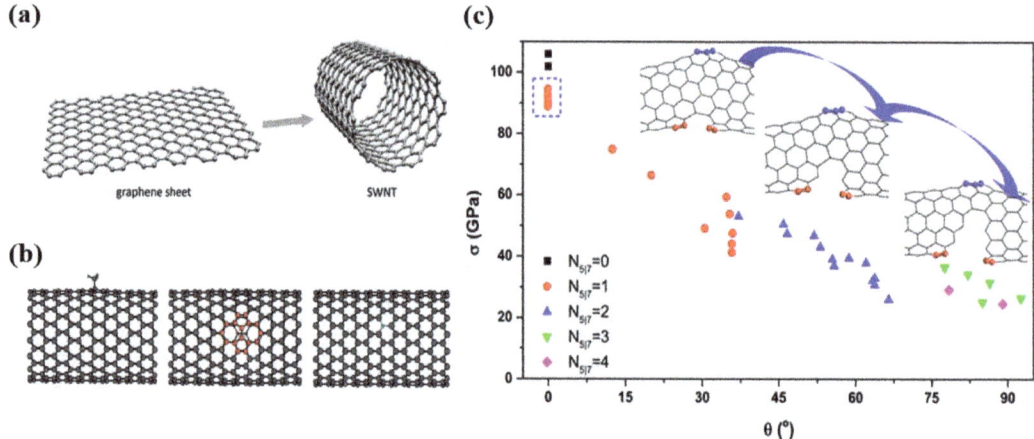

Figure 2. (a) Wrapping of graphene sheet to form an SWCNT. Reproduced with permission from [25]. Copyright 2011, Royal Soc Chemistry. (b) Defects on the SWCNT, from left to right: functionalization defect, SW defect, and vacancy, respectively. Reproduced with permission from [14]. Copyright 2007, Iop Publishing Ltd. (c) The tensile strength of carbon nanotube varies with different numbers of 5 | 7 defects and different chiral angles. Reproduced with permission from [12]. Copyright 2016, Amer Chemical Soc.

2.2. Controllable Preparation of Carbon Nanotubes

Arc discharge [24,32], laser evaporation [33], and chemical vapor deposition (CVD) [34] are the three main methods for preparing CNTs. Compared with the former two methods, the CVD method has the advantages of low temperature, low energy input, and easy control of parameters and is the primary method in academic research and industrial production. The growth of CNTs by the CVD method can be divided into three stages. (i) The catalyst is in a molten state at a high temperature. (ii) The cracked carbon atoms dissolve on the catalyst, precipitate after supersaturation, and (iii) self-assemble to form CNTs [35]. There are many alternative carbon sources for the preparation of CNTs by the CVD method, such as methane, carbon monoxide, ethylene, acetylene, and ethanol. The type of carbon source has a great influence on the structure and quality of the as-grown CNTs. For example, considering the thermal cracking conditions, we thought that methane is the most appropriate carbon source for producing ultralong CNTs with perfect structure [9,36]. The catalyst is another key factor in regulating the structure and quality of CNTs. The catalysts used for producing CNTs are mainly metal catalysts, including magnetic metal catalysts (iron, cobalt, nickel), noble metal catalysts (copper, gold, silver, platinum), as well as molybdenum and tungsten, etc. Iron-based catalysts are the most commonly used and there are many reports on the preparation of CNTs with ferric chloride or ferrocene as catalysts. Ding et al. [31] also reported the role of iron nanoparticles in defect repair, explaining the high efficiency of iron-based catalysts towards CNT growth from the mechanism level. The macroscopic assembly of CNTs requires a large number of CNTs. Therefore, the large-scale production of CNTs, which can be achieved by the CVD method with relatively low cost and good controllability, is the basis of their subsequent assembly into a fiber structure. Early in 1993, Santiesteban et al. [37] first reported the fabrication of CNTs by the CVD method. The fabrication process of high-purity CNTs in large quantities based on CVD developed rapidly in the first decade [38,39]. Our group combined the traditional chemical fluidized-bed technology with the CVD method to realize the large-scale production of CNTs [40–43], which dramatically reduced the cost of CNTs (Figure 3). In addition, ultralong CNTs with macro-scale length can be synthesized by carefully regulating the growth kinetics. These ultralong CNTs possess perfect structure without any defects and can be produced at a wafer scale [9,44]. At the same time, these defect-free CNTs provide an ideal system for analyzing mechanical materials and are expected to yield new results in some branches of solid mechanics. Considering the research paradigm of the bottom-up assembly of CNTs into macrostructures as practical engineering materials, individual CNTs are the most basic structural unit. Obviously, if a large number of CNTs with defect-free or defectless structures can be synthesized and assembled into CNTFs, the excellent intrinsic properties of CNTs can be fully exploited. As a result, developing next-generation high-performance engineering materials can be promoted significantly.

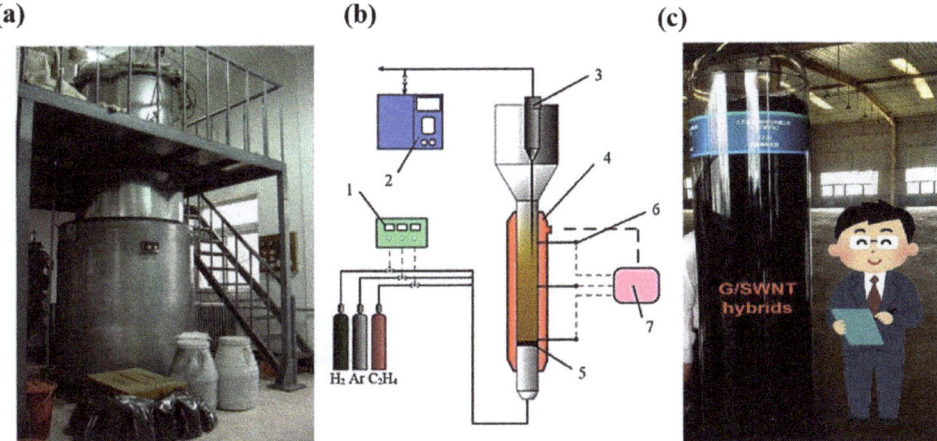

Figure 3. (a,b) Schematic illustration of the large-scale fabrication of CNTs by a fluidized-bed reactor. Reproduced with permission from [41]. Copyright 2008, Elsevier. (c) The photo of collected CNTs and graphene hybrids in mass production. Reproduced with permission from [45]. Copyright 2021, Wiley-Vch Verlag Gmbh.

3. Assembly Technology of Carbon Nanotube Fibers at Different Scales

The CNTFs consist of basic, tightly assembled CNT units. According to the scale of the assembled structure, we divided CNTFs into two categories. One is nanoscale carbon nanotube fibers (nanoscale CNTFs), composed of a small number of CNTs with a diameter up to hundreds of nanometers. The other is microscale carbon nanotube fibers (microscale CNTFs). The number of CNTs assembled can reach hundreds of millions with a micron-scale diameter, far more than the nanoscale CNTFs. It is a prerequisite to develop assembly technologies to combine single CNTs into nanoscale CNTFs or microscale CNTFs and maintain relatively excellent mechanical properties.

3.1. Preparation of Nanoscale Carbon Nanotube Fibers

Carbon nanotube bundles (CNTBs) are typical nanoscale CNTFs. Assembling a small number of CNTs into bundles requires precise nanoscale manipulation. In the early stages, researchers found that bundles with diameters of tens of nanometers can be obtained by controlling conditions using the arc-discharge method. Salvetat et al. [46] prepared SWCNT crystalline ropes by the arc-discharge method. The mean diameter of their nanotubes is about 1.4 nm, and the diameter of the bundles ranges from 3 to 20 nm with a length of several microns. The tubes were arranged in a closely-packed lattice. Espinosa et al. [47] also fabricated bundles using the arc-discharge method. The preparation of nanoscale CNTFs is a process of assembling a small number of CNTs in situ, which can be considered a bottom-up method. In contrast, to obtain CNTBs, it is also effective to assemble CNTs into a macrostructure and cut them into nanoscale CNTFs. Such segmentation from a microscale structure into the nanoscale can be considered an up-down method. Typically,

Yu et al. [48] first fabricated SWCNT "paper" by the laser ablation method, and then they tore the resulting "paper" apart, which caused individual SWCNT bundles to project from the torn edge. The diameter of these nanoscale CNTFs is less than 50 nm. Based on the up-down method, Espinosa et al. [49,50] also obtained CNT bundles with diameters less than 30 nm from DWCNT mats using a mechanical exfoliation technique. From the analysis of the characteristics of these two kinds of processes, the in situ preparation of nanoscale CNTFs without post-treatments has more advantages, as the cutting strategy requires nano-precision mechanical manipulation. More structural defects or impurities will be introduced after mechanical separation, which is not conducive to the subsequent performance research. However, on the other hand, the bottom-up preparation of nanoscale CNTFs by arc discharge has its own limitations, such as high equipment requirements, low controllability, and difficult separation and purification. Therefore, it is necessary to develop a more suitable nanoscale CNTF preparation process. Our group proposed an in situ gas-flow-focusing (GFF) strategy to assemble individual tubes into bundles [10] based on the bottom-up method. As shown in Figure 4a, under the effect of the gas flow, several ultra-long CNTs gradually move towards the center and are assembled by the impact of van der Waals force. The prerequisite for preparing these centimeter-long bundles with a nanoscale diameter is that ultralong CNTs are grown based on a "kite mechanism". Compared with the nanoscale CNTFs directly prepared by the arc-discharge or mechanical stripping method, the core of such an in situ process is the preparation of ultralong CNTs by CVD, so it has the advantages of high controllability, relatively low equipment requirements, and generally perfect fiber structure.

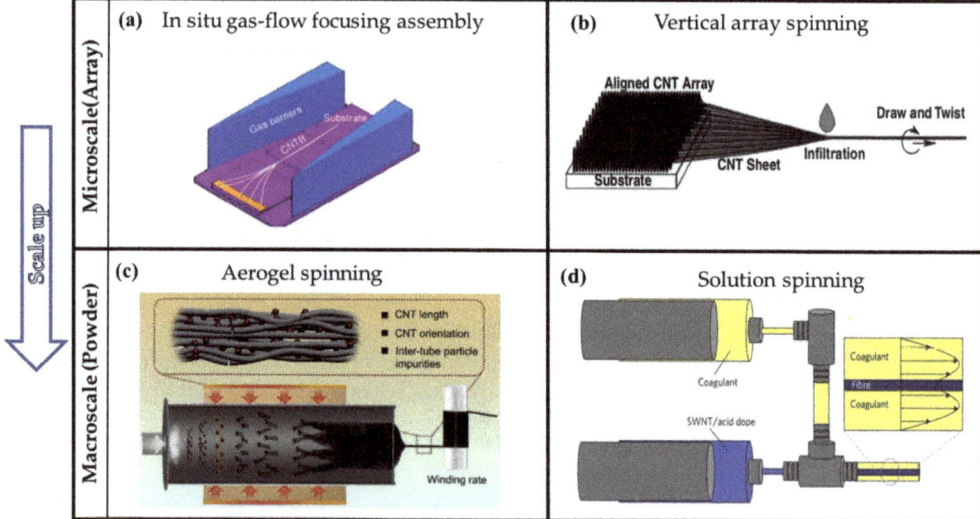

Figure 4. (**a**) Schematic illustration of the in situ fabrication of nanoscale CNTFs by the GFF method. Reproduced with permission from [10]. Copyright 2018, Springer Nature. (**b**) Schematic diagram of vertical-array spinning. Reproduced with permission from [51]. Copyright 2011, Pergamon-Elsevier Science Ltd. (**c**) Schematic diagram of the continuous synthesis of microscale CNTFs by FCCVD. Reproduced with permission from [18]. Copyright 2021, Elsevier Sci Ltd. (**d**) Schematic illustration of solution spinning. Reproduced with permission from [52]. Copyright 2009, Springer Nature.

3.2. Preparation of Microscale Carbon Nanotube Fibers

The spinning techniques for microscale CNTFs are closely related to the traditional process, which can be divided into two main categories, wet spinning [16] and dry spinning [19,22,53]. Solution spinning [54] is the typical wet-spinning method, while vertical-array spinning [55] and CNT aerogel spinning [56] are the two mainstream methods of dry spinning. Understanding the characteristics of different preparation techniques is the basis for better mechanical property optimization.

3.2.1. Solution Spinning

Solution spinning (wet spinning) is a relatively traditional and mature technique that has been widely used to prepare Kevlar and PAN fibers [57,58]. In the past few years, it has also been used to prepare microscale CNTFs or CNTF composites [59,60]. As Figure 4d shows, in the process of solution spinning, the CNT powder is fully dispersed in the solution with the help of a surfactant or superacid as the dispersant. The dispersed solution is injected into the coagulation bath, the solvent is dissolved in the coagulation solution, and the aggregate is precipitated to obtain continuous microscale CNTFs. The most important step for solution spinning is obtaining a highly dispersed and homogeneous CNT dispersion. Until now, a variety of effective solution systems have been developed. In 2000, Vigolo et al. [16] first produced continuous CNTFs with a diameter ranging from a few micrometers to 100 μm through this traditional spinning method. They used sodium dodecyl sulfate (SDS) as a dispersive solvent to obtain homogeneous suspensions at a high SDS concentration. The suspensions were then injected into polyvinyl alcohol using a syringe to produce CNTFs. Although the process was less efficient then, it was the beginning of the preparation of microscale CNTFs. Windle et al. [61] injected ether into the dispersion of CNTs and ethylene glycol, causing the ethylene glycol, ether, and the grown fibers to penetrate each other. Then, the ethylene glycol was heated to remove it, so that the neatly arranged microscale CNTFs were obtained with diameters ranging from 10 to 80 μm. However, such a surfactant–solvent dispersion system makes it difficult to obtain high-concentration and homogeneous CNT dispersions because of the strong van der Waals interaction between tubes. Hence, there is a need to develop more efficient decentralized systems [54]. The DuPont Company has reported the preparation technique of Kevlar fibers by dissolving polymers with concentrated sulfuric acid [57,62]. It has gradually become one of the most efficient systems in the preparation of CNTFs by solution spinning. Smalley and Pasquali et al. have developed superacid dispersion systems in the past two decades [52,54,63–67]. The main mechanism of using superacids to disperse CNTs is to protonate the tube wall so as to achieve efficient dispersion through electrostatic interaction. Smalley et al. dispersed CNTs in fuming sulfuric acid to obtain a CNT dispersion [54]. Pasquali et al. found that CNTs dissolved in chlorosulfonic acid could form a true thermodynamic solution in a liquid crystal phase [52,63]. The liquid crystal phase dispersion used in wet spinning can improve the spinning efficiency and obtain CNTFs with higher orientation. Their mechanical and electrical properties have been greatly improved [64,65]. However, using superacids is not friendly to the environment or equipment. Considering this problem, Pasquali et al. recently proposed a more moderate acid dispersion system. They have proposed a low-corrosive acid solvent system by using methanesulfonic acid or p-toluenesulfonic acid to form a CNT dispersion at concentrations as high as 10 g/L [68]. This system can obtain continuous and high-performance fibers, and it has higher equipment compatibility, which is conducive to mass preparation. In conclusion, the preparation of CNTFs by wet spinning has many mature technical characteristics of other fibers prepared by wet spinning, so it has more advantages in technical reliability, equipment compatibility, and further large-scale production.

3.2.2. Vertical-Array Spinning

As shown in Figure 4b, vertical-array spinning means continuously extracting CNTs connected by van der Waals action from spinnable vertical array CNTs (similar to cocoon spinning) to prepare macroscopic fibers. Vertical-array spinning has strict requirements on the structure of CNT arrays. In order to draw out CNTFs continuously, the vertical arrays of CNTs should be high in height and density while the orientation should be very uniform [69]. In 2002, Fan et al. [55] extracted CNT yarns from a 100 μm-high super-aligned CNT array with tweezers. Superaligned CNT arrays are the ideal raw materials for vertical-array spinning. The prepared continuous microscale CNTFs were 30 cm in length and 200 µm wide. Baughman et al. [70,71] first prepared CNT yarns with stable torque by twisting the yarns drawn out from the vertical arrays. The drawn yarns are very sticky due to their clean surface and extremely high specific surface area. They stick to the surfaces once they touch other objects and cannot be taken off again. This greatly inhibits the development of practical applications for CNT yarns. To solve this problem, Fan et al. [69] designed a new array-spinning process. After they had drawn the yarns out of the super-aligned array, the yarns were pulled through the ethanol droplets, and the centimeters-wide yarns shrank into a fiber structure of 20–30 µm in diameter. In this way, the processed fibers are more convenient for transfer or post-processing, thus making device applications possible. As mentioned above, twisting and solvent contraction are important steps in array spinning. These two improved techniques not only make it easier to produce fibers by array spinning but also can significantly improve the tensile strength and other properties. Li et al. [51] applied the steps of twisting and ethanol infiltration to controllably fabricate continuous microscale CNTFs using a spinning machine (Figure 4b). Compared with the solution, the biggest advantage of array spinning is that there is no need to prepare a high-concentration and homogeneous dispersion, which means that the structural damage of the CNTs can be reduced, and the extraordinary properties of CNTs can be fully exploited. However, the vertical-array spinning technique is relatively immature, it is difficult to achieve industrial scale-up, and the production cost is high. Therefore, this technique is suitable for the fabrication of small multifunctional devices that require microscale CNTFs.

3.2.3. Aerogel Spinning

Aerogel spinning is a continuous spinning process involving three phases. The process is shown in Figure 4c. The CNTs for aerogel spinning are generated by floating catalytic chemical vapor deposition (FCCVD). The catalyst (generally ferrocene) is firstly fed into a high-temperature reactor and reduced by hydrogen. Then, a carbon source such as methane or ethanol is cracked on the catalyst to form a tubular fiber precursor structure. Under gas flow, the precursor is passed through water or other coagulation solution to achieve rapid injection and fibrosis due to the capillary fineness of the liquid. According to the characteristics of the process, it can be considered that the precursor of microscale CNTFs, similar to powder, is in situ assembled into a macro fiber structure. Initially, only the gas–solid phase was involved in aerogel spinning. In 1998, Cheng et al. [38,72] obtained web-like, silver-black, light, and thin materials made of large quantities of bundles ranging from 10 to 40 nm in diameter, mainly SWCNTs. This was the beginning of the direct aerogel spinning of microscale CNTFs by FCCVD. Later, Zhu et al. [73] prepared 20 cm-long CNT yarns with a diameter of 0.3 mm using a similar method. However, the yarns they prepared were isolated and the diameter of the same yarn was not uniform. To collect a large number of continuous fibers, Windle et al. [56] designed different rotating spindles to wrap the aerogel. By drawing CNT aerogel directly from the hot reaction zone, they found that continuous fibers could be collected without length limitations. Similar to vertical-array spinning, aerogel spinning is also improved by solvent shrinkage. Windle et al. [17] densified the fibers by using acetone vapor. Li et al. [74] developed the water-sealing technique to draw fiber precursors into a spindle in water and then collected them in a spindle in the air. Through such a process, yarns with a length of several kilometers can be obtained, and their quality is close to conventional textile yarns. Like vertical-array spinning, aerogel spinning does not need the dispersion step. That means the fibers obtained by this spinning process can also take full advantage of the excellent mechanical properties of CNTs. In addition, wet spinning is based on the traditional spinning process and the technology is relatively mature and easy to achieve industrialization [22,75,76]. The direct spinning method of aerogel has also been capable of large-scale preparation. Li et al. reported that they could fabricate kilometer-level CNTFs with high tensile strength by

FCCVD [18]. Nanocomp, an American company, can fabricate 10 km-long CNT spinning threads and has promoted their practical application in the aerospace industry [77,78]. Therefore, the aerogel direct spinning method based on FCCVD is a very ideal system for studying mechanical property transfer across scales and developing practical applications based on CNTs.

Figure 4 shows the schematic diagram of different assembly processes for CNTFs at different scales. The process of in situ gas-flow focusing assembly or continuously drawing fibers from vertical CNT arrays are precise assembly techniques with nanoscale structure control. Although these microscale assembly techniques have advantages in structure and property control, it is difficult to achieve large-scale preparation, which means a lack of suitable application scenarios. In contrast, solution spinning and aerogel spinning as typical macroscale assembly techniques have more potential for industrial production. Despite that, there are still many significant problems to be solved for the excellent tensile strength transfer to a larger scale. From a bottom-up perspective, it may be effective to apply the methods of fabrication and property optimization of microscale materials to macroscale materials.

4. Tensile Property of Carbon Nanotube Fibers at Different Scales

4.1. Tensile Strength of Single Carbon Nanotubes

A single CNT is a basic structural unit of CNTFs; the mechanical properties determine the macro properties. Tensile strength is a crucial index for evaluating the mechanical properties of materials. For nanoscale single CNTs, precise experiments for tensile strength measurements require delicate nanomanipulation techniques. The theoretical model based on quantum mechanics calculation showed that the tensile strength of a single CNT could be as high as 120 GPa [30]. Several possible defects (SW defects, vacancies, doped atoms, etc. [30,79–82]) significantly influence the mechanical properties of CNTs [83,84]. Many theoretical studies have revealed the excellent intrinsic tensile strength of CNTs and predicted that the presence of defects would greatly reduce the tensile strength, which has been discussed in the former structure section. The tensile strength of CNTs could be reduced by orders of magnitude due to the emergence of a small number of topological defects [12,85]. Experimental studies confirmed these characteristics of the tensile strength of single CNTs. Early research on the mechanical properties of CNTs mainly relied on atomic force microscopy [27,86] or electron microscopy [3,26,28,87,88], and few experiments could observe a single CNT with an ultra-high tensile strength of more than 100 GPa. Using in situ scanning electron microscopy, Yu et al. obtained the tensile strength of multi-walled carbon nanotubes, ranging from 11 to 63 GPa, and the outer diameters ranged from 13 to 36 nm [28]. Chen et al. tested the tensile strength of double-walled carbon nanotubes and triple-walled carbon nanotubes (TWCNTs) with diameters ranging from 1.8 to 3.0 nm and obtained strengths ranging from 13 to 46 GPa [11]. They considered that the poor results were caused by defects in the CNTs. Takakura et al. investigated the tensile strength of SWCNTs with diameters ranging from 1.5 to 3.0 nm. The tensile strength was in the 25~66 GPa range, showing a decreasing trend with the increase in diameter [89]. With the development of controllable preparation technology for CNTs, the length and properties of CNTs have been significantly improved. Espinosa et al. fabricated MWCNTs with a diameter of 15.71 nm and measured a tensile strength of 110 GPa [90]. Zhang et al. fabricated ultra-long CNTs with a length of 55 cm, and their tensile strength was up to 120 GPa [9]. Bai et al. also fabricated centimeter-long and defect-free CNTs with a tensile strength of 118.9 ± 4.5 GPa [15]. Therefore, a single CNT with tensile strength very close to the theoretical limit can be obtained through experiments. From the bottom-up perspective, defect-free single CNTs, as a structural unit of a macrostructure, are a crucial carrier for the study of property transfer across scales. The controllable preparation of defect-free or defectless CNTs is the basis for the assembled CNTFs with high tensile strength.

4.2. Tensile Strength of Carbon Nanotube Fibers at Different Scales

Although single CNTs have shown excellent tensile properties in both theoretical and experimental studies [15,30], it is rather tough to transfer the remarkable properties of single CNTs to larger CNTs across the scale. According to the characteristics of the preparation process, the bundle with tens of nanometers in diameter and the fiber with hundreds of microns can be regarded as the bottom-up assembly of a single CNT. However, the tensile strength decreases significantly when the CNTs are clustered to form macro fibers, ropes, or even bundles.

4.2.1. Nanoscale Carbon Nanotube Fibers

For nanoscale CNTFs such as bundles or yarns, the number of assembled CNTs is far less than that of microscale CNTFs. The bundle formed by the aggregation of several or dozens of CNTs is a crucial bridge from the nanoscale CNTFs to the microscale ones. Based on the controllable preparation of ultra-long CNTs, the author's group fabricated nanoscale CNTFs that are centimeters long with tensile strength close to that of single CNTs [10]. As shown in Figure 4a, CNTBs consisted of only 2–15 CNTs with diameters between 2 and 15 nm. The tensile strength of such CNTBs could be up to 80 GPa, which were the only nanoscale CNTFs with tensile strength close to that of a single CNT at present. Espinosa et al. used an in situ transmission electron microscope to test DWCNT bundles with diameters ranging from 10 to 30 nm, as shown in Figure 5b [50]. They obtained a tensile strength of 17 GPa and a tensile modulus of 0.7 TPa. Cheng et al. prepared SWCNT strands with a diameter of 10–40 nm [38]. Although the diameter of the bundle was very small, the process was developed at a relatively low maturity for fiber assembly. The tensile strength was only 3.6 ± 0.4 GPa in the test [72]. Yu et al. fabricated SWCNT bundles with diameters ranging from 19 to 41 nm and studied their tensile fracture behavior using scanning electron microscopy and atomic force microscopy [48]. The tensile strength of SWCNT bundles ranged from 13 to 52 GPa (mean 30 GPa). The problem of the cross-scale transfer of tensile strength is not solved for most nanoscale CNTFs, though the diameters of these fibers are all less than 100 nm. Nanoscale CNTFs with smaller diameters are more promising to be higher in tensile strength, which should be attributed to their fewer defects and well-controlled tube–tube interactions. Since the diameters of nanoscale CNTFs are lower than 100 nm, the control of defects and tube–tube interactions can still be effective. The tensile strength of nanoscale CNTFs can easily exceed the order of 10 GPa, however, it will be very difficult for microscale CNTFs to maintain such a high level of strength.

4.2.2. Microscale Carbon Nanotube Fibers

For microscale CNTFs, the number of CNTs assembled can reach hundreds of millions, far more than the CNTBs. Since the fiber has a micron diameter, the microscale CNTFs can be considered as a bottom-up assembly of nanoscale CNTFs. Vigolo et al. [16] obtained CNTFs with a tensile strength of only 0.15 MPa. The fiber diameter can be distributed in the range of several microns to 100 microns. Smalley et al. [54] obtained CNTFs with a diameter of less than 1 micron and the tensile modulus and tensile strength reached 120 GPa and 116 MPa, respectively. Pasquali et al. [64] prepared CNTFs with diameters of approximately 9 µm and tensile strength of about 1.3 GPa. Later, they improved the tensile strength of microscale CNTFs to 4.2 GPa after process optimization [65]. Zhu et al. obtained microscale CNTFs with an average diameter of 5 µm and tensile strength up to 3.3 GPa [92]. Windle et al. optimized the post-treatment process and prepared microscale CNTFs of less than 20 µm in diameter with a tensile strength of 9 GPa [17]. Similar to Windle et al., Wang et al. [5] fabricated microscale CNTFs with diameters between 5 and 9 µm and tensile strength ranging from 3.76 to 5.53 GPa. Further, they prepared CNT films with tensile strength of 9.6 GPa [4]. The changes in the tensile strength of microscale CNTFs from a low to a high level are closely related to the scale of fibers and the maturity of the assembly technology. Due to the countless number of CNT units in microscale CNTFs, the efficiency of controlling the defects and tube–tube interactions will be significantly decreased, resulting in a relatively low tensile strength. As shown in Table 1, the tensile strength of CNTs at different scales is listed. The tensile strength of single CNTs and CNTFs with nanoscale diameters has been able to approach the theoretical level. Although significant progress has been made in the controllable preparation and process optimization of microscale CNTFs, their tensile strength is significantly inferior to that of single CNTs and nanoscale CNTFs, which is the typical undesirable performance transfer across different scales.

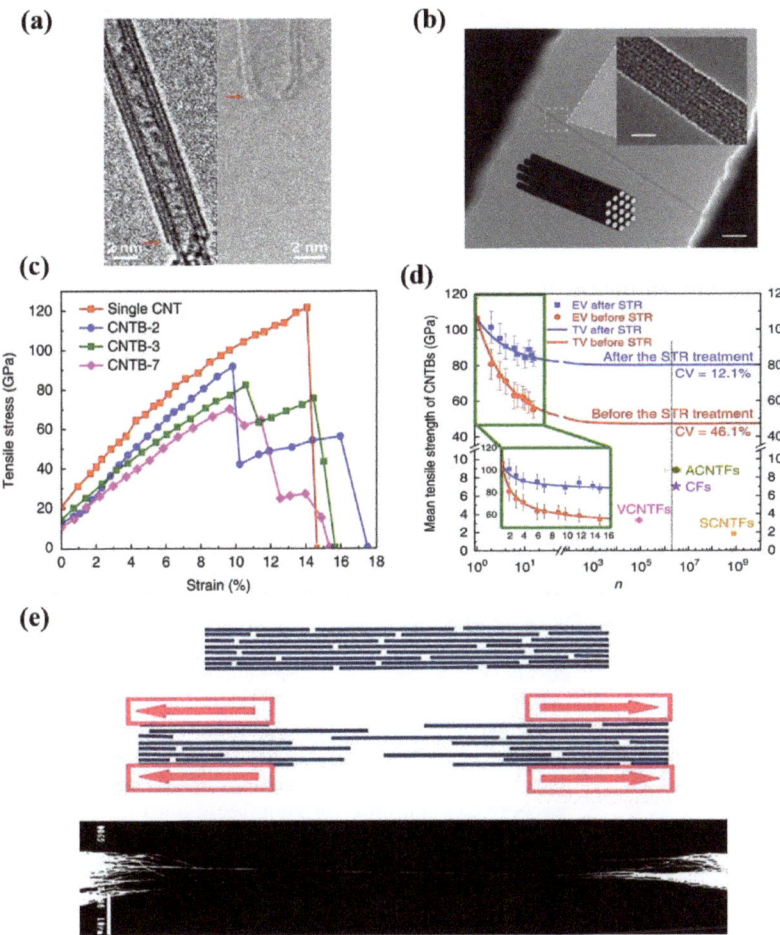

Figure 5. (**a**) TEM images of the flat fracture surfaces (marked by arrows) of the CNTs. Left: static tensile fracture surface; right: fracture surface after cyclic loading. Reproduced with permission from [15]. Copyright 2020, Amer Assoc Advancement Science. (**b**) TEM top view images of a DWCNT bundle (Scale bar: 200 nm). Insert: High-resolution TEM image of the suspended DWCNT bundle (Scale bar: 20 nm). Reproduced with permission from [50]. Copyright 2011, Wiley-Vch Verlag GmbH. (**c**) Stress–strain curves for single CNTs and CNTBs. (**d**) The relationship between the mean tensile strength of CNTBs and their component number before and after STR treatment. (**c**,**d**) Reproduced with permission from [10]. Copyright 2018, Springer Nature. (**e**) Top: Model of CNTFs assembled from individual CNT units. Middle: Model of tensile fracture of CNTFs. Bottom: SEM micrograph of the tensile fracture of actual CNTFs (Scale bar: 10 μm) Reproduced with permission from [91]. Copyright 2011, Amer Chemical Soc.

Table 1. Tensile strength of CNTs at different scales.

Scales	Diameter	Carbon Nanotube	Tensile Strength (GPa)	Ref.
Single tubes	2.0 nm	DWCNTs	118.9 ± 4.5	[15]
	1.0 to 4.0 nm	SWCNTs, DWCNTs, TWCNTs	120	[9]
	15.71 nm	MWCNTs	110	[90]
	1.5 to 3.0 nm	SWCNTs	25 to 66	[89]
	1.8 to 30 nm	DWCNTs, TWCNTs	13 to 46	[11]
	13 to 36 nm	MWCNTs	11 to 63	[28]
Nanoscale CNTFs	10.0 to 25.0 nm	SWCNTs, DWCNTs, TWCNTs	47 to 80	[10]
	19 to 41 nm	SWCNTs	13 to 52	[48]
	10.8 to 27.9 nm	DWCNTs	1.5 to 17.1	[50]
	10 to 40 nm	SWCNTs	3.6 ± 0.4	[38,72]
Microscale CNTFs	/	DWCNTs	9.6	[4]
	7.0 to 20.0 μm	DWCNTs	9	[17]
	8.0 to 9.8 μm	SWCNTs	4.2	[65]
	5.0 to 9.0 μm	DWCNTs	3.76 to 5.53	[5]
	5.0 μm	DWCNTs	3.3	[92]
	15 to 100 μm	SWCNTs	0.15	[16]
	0.2 to 0.6 μm	SWCNTs	0.12	[54]

4.3. Characteristics of Tensile Strength Transfer across Scales

The bottom-up composition of CNTs is similar to that of carbon fiber and cable-stayed bridge wires. Weibull distribution and the Daniel effect can describe the strength distribution of single filaments and carbon fibers [93–95], which have been used to describe the ideal state of performance transfer when CNTs are assembled into macroscopic fibers. Daniel et al. found that the bundles consisting of a large number of monomers with a tensile strength obey the Weibull distribution [96–99], and the average tensile strength $E_\sigma(n)$ is:

$$E_\sigma(n) = \sigma_0[1 - F(\sigma_0)] + c_n/n$$

where $F(\sigma)$ is the breaking probability of a single CNT under stress $\leq \sigma$. Cn is a variable related to the number of components (n). Based on the Weibull distribution and the Daniel effect, Bai et al. proposed a mathematical model to describe the relationship between the tensile strength of nanoscale CNTFs and their component numbers and initial strains [10]. As shown in Figure 5d, the experimental results fit well with the theoretical calculation. With increasing n, there is a quasi-exponential decrease in the mean tensile strength of the nanoscale CNTFs. The mathematical model results also showed that the tensile strength reaches a constant value for a number of constituent tubes larger than a certain value, which is very similar to the characteristics of steel wire in cable-stayed bridges [98]. However, the current prepared microscale CNTFs could not confirm this theoretical prediction, as the maximum tensile strength of microscale CNTFs with macroscopic structure was less than 10 GPa, which was much lower than the tensile strength of single CNTs and nanoscale CNTFs. Analyzing the reasons for this phenomenon is of great significance for the rational design of performance optimization strategies in the furture. The sp^2 hybridized C-C covalent bonds and fewer defects are the structural basis for the ultrahigh tensile strength of CNTs [9,26,100]. As Figure 5a,b show, individual CNTs with perfect structure and well-aligned bundles can make full use of C-C covalent bonds. However, the van der Waals forces are more crucial to be considered when scaling up from the nanoscale to the microscale. For example, the slip between tubes caused by weak tube–tube interactions is an important reason for the low tensile properties of macroscopic fibers. Windle et al. [91] proposed a model of the assembly of CNTFs and their tensile fracture (Figure 5e). Microscale CNTFs are composed of a large number of short CNTs, which means that tube–tube slip asynchronously is more likely to happen before the short CNTs fracture. In other words, the tensile properties of microscale CNTFs depend largely on van der Waals forces rather than C-C covalent bonds. This may be a reason for the unsatisfactory transfer of mechanical properties across scales of CNTs. Therefore, optimizing tube–tube interactions is fundamental as defect control to improve the tensile strength of CNTFs at different scales. Suppose that the macrostructure is composed of numerous ultra-long, defect-free, or defectless CNTs with continuous length, perfect structure, uniform orientation, and uniform initial stress distribution, then there will be a high possibility that its tensile strength can still maintain a rather high level.

5. Optimization Strategies for the Tensile Strength of Carbon Nanotube Fibers at Different Scales

5.1. Defects Control

Defects have a fatal negative effect on the intrinsic properties of CNTs [12,79,101]. In a large number of early experimental studies on the tensile strength of single CNTs, the measured tensile strength was significantly lower than 50 GPa due to the presence of the many structural defects of the CNTs [11,28]. Zhang et al. [9] and Bai et al. [10] fabricated defect-free CNTs with tensile strength close to the theoretical value. Further, Bai et al. assembled these defect-free CNTs into a bundle structure and the tensile strength reached up to 80 GPa, which was also close to the theoretical level. The defect-free structure enabled the desired transfer of tensile strength from single CNTs to nanoscale CNTFs. Due to the bottom-up characteristics of the assembly of CNTFs, the large-scale production of CNTs with fewer defects is the basis of high-strength microscale CNTFs. In the early process of fiber preparation by wet spinning, the raw powder of CNTs had many defects, and the maximum tensile strength of the fibers was less than 1 GPa [16,52]. During the development of wet spinning, Pasquali et al. continuously improved the tensile strength of the fibers, reaching a maximum of 4.2 GPa [65]. This is mainly attributed to the progress in preparing defectless CNTs as raw materials to produce fibers. Recently, Kim et al. have continuously synthesized CNTs with high crystallinity (IG/ID > 60), high aspect ratio (>17,000), and high yield (>6 mg/min) [103]. The high Raman IG/ID signal implied the good quality of the prepared CNTs. The specific strength and modulus of the microscale CNTFs prepared using the products were 2.94 N tex^{-1} and 231 N tex^{-1}, equivalent to the best carbon fibers. This further proves that it is essential to obtain perfect and defect-free CNT raw materials in order to prepare CNT macrostructures with excellent properties.

In addition to the defects introduced by CNTs themselves, the bottom-up assembly process of single tubes into a macroscopic structure also inevitably introduces many defects. More defects will be introduced if vigorous treatments such as superacid and ultrasonic-assisted dispersion are applied [14]. As a result, this will fundamentally reduce the tensile strength of the microscale CNTFs. It is essential to reduce the defects introduced by the assembly process. Smalley et al. [54] prepared CNTFs with very low mechanical strength due to the use of fuming sulfuric acid in the dispersion process, which can easily damage the structure of CNTs and introduce more defects. Pasquali et al. [64] reduced the defects by a liquid crystalline phase system, thus significantly improving the tensile strength of the fibers [52]. The number of defects will be reduced dramatically during dry spinning as it does not require dispersion, a process that inevitably damages the structure. During the vertical-array spinning process, the materials are highly purified vertical CNT arrays, which means that fewer defects will be introduced from both the spinning process and the raw materials themselves. The maximum tensile strength of CNTFs with few defects has reached 3.3 GPa [92]. As a one-step process, aerogel array spinning can assemble CNTs into fibers in situ. Obviously, this process also significantly reduces the introduction of structural defects. Li et al. optimized the process conditions and produced a kilometer-level CNTF with a tensile strength of 3.5 GPa without post-treatment. Dry spinning is a cleaner system that reduces the number of defects in the CNT macrostructure. The tensile strength of CNTFs prepared by dry spinning can be up to 10 GPa after post-treatment [4,17], which is generally higher than that of CNTFs with more structural defects prepared by wet spinning. The difference in defect control caused by different technological processes is a key factor to optimizing the tensile strength.

5.2. Control of Tube–Tube Interactions

Improving the process to achieve good cross-scale transfer is a great challenge. Windle et al. pointed out that the tensile strength of CNTFs depends on the interactions between the tubes [91]. Many factors directly or indirectly affect the interactions between tubes and then affect the tensile strength of the assembled macro fibers. Understanding and controlling the tube–tube interactions are significant.

5.2.1. Initial Strain

The bottom-up assembly of CNTs is similar to the steel wire of a cable-stayed bridge. The main cable of a cable-stayed bridge comprises tens of thousands of untwisted steel wire bundles with a tensile strength of about 2 GPa and a diameter of several millimeters [102–104]. During construction, the steel wire bundles should be divided into hundreds of groups, and the initial strain needs to be as uniform as possible to ensure uniform force. When CNTs are assembled into bundles or fibers from bottom to top, the initial tension of each tube is inevitably different. According to the Daniel effect, the tensile strength of bundle and fiber will decrease rapidly with the increase in the number of tubes if the initial strain is not uniform [96]. Bai et al. studied the initial strain of CNTs in nanoscale CNTFs. The measured tensile strength is shown in Figure 6b. The tensile strength of CNTBs decreases with the increasing number of components, and the breaking process under tensile loading exhibits a multi-stage, one-by-one process. This indicates that the initial stress distribution of CNTs in the nanoscale CNTFs is not uniform, so the CNTs in the fibers cannot bear the load synchronously and equally, and the components break one by one, which leads to the decline of the overall tensile strength. This is an essential point of view elucidating the difficulties during performance transfer across scales. Based on this analysis, the research group proposed a synchronous tightening and relaxing (STR) strategy, as shown in Figure 6a,b, in which the initial stress of the CNTs in the fibers is released to a narrow distribution through nano-manipulation, thus increasing the tensile strength of the nanoscale CNTFs from 47 GPa to more than 80 GPa [10]. Therefore, the tube–tube interaction is an important guarantee of excellent tensile performance, and STR is an effective strategy to make the initial stress of CNT components uniform and improve the tensile strength across scales. Recently, Kim et al. [6] developed a rapid and continuous method to produce highly aligned and densified CNTFs. As Figure 6c shows, the fibers obtained by direct aerogel spinning were dissolved in chlorosulfonic acid, and the alignment of CNTs along the axial was improved after stretching. Such an improvement also made the initial strain of CNTs in microscale CNTFs more uniform. As a result, the tensile strength of CNTFs was increased from 2.1 N tex^{-1} to 4.44 N tex^{-1}, indicating the significance of the uniform initial stress.

5.2.2. Length-Dependent Interactions

CNTs are the basic structural units of CNTFs from bottom to top. When the length of the CNT component increases, the contact area between adjacent CNTs also increases, which improves the local transfer efficiency between tubes and enhances the mechanical properties of fibers. Fibers composed of longer CNTs mean that there are fewer ends. In that case, the force on each part will be more uniform, and the tensile strength will be improved [105,106]. Therefore, increasing the length of CNT units in the macrostructure is an effective strengthening strategy. Pasquali et al. systematically studied the influence of CNT structure on the tensile performance of microscale CNTFs prepared by solution spinning [107]. The results showed that the tensile properties of CNTs were mainly influenced by their aspect ratio rather than other factors such as the number of tube walls, diameter, and crystallinity. In 2013, Pasquali et al. improved the spinning process and used CNTs with an average length of 5 μm as raw materials to prepare high-performance multifunctional fibers with an average tensile strength of 1 GPa [64]. Recently, Pasquali et al. dissolved CNTs with a high aspect ratio (12 μm of average length) in chlorosulphonic acid and obtained microscale CNTFs with a tensile strength of 4.2 GPa, which is far higher than the fibers spun by the CNTs with an average length of 5 μm [65]. By adjusting the growth kinetics, CNT arrays with a higher height can be fabricated, and a macrostructure composed of longer CNT units can be obtained by spinning. Zhu et al. studied the effect of array height on the tensile strength of spun fibers [108]. They fabricated 300 μm-, 500 μm-, and 650 μm-high vertical arrays and spun them into fibers. These fibers' tensile strengths are 0.32, 0.56, and 0.85 GPa, respectively. Further, Zhu et al., fabricated fibers with longer (1 mm) individual CNTs. The tensile strength of the fabricated fibers can be up to 3.3 GPa. Windle et al. [17] and Li et al. [18] fabricated fibers based on the FCCVD method, and also proved that with the increase in single CNT length, the tensile strength of the macro structure of CNTs would be higher. The enhancement of the interaction force between tubes results from an increase in the length of constituent CNTs, which can reduce the slip between tubes and make full use of the mechanical properties of the CNTs.

5.2.3. Packing Density

The increased packing density can reduce the fiber cross-sectional area, which will effectively reduce the distance between tubes and improve the tube–tube interactions so that the load transfer capacity will be increased [55,109,110]. Xie et al. fabricated an MWCNT rope with a tube spacing of about 100 nm, and the tensile strength was measured to be 4 GPa [111]. In contrast, Bai et al. assembled tubes into bundles tightly without pores, and the tensile strength of such dense nanoscale CNTFs can be improved to as high as 47 GPa without any post-treatments [10]. For microscale CNTFs, there are many effective methods to increase the packing density [112,113]. Baughman et al. introduced the twisting step in this spinning method [70], which increased the fiber density to 0.8 g/cm^3 and the tensile strength to 150~300 MPa, as shown in Figure 6e. Li et al. [56] introduced a twisting and collecting device at the end of the furnace, resulting in the continuous in situ preparation of microscale CNTFs with a tensile strength of 3.24 GPa (1 mm pinch) [5,17]. Physical and chemical crosslinking treatment [114,115] and densification solvent [116,117] are also effective methods for fiber densification. Polymers such as polyvinyl alcohol and epoxy resin are good densifying solvents, which will contribute to enhancing the load transfer capacity and effectively improving composite fibers' mechanical properties [118,119]. Jiang et al. densified the spun fibers with PVA solution to obtain CNT composite fibers with a tensile strength of 2 GPa [117]. Dalton et al. prepared a CNT/polyvinyl alcohol composite by wet spinning [59,60], and an ultimate tensile strength of 1.8 GPa was obtained in the test. For aerogel spinning, there are many holes in the prepared fibers. Li et al. [122] reported the densification of different solvents and found that densification using non-volatile solvents with high polarity could significantly enhance the strength. Li et al. [123] were also inspired by straw bundling and used the character of self-contraction of silk fibroin to densify CNTFs locally. The tensile strength of CNTFs was enhanced from 355 MPa to 960 MPa. Mechanical densification is a simpler and more effective post-treatment densification technique. Figure 6f shows that by collecting CNT cylinders and introducing the rolling strategy for densification, Wang et al. [6] obtained a CNT macrostructure with an average tensile strength of 4.34 GPa, which is 12 times higher than that of unrolled fibers. After rolling treatment and the optimization of the winding rate, the density of fibers was significantly improved from 0.53 g cm^{-3} to 1.85 g cm^{-3}, and the tensile strength was tested as high as 9.6 GPa, which is the highest value among macrostructures of CNT until now [4]. These indicated that rolling is an effective strategy to improve the mechanical properties of microscale CNTFs. The above results have addressed the significance of higher packing density for the improvement of tensile strength.

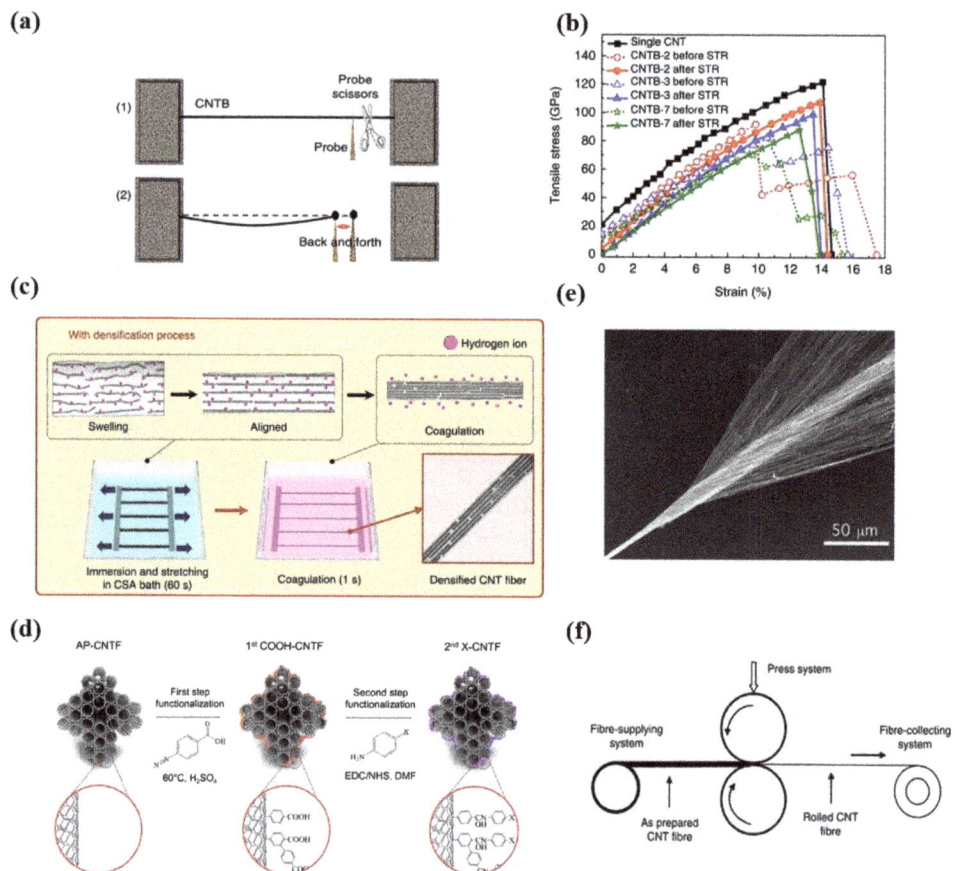

Figure 6. (**a**) Schematic diagram to make the initial stress of each tube more uniform. (**b**) Stress–strain curves for a single CNT and for CNTBs before and after the stress uniformity treatment. (**a**,**b**) Reproduced with permission from [10]. Copyright 2018, Springer Nature. (**c**) Schematic of the stretching and densification of CNTFs. Reproduced with permission from [6]. Copyright 2019, Springer Nature. (**d**) Schematic of the two-step chemical molecule cross-linking. Reproduced with permission from [120]. Copyright 2021, Elsevier Sci Ltd. (**e**) SEM images of a twisted CNT yarn. Reproduced with permission from [70]. Copyright 2004, Amer Assoc Advancement Science. (**f**) A schematic of the system for rolling CNTFs. Reproduced with permission from [5]. Copyright 2014, Springer Nature.

6. Prospects for the Application of High-Strength Carbon Nanotube Fibers

6.1. Structural Reinforcing Material

Due to their outstanding intrinsic mechanical properties, CNTs are ideal candidates for structural reinforcement materials. Although the tensile strength of CNTFs is far from the theoretical value, there have been many CNTFs with a tensile strength higher than T1000, one of the strongest commercial carbon fibers [4,120]. As structural reinforcement materials, CNTFs can greatly improve the strength of materials such as polymers [119,121,122], metals [123,124], and ceramics [125,126], thus playing an important role in many fields such as aerospace and building materials. NASA has identified CNTFs as an alternative reinforcement material [122]. They used continuous CNT yarns reinforced Ultem to build the frame of a quadcopter. They also demonstrated the application value of CNT yarns in coating materials (Figure 7a,b). The breaking load of a bare aluminum ring was about 5000 N while the breaking load of an aluminum ring coated with Epon 828/CNT yarn could be higher than 11,000 N [123]. With lightweight and high-strength mechanical properties, CNTFs are ideal materials for various kinds of armor, especially for body armor applications [127,128]. Lee et al. [128] studied the dynamic strengthening phenomenon of CNTFs under extreme mechanical impulses. They found that the kinetic energy absorption properties of CNTFs are superior to other high-performance fibers such as nylon, Kevlar, and aluminum monofilament. Structural reinforcement is one of the most important applications of CNTs. With the breakthrough of high-strength CNTFs, more achievements can be realized in high-end fields. Even the space elevator [1] can be possible.

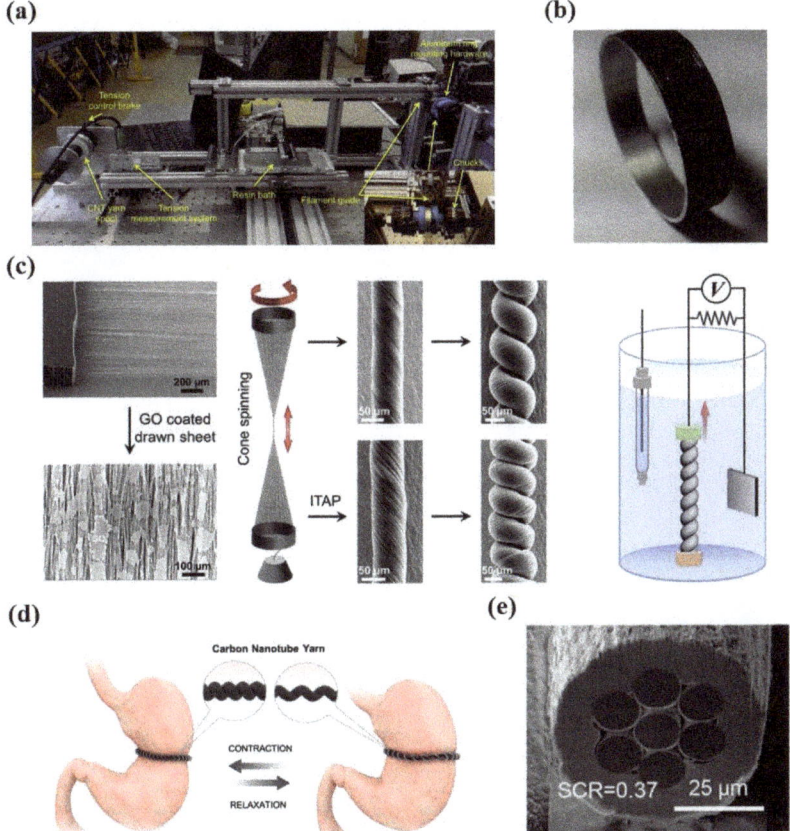

Figure 7. (**a**) Equipment in which Al rings are overwrapped with CNTFs. (**b**) Al ring overwrapped with Epon 828/CNT yarn. (**a**,**b**) Reproduced with permission from [123]. Copyright 2016, Elsevier Sci Ltd. (**c**) SEM images and illustration of cone spinning for fabricating twisted and coiled neat CNT yarns from forest-drawn CNT sheets and its designed application device. Reproduced with permission from [129]. Copyright 2022, Wiley-Vch Verlag Gmbh. (**d**) Coiled CNT yarns for stomach sensors. Reproduced with permission from [130]. Copyright 2019, Amer Chemical Soc. (**e**) Cross-sectional SEM images for sheath-driven PI_PDMS@CNT muscles. Reproduced with permission from [131]. Copyright 2022, Wiley-Vch Verlag Gmbh.

6.2. Energy Storage

In recent years, new energy development has attracted attention to energy storage. Energy storage devices are an important emerging application field of CNTFs, and some advances have been made in device design and fabrication. CNTFs with high mechanical properties are important materials for mechanoelectrical energy conversion [21,132,133]. Baughman et al. [134] designed a CNT yarn harvester that can convert tensile or torsional mechanical energy into electrical energy. The peak electrical energy generated by the coiled fibers was 250 W/kg as the stretching cycle frequency reached 30 Hz. At mechanical frequencies between 6 and 600 Hz, the peak electrical energy generated per harvester weight was higher than that generated by any prior-art mechanical energy harvesters. However, the conversion efficiency of the twistron fibers was not high enough. Recently, Baughman and Kim et al. [129] proposed some effective strategies to improve energy conversion efficiency. As shown in Figure 7c, these improvement strategies include optimizing the orientation, applying tensile stress, electrothermal pulse annealing under stress, and loading graphene nanosheets [129,135]. The optimized twistrons reached a peak power output of 12-fold that of the mechanical energy collector in the prior art at a frequency of 30 Hz. These electromechanical conversion devices are expected to be applied in wave energy generation, using human motion to power sensors and energy storage, and other emerging fields of energy storage [130,136–138]. Figure 7d shows that Kim et al. [130] designed a sensor for detecting gastric deformations based on coiled CNT yarn.

6.3. Artificial Muscle

Many studies have reported that the prepared CNTFs possess high mechanical and electrical properties. Pasquali et al. [65] reported that fibers prepared by wet spinning had a tensile strength of 4.2 GPa and electrical conductivity of 10.9 MS·m^{-1}, comparable to copper. Artificial muscle materials need excellent mechanical and electrical properties. Therefore, the research on artificial muscle based on CNTFs has attracted much attention. Kim and Baughman et al. [139,140] fabricated artificial muscles by spinning twisted CNTs, exhibiting the contractile motion of an all-solid-state stretching muscle. Further study showed that the tension and contraction of two-ply coiled CNT yarn were as high as 16.5% under voltage driving, which was about 30 times higher than natural muscle [141]. They increased the capacitance by introducing graphene, and the tensile stroke of artificial muscle made with CNT yarns was increased two-fold [139]. Recently, they designed an artificial muscle structure with a polymer core and CNT sheath (Figure 7e). This structure can achieve an average power density of 12 kW/kg, which is 42-times that of human skeleton muscle [131]. Fibrous artificial muscles are being investigated for applications such as robotics, prostheses, and exoskeletons. The design and manufacture of related devices have been relatively mature. CNTF-based artificial muscles are also more multifunctional.

7. Conclusions

Carbon nanotubes are a kind of ultra-light and high-strength material. The fiber structure composed of CNTs is expected to be a significant breakthrough for the next generation of high-strength mechanical materials. However, the mechanical properties such as the tensile strength of CNTFs are still far lower than the theoretical levels. Further research on the preparation and assembly of CNTs is still urgently needed for the revolutionary development of nanomaterials. In this review, we have analyzed the controllable preparation and tensile strength of CNTs at different scales. Based on recent years' research on the controllable preparation of CNTFs, the significance of defect control and efficient treatment process to control tube–tube interactions is emphasized towards ultra-high-strength CNTFs. We introduce the unique structure and large-scale preparation of CNTs, which is the basic unit of nanoscale assembly. The significance of defect control for high-strength CNTFs is emphasized. In addition, the mechanical strength and toughness can be significantly improved by eliminating the tube–tube non-uniform interactions or other post-treatment interactions. Based on the bottom-up method, the mass production of defect-free CNTs and accurate assembly into the macro-scale with fewer defects, fine alignment, and higher density is the priority to obtain macro-scale assemblies with excellent tensile strength. In addition to tensile strength, mechanical properties such as fatigue, bending, and torsion, as well as electrical and thermal properties, face a similar problem of performance transfer across scales. With further analysis of the influencing factors and optimization strategies for the cross-scale transfer of other properties, the excellent intrinsic properties of CNTs can be fully utilized, and CNTs can play a more significant role in many fields.

However, there are still many challenges to be solved. The first is the definition of high-end application scenarios for these super strong, super tough materials. There are ideal concepts, such as space elevators and flywheels, but they all require cross-disciplinary assistance that is not available with current technology. The second is that the continuous length of a single tube can be only meter-level, while tens of billions of tubes are needed to make a macroscopic fiber even with a kilometer in length. Such non-damage welding and inter-tube orientation of nanoscale fibers are challenges to science, engineering, and material manufacturing. Therefore, tremendous breakthroughs at the technical level are still needed. Finally, in order to further improve the strength and achieve practical application, the large-scale preparation of CNTs with defect-free or defectless structures is crucial. However, without engineering input in application scenarios, the development of the large-scale preparation of such nanomaterials will be an unspontaneous and challenging process. Despite of that, we believe that the continuous attention from researchers in the fields of emerging energy and materials can finally find solutions to those tough but significant problems, such as the preparation and performance optimization of cutting-edge fiber materials like CNTFs.

Author Contributions: Conceptualization, Y.Z. and H.Y.; methodology, F.W., Z.Z. and Y.B.; validation, F.W., Z.Z. and Y.B.; formal analysis, Y.Z. and Z.Z.; writing—original draft preparation, Y.Z. and Z.Z.; writing—review and editing, Y.Z., Z.Z., H.Y. and M.J.A.; supervision, F.W. and Z.Z.; project administration, Z.Z.; funding acquisition, Z.Z. and F.W. All authors have read and agreed to the published version of the manuscript.

Funding: This research was funded by the National Natural Science Foundation of China (grant 22108155) and the China Postdoctoral Science Foundation (grant 2021M691764 and 2021T140367).

Institutional Review Board Statement: Not applicable.

Informed Consent Statement: Not applicable.

Data Availability Statement: Not applicable.

Acknowledgments: The authors gratefully acknowledge the technical and financial support from Tsinghua University (THU).

Conflicts of Interest: The authors declare no conflict of interest.

References

1. Yakobson, B.I.; Smalley, R.E. Fullerene Nanotubes: C 1,000,000 and Beyond. *Am. Sci.* **1997**, *85*, 324–337.
2. Coulson, C.A. *Valence*, 2nd ed.; Oxford University Press: London, UK, 1952; pp. 350.
3. Demczyk, B.G.; Wang, Y.M.; Cumings, J.; Hetman, M.; Han, W.; Zettl, A.; Ritchie, R.O. Direct Mechanical Measurement of the Tensile Strength and Elastic Modulus of Multiwalled Carbon Nanotubes. *Mater. Sci. Eng. A Struct. Mater. Prop. Microstruct. Process.* **2002**, *334*, 173–178. [CrossRef]
4. Xu, W.; Chen, Y.; Zhan, H.; Wang, J.N. High-Strength Carbon Nanotube Film from Improving Alignment and Densification. *Nano Lett.* **2016**, *16*, 946–952. [CrossRef] [PubMed]
5. Wang, J.N.; Luo, X.G.; Wu, T.; Chen, Y. High-Strength Carbon Nanotube Fibre-Like Ribbon with High Ductility and High Electrical Conductivity. *Nat. Commun.* **2014**, *5*, 1–8. [CrossRef]
6. Lee, J.; Lee, D.M.; Jung, Y.; Park, J.; Lee, H.S.; Kim, Y.K.; Park, C.R.; Jeong, H.S.; Kim, S.M. Direct Spinning and Densification Method for High-Performance Carbon Nanotube Fibers. *Nat. Commun.* **2019**, *10*, 1–10. [CrossRef]
7. Gao, E.L.; Lu, W.B.; Xu, Z.P. Strength Loss of Carbon Nanotube Fibers Explained in a Three-Level Hierarchical Model. *Carbon* **2018**, *138*, 134–142. [CrossRef]
8. Salvetat, J.P.; Bonard, J.M.; Thomson, N.H.; Kulik, A.J.; Forro, L.; Benoit, W.; Zuppiroli, L. Mechanical Properties of Carbon Nanotubes. *Appl. Phys. A Mater. Sci. Process.* **1999**, *69*, 255–260. [CrossRef]
9. Zhang, R.; Zhang, Y.; Zhang, Q.; Xie, H.; Qian, W.; Wei, F. Growth of Half-Meter Long Carbon Nanotubes Based on Schulz-Flory Distribution. *ACS Nano* **2013**, *7*, 6156–6161. [CrossRef]

10. Bai, Y.; Zhang, R.; Ye, X.; Zhu, Z.; Xie, H.; Shen, B.; Cai, D.; Liu, B.; Zhang, C.; Jia, Z.; et al. Carbon Nanotube Bundles with Tensile Strength over 80 Gpa. *Nat. Nanotechnol.* **2018**, *13*, 589–595. [CrossRef]
11. Wei, X.L.; Chen, Q.; Peng, L.M.; Cui, R.L.; Li, Y. Tensile Loading of Double-Walled and Triple-Walled Carbon Nanotubes and Their Mechanical Properties. *J. Phys. Chem. C* **2009**, *113*, 17002–17005. [CrossRef]
12. Zhu, L.Y.; Wang, J.L.; Ding, F. The Great Reduction of a Carbon Nanotube's Mechanical Performance by a Few Topological Defects. *ACS Nano* **2016**, *10*, 6410–6415. [CrossRef] [PubMed]
13. Tagawa, T.; Miyata, T. Size Effect on Tensile Strength of Carbon Fibers. *Mater. Sci. Eng. A Struct. Mater. Prop. Microstruct. Process.* **1997**, *238*, 336–342. [CrossRef]
14. Yang, M.; Koutsos, V.; Zaiser, M. Size Effect in the Tensile Fracture of Single-Walled Carbon Nanotubes with Defects. *Nanotechnology* **2007**, *18*, 155708. [CrossRef]
15. Bai, Y.; Yue, H.; Wang, J.; Shen, B.; Sun, S.; Wang, S.; Wang, H.; Li, X.; Xu, Z.; Zhang, R.; et al. Super-Durable Ultralong Carbon Nanotubes. *Science* **2020**, *369*, 1104–1106. [CrossRef] [PubMed]
16. Vigolo, B.; Penicaud, A.; Coulon, C.; Sauder, C.; Pailler, R.; Journet, C.; Bernier, P.; Poulin, P. Macroscopic Fibers and Ribbons of Oriented Carbon Nanotubes. *Science* **2000**, *290*, 1331–1334. [CrossRef]
17. Koziol, K.; Vilatela, J.; Moisala, A.; Motta, M.; Cunniff, P.; Sennett, M.; Windle, A. High-Performance Carbon Nanotube Fiber. *Science* **2007**, *318*, 1892–1895. [CrossRef]
18. Zhou, T.; Niu, Y.T.; Li, Z.; Li, H.F.; Yong, Z.Z.; Wu, K.J.; Zhang, Y.Y.; Li, Q.W. The Synergetic Relationship between the Length and Orientation of Carbon Nanotubes in Direct Spinning of High-Strength Carbon Nanotube Fibers. *Mater. Des.* **2021**, *203*, 109557. [CrossRef]
19. Zhang, X.H.; Lu, W.B.; Zhou, G.H.; Li, Q.W. Understanding the Mechanical and Conductive Properties of Carbon Nanotube Fibers for Smart Electronics. *Adv. Mater.* **2020**, *32*, 1902028. [CrossRef] [PubMed]
20. Yadav, M.D.; Dasgupta, K.; Patwardhan, A.W.; Joshi, J.B. High Performance Fibers from Carbon Nanotubes: Synthesis, Characterization, and Applications in Composites—A Review. *Ind. Eng. Chem. Res.* **2017**, *56*, 12407–12437. [CrossRef]
21. Li, K.; Ni, X.; Wu, Q.; Yuan, C.; Li, C.; Li, D.; Chen, H.; Lv, Y.; Ju, A. Carbon-Based Fibers: Fabrication, Characterization and Application. *Adv. Fiber Mater.* **2022**, *4*, 631–682. [CrossRef]
22. Lu, W.; Zu, M.; Byun, J.-H.; Kim, B.-S.; Chou, T.-W. State of the Art of Carbon Nanotube Fibers: Opportunities and Challenges. *Adv. Mater.* **2012**, *24*, 1805–1833. [CrossRef] [PubMed]
23. Mikhalchan, A.; Jose Vilatela, J. A Perspective on High-Performance Cnt Fibres for Structural Composites. *Carbon* **2019**, *150*, 191–215. [CrossRef]
24. Iijima, S. Helical Microtubules of Graphitic Carbon. *Nature* **1991**, *354*, 56–58. [CrossRef]
25. Prasek, J.; Drbohlavova, J.; Chomoucka, J.; Hubalek, J.; Jasek, O.; Adam, V.; Kizek, R. Methods for Carbon Nanotubes Synthesis-Review. *J. Mater. Chem.* **2011**, *21*, 15872–15884. [CrossRef]
26. Treacy, M.M.J.; Ebbesen, T.W.; Gibson, J.M. Exceptionally High Young's Modulus Observed for Individual Carbon Nanotubes. *Nature* **1996**, *381*, 678–680. [CrossRef]
27. Salvetat, J.P.; Kulik, A.J.; Bonard, J.M.; Briggs, G.A.D.; Stockli, T.; Metenier, K.; Bonnamy, S.; Beguin, F.; Burnham, N.A.; Forro, L. Elastic Modulus of Ordered and Disordered Multiwalled Carbon Nanotubes. *Adv. Mater.* **1999**, *11*, 161–165. [CrossRef]
28. Yu, M.F.; Lourie, O.; Dyer, M.J.; Moloni, K.; Kelly, T.F.; Ruoff, R.S. Strength and Breaking Mechanism of Multiwalled Carbon Nanotubes under Tensile Load. *Science* **2000**, *287*, 637–640. [CrossRef]
29. Wei, C.Y.; Cho, K.J.; Srivastava, D. Tensile Strength of Carbon Nanotubes under Realistic Temperature and Strain Rate. *Phys. Rev. B* **2003**, *67*, 115407. [CrossRef]
30. Mielke, S.L.; Troya, D.; Zhang, S.; Li, J.L.; Xiao, S.P.; Car, R.; Ruoff, R.S.; Schatz, G.C.; Belytschko, T. The Role of Vacancy Defects and Holes in the Fracture of Carbon Nanotubes. *Chem. Phys. Lett.* **2004**, *390*, 413–420. [CrossRef]
31. Yuan, Q.; Xu, Z.; Yakobson, B.I.; Ding, F. Efficient Defect Healing in Catalytic Carbon Nanotube Growth. *Phys. Rev. Lett.* **2012**, *108*, 245505. [CrossRef]
32. Sun, X.; Bao, W.; Lv, Y.; Deng, J.; Wang, X. Synthesis of High Quality Single-Walled Carbon Nanotubes by Arc Discharge Method in Large Scalele. *Mater. Lett.* **2007**, *61*, 3956–3958. [CrossRef]
33. Maser, W.K.; Munoz, E.; Benito, A.M.; Martinez, M.T.; de la Fuente, G.F.; Maniette, Y.; Anglaret, E.; Sauvajol, J.L. Production of High-Density Single-Walled Nanotube Material by a Simple Laser-Ablation Method. *Chem. Phys. Lett.* **1998**, *292*, 587–593. [CrossRef]
34. Huang, S.M.; Maynor, B.; Cai, X.Y.; Liu, J. Ultralong, Well-Aligned Single-Walled Carbon Nanotube Architectures on Surfaces. *Adv. Mater.* **2003**, *15*, 1651–1655. [CrossRef]
35. Kong, J.; Soh, H.T.; Cassell, A.M.; Quate, C.F.; Dai, H.J. Synthesis of Individual Single-Walled Carbon Nanotubes on Patterned Silicon Wafers. *Nature* **1998**, *395*, 878–881. [CrossRef]
36. Wen, Q.; Qian, W.; Nie, J.; Cao, A.; Ning, G.; Wang, Y.; Hu, L.; Zhang, Q.; Huang, J.; Wei, F. 100 Mm Long, Semiconducting Triple-Walled Carbon Nanotubes. *Adv. Mater.* **2010**, *22*, 1867–1871. [CrossRef] [PubMed]
37. Joseyacaman, M.; Mikiyoshida, M.; Rendon, L.; Santiesteban, J.G. Catalytic Growth of Carbon Microtubules with Fullerene Structure. *Appl. Phys. Lett.* **1993**, *62*, 202–204. [CrossRef]
38. Cheng, H.M.; Li, F.; Su, G.; Pan, H.Y.; He, L.L.; Sun, X.; Dresselhaus, M.S. Large-Scale and Low-Cost Synthesis of Single-Walled Carbon Nanotubes by the Catalytic Pyrolysis of Hydrocarbons. *Appl. Phys. Lett.* **1998**, *72*, 3282–3284. [CrossRef]

39. Cheng, H.M.; Li, F.; Sun, X.; Brown, S.D.M.; Pimenta, M.A.; Marucci, A.; Dresselhaus, G.; Dresselhaus, M.S. Bulk Morphology and Diameter Distribution of Single-Walled Carbon Nanotubes Synthesized by Catalytic Decomposition of Hydrocarbons. *Chem. Phys. Lett.* **1998**, *289*, 602–610. [CrossRef]
40. Wang, Y.; Wei, F.; Luo, G.H.; Yu, H.; Gu, G.S. The Large-Scale Production of Carbon Nanotubes in a Nano-Agglomerate Fluidized-Bed Reactor. *Chem. Phys. Lett.* **2002**, *364*, 568–572. [CrossRef]
41. Wei, F.; Zhang, Q.; Qian, W.-Z.; Yu, H.; Wang, Y.; Luo, G.-H.; Xu, G.-H.; Wang, D.-Z. The Mass Production of Carbon Nanotubes Using a Nano-Agglomerate Fluidized Bed Reactor: A Multiscale Space-Time Analysis. *Powder Technol.* **2008**, *183*, 10–20. [CrossRef]
42. Wang, Q.X.; Ning, G.Q.; Wei, F.; Luo, G.H. Production of High Quality Single-Walled Carbon Nanotubes in a Nano-Agglomerated Fluidized Bed Reactor. In Proceedings of the Symposium on Materials and Devices for Smart Systems Held at the 2003 MRS Fall Meeting, Boston, MA, USA, 1–5 December 2003; pp. 313–318.
43. Zhang, Q.; Yu, H.; Liu, Y.; Qian, W.; Wang, Y.; Luo, G.; Wei, F. Few Walled Carbon Nanotube Production in Large-Scale by Nano-Agglomerated Fluidized-Bed Process. *Nano* **2008**, *3*, 45–50. [CrossRef]
44. Zhu, Z.; Wei, N.; Cheng, W.; Shen, B.; Sun, S.; Gao, J.; Wen, Q.; Zhang, R.; Xu, J.; Wang, Y.; et al. Rate-Selected Growth of Ultrapure Semiconducting Carbon Nanotube Arrays. *Nat. Commun.* **2019**, *10*, 1–8. [CrossRef] [PubMed]
45. Zhu, Z.; Cui, C.; Bai, Y.; Gao, J.; Jiang, Y.; Li, B.; Wang, Y.; Zhang, Q.; Qian, W.; Wei, F. Advances in Precise Structure Control and Assembly toward the Carbon Nanotube Industry. *Adv. Funct. Mater.* **2021**, *32*, 2109401. [CrossRef]
46. Salvetat, J.P.; Briggs, G.A.D.; Bonard, J.M.; Bacsa, R.R.; Kulik, A.J.; Stockli, T.; Burnham, N.A.; Forro, L. Elastic and Shear Moduli of Single-Walled Carbon Nanotube Ropes. *Phys. Rev. Lett.* **1999**, *82*, 944–947. [CrossRef]
47. Kis, A.; Csanyi, G.; Salvetat, J.P.; Lee, T.N.; Couteau, E.; Kulik, A.J.; Benoit, W.; Brugger, J.; Forro, L. Reinforcement of Single-Walled Carbon Nanotube Bundles by Intertube Bridging. *Nat. Mater.* **2004**, *3*, 153–157. [CrossRef]
48. Yu, M.F.; Files, B.S.; Arepalli, S.; Ruoff, R.S. Tensile Loading of Ropes of Single Wall Carbon Nanotubes and Their Mechanical Properties. *Phys. Rev. Lett.* **2000**, *84*, 5552–5555. [CrossRef]
49. Naraghi, M.; Filleter, T.; Moravsky, A.; Locascio, M.; Loutfy, R.O.; Espinosa, H.D. A Multiscale Study of High Performance Double-Walled Nanotube-Polymer Fibers. *ACS Nano* **2010**, *4*, 6463–6476. [CrossRef]
50. Filleter, T.; Bernal, R.; Li, S.; Espinosa, H.D. Ultrahigh Strength and Stiffness in Cross-Linked Hierarchical Carbon Nanotube Bundles. *Adv. Mater.* **2011**, *23*, 2855–2860. [CrossRef]
51. Jia, J.; Zhao, J.; Xu, G.; Di, J.; Yong, Z.; Tao, Y.; Fang, C.; Zhang, Z.; Zhang, X.; Zheng, L.; et al. A Comparison of the Mechanical Properties of Fibers Spun from Different Carbon Nanotubes. *Carbon* **2011**, *49*, 1333–1339. [CrossRef]
52. Davis, V.A.; Parra-Vasquez, A.N.G.; Green, M.J.; Rai, P.K.; Behabtu, N.; Prieto, V.; Booker, R.D.; Schmidt, J.; Kesselman, E.; Zhou, W.; et al. True Solutions of Single-Walled Carbon Nanotubes for Assembly into Macroscopic Materials. *Nat. Nanotechnol.* **2009**, *4*, 830–834. [CrossRef]
53. Jung, Y.; Cho, Y.S.; Lee, J.W.; Oh, J.Y.; Park, C.R. How Can We Make Carbon Nanotube Yarn Stronger? *Compos. Sci. Technol.* **2018**, *166*, 95–108. [CrossRef]
54. Ericson, L.M.; Fan, H.; Peng, H.Q.; Davis, V.A.; Zhou, W.; Sulpizio, J.; Wang, Y.H.; Booker, R.; Vavro, J.; Guthy, C.; et al. Macroscopic, Neat, Single-Walled Carbon Nanotube Fibers. *Science* **2004**, *305*, 1447–1450. [CrossRef] [PubMed]
55. Jiang, K.; Li, Q.; Fan, S. Nanotechnology: Spinning Continuous Carbon Nanotube Yarns. *Nature* **2002**, *419*, 801. [CrossRef] [PubMed]
56. Li, Y.L.; Kinloch, I.A.; Windle, A.H. Direct Spinning of Carbon Nanotube Fibers from Chemical Vapor Deposition Synthesis. *Science* **2004**, *304*, 276–278. [CrossRef]
57. Jassal, M.; Ghosh, S. Aramid Fibres—An Overview. *Indian J. Fibre Text. Res.* **2002**, *27*, 290–306.
58. Khayyam, H.; Jazar, R.N.; Nunna, S.; Golkarnarenji, G.; Badii, K.; Fakhrhoseini, S.M.; Kumar, S.; Naebe, M. Pan Precursor Fabrication, Applications and Thermal Stabilization Process in Carbon Fiber Production: Experimental and Mathematical Modelling. *Prog. Mater Sci.* **2020**, *107*, 100575. [CrossRef]
59. Dalton, A.B.; Collins, S.; Munoz, E.; Razal, J.M.; Ebron, V.H.; Ferraris, J.P.; Coleman, J.N.; Kim, B.G.; Baughman, R.H. Super-Tough Carbon-Nanotube Fibres. *Nature* **2003**, *423*, 703. [CrossRef]
60. Dalton, A.B.; Collins, S.; Razal, J.; Munoz, E.; Ebron, V.H.; Kim, B.G.; Coleman, J.N.; Ferraris, J.P.; Baughman, R.H. Continuous Carbon Nanotube Composite Fibers: Properties, Potential Applications, and Problems. *J. Mater. Chem.* **2004**, *14*, 1–3. [CrossRef]
61. Zhang, S.; Koziol, K.K.K.; Kinloch, I.A.; Windle, A.H. Macroscopic Fibers of Well-Aligned Carbon Nanotubes by Wet Spinning. *Small* **2008**, *4*, 1217–1222. [CrossRef]
62. Wong, C.P.; Ohnuma, H.; Berry, G.C. Properties of Some Rodlike Polymers in Solution. *J. Polym. Sci. Part C Polym. Symp.* **1978**, *65*, 173–192. [CrossRef]
63. Parra-Vasquez, A.N.G.; Behabtu, N.; Green, M.J.; Pint, C.L.; Young, C.C.; Schmidt, J.; Kesselman, E.; Goyal, A.; Ajayan, P.M.; Cohen, Y.; et al. Spontaneous Dissolution of Ultralong Single- and Multiwalled Carbon Nanotubes. *ACS Nano* **2010**, *4*, 3969–3978. [CrossRef] [PubMed]
64. Behabtu, N.; Young, C.C.; Tsentalovich, D.E.; Kleinerman, O.; Wang, X.; Ma, A.W.K.; Bengio, E.A.; ter Waarbeek, R.F.; de Jong, J.J.; Hoogerwerf, R.E.; et al. Strong, Light, Multifunctional Fibers of Carbon Nanotubes with Ultrahigh Conductivity. *Science* **2013**, *339*, 182–186. [CrossRef] [PubMed]
65. Taylor, L.W.; Dewey, O.S.; Headrick, R.J.; Komatsu, N.; Peraca, N.M.; Wehmeyer, G.; Kono, J.; Pasquali, M. Improved Properties, Increased Production, and the Path to Broad Adoption of Carbon Nanotube Fibers. *Carbon* **2021**, *171*, 689–694. [CrossRef]

66. Headrick, R.J.; Tsentalovich, D.E.; Berdegue, J.; Bengio, E.A.; Liberman, L.; Kleinerman, O.; Lucas, M.S.; Talmon, Y.; Pasquali, M. Structure-Property Relations in Carbon Nanotube Fibers by Downscaling Solution Processing. *Adv. Mater.* **2018**, *30*, 1704482. [CrossRef] [PubMed]
67. Tsentalovich, D.E.; Ma, A.W.K.; Lee, J.A.; Behabtu, N.; Bengio, E.A.; Choi, A.; Hao, J.; Luo, Y.; Headrick, R.J.; Green, M.J.; et al. Relationship of Extensional Viscosity and Liquid Crystalline Transition to Length Distribution in Carbon Nanotube Solutions. *Macromolecules* **2016**, *49*, 681–689. [CrossRef]
68. Headrick, R.J.; Williams, S.M.; Owens, C.E.; Taylor, L.W.; Dewey, O.S.; Ginestra, C.J.; Liberman, L.; Ya'akobi, A.M.; Talmon, Y.; Maruyama, B.; et al. Versatile Acid Solvents for Pristine Carbon Nanotube Assembly. *Sci. Adv.* **2022**, *8*, eabm3285. [CrossRef] [PubMed]
69. Zhang, X.B.; Jiang, K.L.; Teng, C.; Liu, P.; Zhang, L.; Kong, J.; Zhang, T.H.; Li, Q.Q.; Fan, S.S. Spinning and Processing Continuous Yarns from 4-Inch Wafer Scale Super-Aligned Carbon Nanotube Arrays. *Adv. Mater.* **2006**, *18*, 1505–1510. [CrossRef]
70. Zhang, M.; Atkinson, K.R.; Baughman, R.H. Multifunctional Carbon Nanotube Yarns by Downsizing an Ancient Technology. *Science* **2004**, *306*, 1358–1361. [CrossRef]
71. Zhang, M.; Fang, S.L.; Zakhidov, A.A.; Lee, S.B.; Aliev, A.E.; Williams, C.D.; Atkinson, K.R.; Baughman, R.H. Strong, Transparent, Multifunctional, Carbon Nanotube Sheets. *Science* **2005**, *309*, 1215–1219. [CrossRef]
72. Li, F.; Cheng, H.M.; Bai, S.; Su, G.; Dresselhaus, M.S. Tensile Strength of Single-Walled Carbon Nanotubes Directly Measured from Their Macroscopic Ropes. *Appl. Phys. Lett.* **2000**, *77*, 3161–3163. [CrossRef]
73. Zhu, H.W.; Xu, C.L.; Wu, D.H.; Wei, B.Q.; Vajtai, R.; Ajayan, P.M. Direct Synthesis of Long Single-Walled Carbon Nanotube Strands. *Science* **2002**, *296*, 884–886. [CrossRef]
74. Zhong, X.-H.; Li, Y.-L.; Liu, Y.-K.; Qiao, X.-H.; Feng, Y.; Liang, J.; Jin, J.; Zhu, L.; Hou, F.; Li, J.-Y. Continuous Multilayered Carbon Nanotube Yarns. *Adv. Mater.* **2010**, *22*, 692–696. [CrossRef] [PubMed]
75. Steinmetz, J.; Glerup, M.; Paillet, M.; Bernier, P.; Holzinger, M. Production of Pure Nanotube Fibers Using a Modified Wet-Spinning Method. *Carbon* **2005**, *43*, 2397–2400. [CrossRef]
76. He, Z.; Zhou, G.; Byun, J.-H.; Lee, S.-K.; Um, M.-K.; Park, B.; Kim, T.; Lee, S.B.; Chou, T.-W. Highly Stretchable Multi-Walled Carbon Nanotube/Thermoplastic Polyurethane Composite Fibers for Ultrasensitive, Wearable Strain Sensors. *Nanoscale* **2019**, *11*, 5884–5890. [CrossRef] [PubMed]
77. Klammer, R. Huntsman Acquires Nanocomp Technologies. *Surf. Coat. Int.* **2018**, *101*, 144.
78. Nanocomp Technologies. Nanocomp Technologies Awarded Air Force Contract. *Adv. Mater. Process.* **2010**, *168*, 12.
79. Sammalkorpi, M.; Krasheninnikov, A.; Kuronen, A.; Nordlund, K.; Kaski, K. Mechanical Properties of Carbon Nanotubes with Vacancies and Related Defects. *Phys. Rev. B* **2004**, *70*, 245416. [CrossRef]
80. Zhang, S.L.; Mielke, S.L.; Khare, R.; Troya, D.; Ruoff, R.S.; Schatz, G.C.; Belytschko, T. Mechanics of Defects in Carbon Nanotubes: Atomistic and Multiscale Simulations. *Phys. Rev. B* **2005**, *71*, 115403. [CrossRef]
81. Ebbesen, T.W.; Takada, T. Topological and Sp3 Defect Structures in Nanotubes. *Carbon* **1995**, *33*, 973–978. [CrossRef]
82. Samsonidze, G.G.; Samsonidze, G.G.; Yakobson, B.I. Kinetic Theory of Symmetry-Dependent Strength in Carbon Nanotubes. *Phys. Rev. Lett.* **2002**, *88*, 065501. [CrossRef]
83. Yakobson, B.I. Mechanical Relaxation and "Intramolecular Plasticity" in Carbon Nanotubes. *Appl. Phys. Lett.* **1998**, *72*, 918–920. [CrossRef]
84. Nardelli, M.B.; Yakobson, B.I.; Bernholc, J. Brittle and Ductile Behavior in Carbon Nanotubes. *Phys. Rev. Lett.* **1998**, *81*, 4656–4659. [CrossRef]
85. Yuan, Q.; Li, L.; Li, Q.; Ding, F. Effect of Metal Impurities on the Tensile Strength of Carbon Nanotubes: A Theoretical Study. *J. Phys. Chem. C* **2013**, *117*, 5470–5474. [CrossRef]
86. Wong, E.W.; Sheehan, P.E.; Lieber, C.M. Nanobeam Mechanics: Elasticity, Strength, and Toughness of Nanorods and Nanotubes. *Science* **1997**, *277*, 1971–1975. [CrossRef]
87. Wei, X.; Chen, Q.; Xu, S.; Peng, L.; Zuo, J. Beam to String Transition of Vibrating Carbon Nanotubes under Axial Tension. *Adv. Funct. Mater.* **2009**, *19*, 1753–1758. [CrossRef]
88. Ganesan, Y.; Peng, C.; Lu, Y.; Ci, L.; Srivastava, A.; Ajayan, P.M.; Lou, J. Effect of Nitrogen Doping on the Mechanical Properties of Carbon Nanotubes. *ACS Nano* **2010**, *4*, 7637–7643. [CrossRef] [PubMed]
89. Takakura, A.; Beppu, K.; Nishihara, T.; Fukui, A.; Kozeki, T.; Namazu, T.; Miyauchi, Y.; Itami, K. Strength of Carbon Nanotubes Depends on Their Chemical Structures. *Nat. Commun.* **2019**, *10*, 1–7. [CrossRef]
90. Peng, B.; Locascio, M.; Zapol, P.; Li, S.; Mielke, S.L.; Schatz, G.C.; Espinosa, H.D. Measurements of near-Ultimate Strength for Multiwalled Carbon Nanotubes and Irradiation-Induced Crosslinking Improvements. *Nat. Nanotechnol.* **2008**, *3*, 626–631. [CrossRef]
91. Vilatela, J.J.; Elliott, J.A.; Windle, A.H. A Model for the Strength of Yarn-Like Carbon Nanotube Fibers. *ACS Nano* **2011**, *5*, 1921–1927. [CrossRef]
92. Zhang, X.; Li, Q.; Holesinger, T.G.; Arendt, P.N.; Huang, J.; Kirven, P.D.; Clapp, T.G.; DePaula, R.F.; Liao, X.; Zhao, Y.; et al. Ultrastrong, Stiff, and Lightweight Carbon-Nanotube Fibers. *Adv. Mater.* **2007**, *19*, 4198–4201. [CrossRef]
93. Weibull, W. A Statistical Distribution Function of Wide Applicability. *J. Appl. Mech. Trans. Asme* **1951**, *18*, 293–297. [CrossRef]
94. Nakagawa, T.; Osaki, S. Discrete Weibull Distribution. *IEEE Trans. Reliab.* **1975**, *24*, 300–301. [CrossRef]

95. Zhou, Y.; Wang, Y.; Xia, Y.; Jeelani, S. Tensile Behavior of Carbon Fiber Bundles at Different Strain Rates. *Mater. Lett.* **2010**, *64*, 246–248. [CrossRef]
96. Daniels, H.E. The Statistical Theory of the Strength of Bundles of Threads. I. *Proc. R. Soc. Lond. Ser. A Math. Phys. Sci.* **1945**, *183*, 405–435.
97. Gollwitzer, S.; Rackwitz, R. On the Reliability of Daniels Systems. *Struct. Saf.* **1990**, *7*, 229–243. [CrossRef]
98. Lan, C.; Li, H.; Ju, Y. Bearing Capacity Assessment for Parallel Wire Cables. *China Civ. Eng. J.* **2013**, *46*, 31–38.
99. Chi, Z.F.; Chou, T.W.; Shen, G.Y. Determination of Single Fiber Strength Distribution from Fiber Bundle Testings. *J. Mater. Sci.* **1984**, *19*, 3319–3324. [CrossRef]
100. Gao, G.; Cagin, T.; Goddard, W.A. Energetics, Structure, Mechanical and Vibrational Properties of Single-Walled Carbon Nanotubes. *Nanotechnology* **1999**, *9*, 184–191. [CrossRef]
101. Belytschko, T.; Xiao, S.P.; Schatz, G.C.; Ruoff, R.S. Atomistic Simulations of Nanotube Fracture. *Phys. Rev. B* **2002**, *65*, 235430. [CrossRef]
102. Hu, J.; Zheng, Q. Application of 2000 mpa Parallel Wire Cables to Cable-Stayed Rail-Cum-Road Bridge with Main Span Length over 1000 m. *Bridge Constr.* **2019**, *49*, 48–53.
103. Li, S.; Xu, Y.; Zhu, S.; Guan, X.; Bao, Y. Probabilistic Deterioration Model of High-Strength Steel Wires and Its Application to Bridge Cables. *Struct. Infrastruct. Eng.* **2015**, *11*, 1240–1249. [CrossRef]
104. Fang, I.K.; Chen, C.R.; Chang, I.S. Field Static Load Test on Kao-Ping-Hsi Cable-Stayed Bridge. *J. Bridge Eng.* **2004**, *9*, 531–540. [CrossRef]
105. Li, Q.; Zhang, X.; DePaula, R.F.; Zheng, L.; Zhao, Y.; Stan, L.; Holesinger, T.G.; Arendt, P.N.; Peterson, D.E.; Zhu, Y.T. Sustained Growth of Ultralong Carbon Nanotube Arrays for Fiber Spinning. *Adv. Mater.* **2006**, *18*, 3160–3163. [CrossRef]
106. Li, Q.; Li, Y.; Zhang, X.; Chikkannanavar, S.B.; Zhao, Y.; Dangelewicz, A.M.; Zheng, L.; Doorn, S.K.; Jia, Q.; Peterson, D.E.; et al. Structure-Dependent Electrical Properties of Carbon Nanotube Fibers. *Adv. Mater.* **2007**, *19*, 3358–3363. [CrossRef]
107. Tsentalovich, D.E.; Headrick, R.J.; Mirri, F.; Hao, J.; Behabtu, N.; Young, C.C.; Pasquali, M. Influence of Carbon Nanotube Characteristics on Macroscopic Fiber Properties. *ACS Appl. Mater. Interfaces* **2017**, *9*, 36189–36198. [CrossRef]
108. Zhang, X.F.; Li, Q.W.; Tu, Y.; Li, Y.A.; Coulter, J.Y.; Zheng, L.X.; Zhao, Y.H.; Jia, Q.X.; Peterson, D.E.; Zhu, Y.T. Strong Carbon-Nanotube Fibers Spun from Long Carbon-Nanotube Arrays. *Small* **2007**, *3*, 244–248. [CrossRef]
109. Zhang, Y.; Zou, G.; Doorn, S.K.; Htoon, H.; Stan, L.; Hawley, M.E.; Sheehan, C.J.; Zhu, Y.; Jia, Q. Tailoring the Morphology of Carbon Nanotube Arrays: From Spinnable Forests to Undulating Foams. *ACS Nano* **2009**, *3*, 2157–2162. [CrossRef]
110. Huynh, C.P.; Hawkins, S.C. Understanding the Synthesis of Directly Spinnable Carbon Nanotube Forests. *Carbon* **2010**, *48*, 1105–1115. [CrossRef]
111. Xie, S.S.; Li, W.Z.; Pan, Z.W.; Chang, B.H.; Sun, L.F. Mechanical and Physical Properties on Carbon Nanotube. *J. Phys. Chem. Solids* **2000**, *61*, 1153–1158. [CrossRef]
112. Fang, S.; Zhang, M.; Zakhidov, A.A.; Baughman, R.H. Structure and Process-Dependent Properties of Solid-State Spun Carbon Nanotube Yarns. *J. Phys. Condens. Matter* **2010**, *22*, 334221. [CrossRef]
113. Kuznetsov, A.A.; Fonseca, A.F.; Baughman, R.H.; Zakhidov, A.A. Structural Model for Dry-Drawing of Sheets and Yarns from Carbon Nanotube Forests. *ACS Nano* **2011**, *5*, 985–993. [CrossRef] [PubMed]
114. Miao, M.; Hawkins, S.C.; Cai, J.Y.; Gengenbach, T.R.; Knott, R.; Huynh, C.P. Effect of Gamma-Irradiation on the Mechanical Properties of Carbon Nanotube Yarns. *Carbon* **2011**, *49*, 4940–4947. [CrossRef]
115. Cai, J.Y.; Min, J.; McDonnell, J.; Church, J.S.; Easton, C.D.; Humphries, W.; Lucas, S.; Woodhead, A.L. An Improved Method for Functionalisation of Carbon Nanotube Spun Yarns with Aryldiazonium Compounds. *Carbon* **2012**, *50*, 4655–4662. [CrossRef]
116. Li, S.; Zhang, X.; Zhao, J.; Meng, F.; Xu, G.; Yong, Z.; Jia, J.; Zhang, Z.; Li, Q. Enhancement of Carbon Nanotube Fibres Using Different Solvents and Polymers. *Compos. Sci. Technol.* **2012**, *72*, 1402–1407. [CrossRef]
117. Liu, K.; Sun, Y.; Lin, X.; Zhou, R.; Wang, J.; Fan, S.; Jiang, K. Scratch-Resistant, Highly Conductive, and High-Strength Carbon Nanotube-Based Composite Yarns. *ACS Nano* **2010**, *4*, 5827–5834. [CrossRef] [PubMed]
118. Kozlov, M.E.; Capps, R.C.; Sampson, W.M.; Ebron, V.H.; Ferraris, J.P.; Baughman, R.H. Spinning Solid and Hollow Polymer-Free Carbon Nanotube Fibers. *Adv. Mater.* **2005**, *17*, 614–617. [CrossRef]
119. Nurazzi, N.M.; Sabaruddin, F.A.; Harussani, M.M.; Kamarudin, S.H.; Rayung, M.; Asyraf, M.R.M.; Aisyah, H.A.; Norrrahim, M.N.F.; Ilyas, R.A.; Abdullah, N.; et al. Mechanical Performance and Applications of Cnts Reinforced Polymer Composites—A Review. *Nanomaterials* **2021**, *11*, 2186. [CrossRef]
120. Kim, T.; Shin, J.; Lee, K.; Jung, Y.; Lee, S.B.; Yang, S.J. A Universal Surface Modification Method of Carbon Nanotube Fibers with Enhanced Tensile Strength. *Compos. Part A Appl. Sci. Manuf.* **2021**, *140*, 106182. [CrossRef]
121. Cetin, M.E. Investigation of Carbon Nanotube Reinforcement to Polyurethane Adhesive for Improving Impact Performance of Carbon Fiber Composite Sandwich Panels. *Int. J. Adhes. Adhes.* **2022**, *112*, 103002. [CrossRef]
122. Gardner, J.M.; Sauti, G.; Kim, J.W.; Cano, R.J.; Wincheski, R.A.; Stelter, C.J.; Grimsley, B.W.; Working, D.C.; Siochi, E.J. 3-D Printing of Multifunctional Carbon Nanotube Yarn Reinforced Components. *Addit. Manuf.* **2016**, *12*, 38–44. [CrossRef]
123. Kim, J.W.; Sauti, G.; Cano, R.J.; Wincheski, R.A.; Ratcliffe, J.G.; Czabaj, M.; Gardner, N.W.; Siochi, E.J. Assessment of Carbon Nanotube Yarns as Reinforcement for Composite Overwrapped Pressure Vessels. *Compos. Part A Appl. Sci. Manuf.* **2016**, *84*, 256–265. [CrossRef]

124. Han, B.; Guo, E.; Xue, X.; Zhao, Z.; Luo, L.; Qu, H.; Niu, T.; Xu, Y.; Hou, H. Fabrication and Densification of High Performance Carbon Nanotube/Copper Composite Fibers. *Carbon* **2017**, *123*, 593–604. [CrossRef]
125. Zhu, P.; Wang, C.; Wang, X.; Yang, H.; Xu, G.; Qin, C.; Xiong, L.; Wen, G.; Xia, L. Superior Mechanical Properties of Lithium Aluminosilicate Composites by Pyrolytic Carbon Intercalated Carbon Fibers/Carbon Nanotubes Multi-Scale Reinforcements. *Ceram. Int.* **2021**, *47*, 32837–32846. [CrossRef]
126. Sadeghi, A.M.; Esmaeili, J. Hybrid-Fibre-Reinforced Concrete Containing Multi-Wall Carbon Nanotubes. *Proc. Inst. Civ. Eng. Struct. Build.* **2020**, *173*, 646–654. [CrossRef]
127. Benzait, Z.; Trabzon, L. A Review of Recent Research on Materials Used in Polymer-Matrix Composites for Body Armor Application. *J. Compos. Mater.* **2018**, *52*, 3241–3263. [CrossRef]
128. Xie, W.T.; Zhang, R.Y.; Headrick, R.J.; Taylor, L.W.; Kooi, S.; Pasquali, M.; Muftu, S.; Lee, J.H. Dynamic Strengthening of Carbon Nanotube Fibers under Extreme Mechanical Impulses. *Nano Lett.* **2019**, *19*, 3519–3526. [CrossRef]
129. Wang, Z.; Mun, T.J.; Machado, F.M.; Moon, J.H.; Fang, S.; Aliev, A.E.; Zhang, M.; Cai, W.; Mu, J.; Hyeon, J.S.; et al. More Powerful Twistron Carbon Nanotube Yarn Mechanical Energy Harvesters. *Adv. Mater.* **2022**, *34*, 2201826. [CrossRef]
130. Jang, Y.; Kim, S.M.; Kim, K.J.; Sim, H.J.; Kim, B.J.; Park, J.W.; Baughman, R.H.; Ruhparwar, A.; Kim, S.J. Self-Powered Coiled Carbon-Nanotube Yarn Sensor for Gastric Electronics. *ACS Sens.* **2019**, *4*, 2893–2899. [CrossRef]
131. Hu, X.; Jia, J.; Wang, Y.; Tang, X.; Fang, S.; Wang, Y.; Baughman, R.H.; Ding, J. Fast Large-Stroke Sheath-Driven Electrothermal Artificial Muscles with High Power Densities. *Adv. Funct. Mater.* **2022**, *32*, 2200591. [CrossRef]
132. Chen, S.H.; Qiu, L.; Cheng, H.M. Carbon-Based Fibers for Advanced Electrochemical Energy Storage Devices. *Chem. Rev.* **2020**, *120*, 2811–2878. [CrossRef]
133. Zhou, X.; Fang, S.; Leng, X.; Liu, Z.; Baughman, R.H. The Power of Fiber Twist. *Acc. Chem. Res.* **2021**, *54*, 2624–2636. [CrossRef] [PubMed]
134. Kim, S.H.; Haines, C.S.; Li, N.; Kim, K.J.; Mun, T.J.; Choi, C.; Di, J.T.; Oh, Y.J.; Oviedo, J.P.; Bykova, J.; et al. Harvesting Electrical Energy from Carbon Nanotube Yarn Twist. *Science* **2017**, *357*, 773–778. [CrossRef]
135. Di, J.; Fang, S.; Moura, F.A.; Galvao, D.S.; Bykova, J.; Aliev, A.; de Andrade, M.J.; Lepro, X.; Li, N.; Haines, C.; et al. Strong, Twist-Stable Carbon Nanotube Yarns and Muscles by Tension Annealing at Extreme Temperatures. *Adv. Mater.* **2016**, *28*, 6598–6605. [CrossRef] [PubMed]
136. Vallem, V.; Sargolzaeiaval, Y.; Ozturk, M.; Lai, Y.-C.; Dickey, M.D. Energy Harvesting and Storage with Soft and Stretchable Materials. *Adv. Mater.* **2021**, *33*, 2004832. [CrossRef] [PubMed]
137. Mun, T.J.; Kim, S.H.; Park, J.W.; Moon, J.H.; Jang, Y.; Huynh, C.; Baughman, R.H.; Kim, S.J. Wearable Energy Generating and Storing Textile Based on Carbon Nanotube Yarns. *Adv. Funct. Mater.* **2020**, *30*, 2000411. [CrossRef]
138. Jang, Y.; Kim, S.M.; Spinks, G.M.; Kim, S.J. Carbon Nanotube Yarn for Fiber-Shaped Electrical Sensors, Actuators, and Energy Storage for Smart Systems. *Adv. Mater.* **2020**, *32*, 1902670. [CrossRef]
139. Hyeon, J.S.; Park, J.W.; Baughman, R.H.; Kim, S.J. Electrochemical Graphene/Carbon Nanotube Yarn Artificial Muscles. *Sens. Actuators B—Chem.* **2019**, *286*, 237–242. [CrossRef]
140. Foroughi, J.; Spinks, G.M.; Wallace, G.G.; Oh, J.; Kozlov, M.E.; Fang, S.L.; Mirfakhrai, T.; Madden, J.D.W.; Shin, M.K.; Kim, S.J.; et al. Torsional Carbon Nanotube Artificial Muscles. *Science* **2011**, *334*, 494–497. [CrossRef]
141. Lee, J.A.; Li, N.; Haines, C.S.; Kim, K.J.; Lepro, X.; Ovalle-Robles, R.; Kim, S.J.; Baughman, R.H. Electrochemically Powered, Energy-Conserving Carbon Nanotube Artificial Muscles. *Adv. Mater.* **2017**, *29*, 1700870. [CrossRef]

Article

Single-Walled Carbon Nanotubes with Red Phosphorus in Lithium-Ion Batteries: Effect of Surface and Encapsulated Phosphorus

Anna A. Vorfolomeeva [1], Svetlana G. Stolyarova [1], Igor P. Asanov [1], Elena V. Shlyakhova [1], Pavel E. Plyusnin [1], Evgeny A. Maksimovskiy [1], Evgeny Yu. Gerasimov [2], Andrey L. Chuvilin [3,4], Alexander V. Okotrub [1] and Lyubov G. Bulusheva [1,*]

[1] Nikolaev Institute of Inorganic Chemistry SB RAS, 3 Acad. Lavrentiev Ave., 630090 Novosibirsk, Russia
[2] Boreskov Institute of Catalysis, SB RAS, 5 Acad. Lavrentiv Ave., 630090 Novosibirsk, Russia
[3] CIC NanoGUNE BRTA, Tolosa Hiribidea 76, E-20018 Donostia-San Sebastián, Spain
[4] IKERBASQUE, Basque Foundation of Science, Maria Diaz de Haro 3, E-48013 Bilbao, Spain
* Correspondence: bul@niic.nsc.ru; Tel.: +73-8333-053-52

Abstract: Single-walled carbon nanotubes (SWCNTs) with their high surface area, electrical conductivity, mechanical strength and elasticity are an ideal component for the development of composite electrode materials for batteries. Red phosphorus has a very high theoretical capacity with respect to lithium, but has poor conductivity and expends considerably as a result of the reaction with lithium ions. In this work, we compare the electrochemical performance of commercial SWCNTs with red phosphorus deposited on the outer surface of nanotubes and/or encapsulated in internal channels of nanotubes in lithium-ion batteries. External phosphorus, condensed from vapors, is easily oxidized upon contact with the environment and only the un-oxidized phosphorus cores participate in electrochemical reactions. The support of the SWCNT network ensures a stable long-term cycling for these phosphorus particles. The tubular space inside the SWCNTs stimulate the formation of chain phosphorus structures. The chains reversibly interact with lithium ions and provide a specific capacity of 1545 mAh·g^{-1} (calculated on the mass of phosphorus in the sample) at a current density of 0.1 A·g^{-1}. As compared to the sample containing external phosphorus, SWCNTs with encapsulated phosphorus demonstrate higher reaction rates and a slight loss of initial capacity (~7%) on the 1000th cycle at 5 A·g^{-1}.

Keywords: carbon nanotubes; phosphorus; encapsulation; lithium-ion batteries

1. Introduction

Elementary phosphorus (P) has attracted considerable attention as a promising anode material for lithium-ion batteries (LIBs) due to the strong affinity between P and the metal ion [1]. Phosphorus in the course of electrochemical reactions is able to accept three lithium ions (Li$_3$P) [2], which provide a theoretical capacity of 2596 mAh·g^{-1} [3]. Among all allotropes, red phosphorus is commercially available, chemically stabile and environmentally friendly [4].

Limitations in the use of red phosphorus in LIBs are associated with its low electrical conductivity (~10^{-14} S·cm^{-1}) [5,6], which leads to slow electrochemical redox reactions and poor rate performance. In addition, the high initial capacity of phosphorus drops rapidly during charge-discharge due to volume expansion (>300%) with the attachment of lithium [7]. As a result, red phosphorus anodes exhibit a rapid capacity loss, low Coulombic efficiency and electrode deterioration during cycling [8–10].

To alleviate these problems, a carbon matrix is used [1]. Allotropes of carbon are generally hydrophobic and resistant to chemicals under normal conditions. The combination of phosphorus and carbon is an opportunity to obtain a conductive material in which the carbon protects and/or stabilizes phosphorus during operation of the device [11]. Most of these combinations are obtained either by deposition of phosphorus on carbon using the evaporation-condensation technique [4,11–16], or by simple mixing of the components in a ball mill [2,17–23]. However, in this case, the carbon matrix is able to prevent material degradation only for a short time of LIB operation (less than 100 cycles) [24,25].

At present, it is generally accepted that phosphorus-based anodes should be made in the form of nanocomposites, in which the phosphorus component is finely dispersed in carbon [21]. Such materials are porous (micro- or mesoporous) carbon [3,4,13], modified graphene [6,26,27], carbon nanofibers [28] and carbon nanotubes (CNTs) [2,20,29]. Besides, it is very important to maintain a good electrical connection between the phosphorus and the conductive matrix during the lithiation processes.

The combination of phosphorus with porous carbon has been shown to provide good cycling stability of the composite electrode due to pores that prevent phosphorus sputtering [30]. Channels between CNTs and inside them can also perform this phosphorus-stabilizing role [31], preventing the contact of phosphorus with environmental molecules. CNTs are characterized by high conductivity ($>10^2$ S·cm^{-1}) [32], mechanical strength (Young's modulus < 1 TPa) [33–35], tensile strength < 70 GPa [35,36]) and flexibility [37,38]. Currently, several studies are reported using red phosphorus/CNT composites to improve electrochemical performance in lithiation/delithiation processes [17,39,40]. A high Coulombic efficiency in the fiftieth cycle was achieved for phosphorus-encapsulated CNTs; however, the specific capacity during cycling significantly decreased [41].

It can be assumed that the structure of phosphorus affects its interaction with lithium ions, although there are no clear studies in this area. The arrangement of phosphorus atoms inside the nanotube depends on the size of the internal space [42–45] and the synthesis conditions. The vaporization-condensation process is commonly used to fill CNTs with phosphorus [29,41,44]. This process also leads to the deposition of phosphorus on the outer surface of CNTs [46–48]. Phosphorus oxides, formed when the deposit comes into contact with laboratory air, are easily washed off with ethanol [43] or CS_2 [5,48], but the core of red phosphorus nanoparticles is quite stable. During LIB operation, this external phosphorus can detach from the conductive carbon surface and expand uncontrollably [44]. How to mitigate or suppress these negative effects is the key to improving battery performance [39].

Here, we study the effect of phosphorus located on the surface of SWCNTs and/or inside the channels of SWCNTs on the performance of the composite electrode in LIBs. Commercial SWCNTs (trademark Tuball) 1.6–2.9 nm in diameter with closed or open ends were used as a substrate for red phosphorus condensation at 800 °C. The structure and composition of the synthesized nanomaterials were determined by scanning electron microscopy (SEM) combined with elemental mapping by energy-dispersive X-ray spectroscopy (EDS), high-angle annular dark-field scanning transmission electron microscopy (HAADF-STEM), powder X-ray diffraction (XRD), Raman spectroscopy, thermogravimetric analysis (TGA) and X-ray photoelectron spectroscopy (XPS). The nanomaterials were comparatively studied in LIBs using galvanostatic discharge-charge (GDC) and cyclic voltammetry (CV) measurements. SWCNTs with the encapsulated phosphorus showed the best rate and cycling performance as well as faster kinetics of redox reactions as compared to SWCNTs with external phosphorus deposits.

2. Materials and Methods

2.1. Synthesis

SWCNTs were purchased from OCSiAl (Novosibirsk, Russia). Treatment with concentrated HCl was used to remove the accessible iron catalyst from the sample and the resulting SWCNTs served as a substrate to form the external phosphorus deposits (sample SWCNTs/P). To open the caps of the nanotubes, the original sample was heated in air at 500 °C for 1 h and then washed with concentrated HCl and dried in air. The obtained sample is labeled as o-SWCNTs and used as a reference.

Synthesis of composite nanomaterials was carried out in an H-shaped quartz ampoule by the vaporization-condensation method using amorphous red phosphorus (Prime Chemicals Group, >98%). A phosphorus sample (60 mg) and a nanotube sample (30 mg) were placed in different parts of an H-shaped quartz ampoule, the ampoule was pumped out, sealed, heated in a muffle furnace to 800 °C with a rate of ~7 °C·min^{-1} and kept at the final temperature for 48 h. After that, the ampoule was naturally cooled to room temperature and carefully opened in air. The product obtained with o-SWCNTs is designated as P@SWCNTs/P. Treatment with an aqueous NaOH solution (2.5 M) was used to remove external phosphorus deposits [46]. Briefly, a P@SWCNTs/P sample was vigorously stirred at 60 °C for 5 h, filtered, washed to neutral pH and dried in an oven. This sample is designated P@SWCNTs. A reference sample of red phosphorus P_{re} was synthesized by recrystallization of a commercial amorphous phosphorus sample under the conditions used for the synthesis of composite nanomaterials.

2.2. Instrumental Methods

The sample morphology was characterized by SEM on a JEOL JSM 6700 F (JEOL LTD., Tokyo, Japan) microscope. Elemental analysis was carried out by EDS on a Bruker XFlash 6 spectrometer. HAADF-STEM and EDS mapping were carried out at 80 kV on a Themis Z microscope (Thermo Fisher Scientific Inc., Eindhoven, The Netherlands) equipped with x-FEG monochromator, dual side C_s spherical aberration correction and Super-X EDS detector. High-resolution TEM (HRTEM) images were obtained using a Titan 60–300 microscope (FEI, Eindhoven, The Netherlands) at an accelerating voltage of 80 kV.

Raman spectra were collected on a LabRAM HR Evolution Horiba spectrometer (Horiba, Kyoto, Japan) using excitation from an Ar$^+$ laser at 514 nm at a power of 1 mW. The spectra at room temperature were obtained in the backscattering geometry. The laser beam was focused to a diameter of 2 μm using an LMPlan FL 50×/0.50 Olympus objective. The spectral resolution was 3.0 cm^{-1}. The size of the laser spot was 1 μm. The XRD patterns of the samples were obtained on a Shimadzu XRD-7000 diffractometer (Shimadzu Europa GmbH, Duisburg, Germany) using Cu Kα radiation and Ni filter on the reflected beam.

Simultaneous thermal analysis (STA) includes thermogravimetric (TG) and evolved gas analysis using mass spectrometry. Measurements were performed on a NETZSCH STA 449F1 Jupiter instrument (Selb/Bayern, Germany). The sample was placed in an open Al$_2$O$_3$ crucible and heated in a helium flow (20 mL·min^{-1}) from room temperature to 1000 °C at a rate of 10 °C·min^{-1}. An electron impact ionizer operated at an energy of 70 eV. The ion currents of the selected mass/charge (m/z) numbers were monitored in multiple-ion detection mode with a collection time of 0.1 s for each channel.

XPS measurements were carried out on an X-ray photoelectron spectrometer FleXPS (Specs GmbH, Berlin, Germany) via an electron energy analyzer Phoibos 150 with a delay line detector (DLD) electron detector. The spectra were excited by Al Kα radiation (1486.7 eV). The transmission energy of the analyzer was 20 eV. The vacuum in an analytical chamber was ~10^{-9} mbar. Elemental composition was evaluated from the ratio of the areas under corresponding core-level peaks of the survey spectra, taking into account the photoionization cross sections at a given photon energy. Data processing was carried out using the CasaXPS package (Casa Software Ltd., Teignmouth, UK). The fitting of the spectra was performed using symmetric lines as a product of the Gaussian and Lorentzian components. The XPS P 2p spectrum was fitted by two spin–orbit P 2p$_{3/2}$–P 2p$_{1/2}$ doublets with a ratio of the components of 2:1 and separation of 0.84 eV after subtraction of the background signal by Shirley's method.

2.3. Electrochemical Measurements

To prepare the electrode material, a portion of the testing sample (~30 mg) was mixed with polyvinylidene fluoride (PVDF-2, 10 wt%), necessary for better adhesion to the copper foil surface, conductive carbon additive super P (10 wt%) and 2–3 mL of N-methyl-pyrrolidone solvent. The mixture was stirred with steel balls on a vibrating mixer for 1 h.

The resulting suspension was distributed over a copper foil and dried at 80 °C for 12 h under vacuum. Electrodes with a diameter of 10 mm were obtained using a cylindrical cutter. The mass of the active electrode material was determined from the mass difference between the initial copper substrate and the substrate with the deposited material. The weight of each electrode material was ca. 0.4–0.5 mg. The electrochemical cells were assembled in a glove box filled with argon with the water and oxygen content less than 0.1 ppm. The CR2032 coin cells were assembled with lithium metal as the counter electrode. The electrolyte was a 1 M solution of $LiPF_6$ in a mixture of ethylene carbonate and dimethyl carbonate (1:1 by volume) from Merck Co. GDC tests were performed on a NEWARE CT-3008 station (Neware Technology Ltd., Shenzhen, China) in the voltage range from 0.01 to 2.50 V vs. Li/Li^+ at current densities from 0.1 $A \cdot g^{-1}$ to 5 $A \, g^{-1}$. CV measurements were carried out using a Bio-Logic SP-300 station (Bio-Logic Science Instrument, Seyssinet-Parist, France) from 0.01 to 2.5 V at scan rates varied from 0.1 to 1 $mV \cdot s^{-1}$.

3. Results

3.1. Characterization of Nanomaterials

The synthesis of composite nanomaterials is illustrated in Figure 1. At high temperatures, phosphorus evaporates and moves into the part containing carbon nanotubes. Vapors penetrate into the cavities of nanotubes when their caps are opened (o-SWCNTs sample). Upon cooling, phosphorus species condense on accessible substrates, in particularly, on the inner and outer surfaces of nanotubes and the ampoule walls. Therefore, the P@SWCNTs/P sample contains encapsulated phosphorus and phosphorus covering the nanotubes. Treatment of this sample with NaOH removes external phosphorus [46] and the resulting P@SWCNTs are nanotubes filled with phosphorus. In the SWCNTs/P sample, phosphorus should be present only on the outer surface of the nanotubes, because their ends are closed.

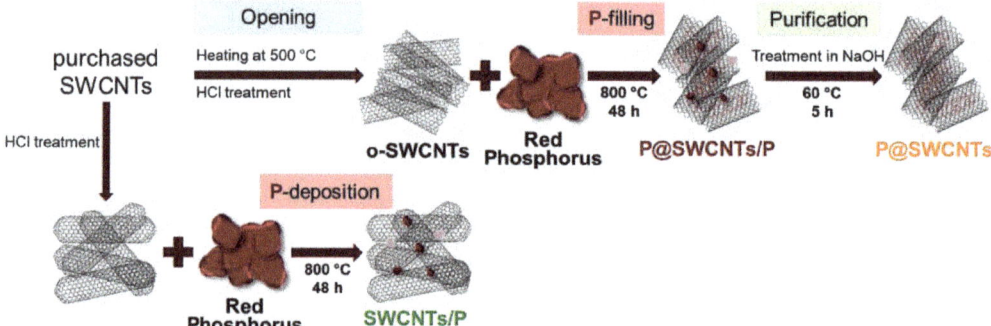

Figure 1. Schematic illustration of the synthesis of composite nanomaterials from SWCNTs and red phosphorus.

The SEM study shows that the P@SWCNTs/P sample has a fibrous structure; however, individual bundles of nanotubes can be distinguished only at breaks (Figure 2a). This is due to the coating of nanotubes with phosphorus. The bundles of nanotubes are clearly visible in the P@SWCNTs image (Figure 2b), which confirms effectiveness of the purification procedure used. EDS mapping shows differences between P@SWCNTs/P (Figure 2c) and P@SWCNTs (Figure 2d). In the former image, there are areas of phosphorus accumulation, which may correspond to surface phosphorus nanoparticles. In the latter image, phosphorus is evenly distributed throughout the sample.

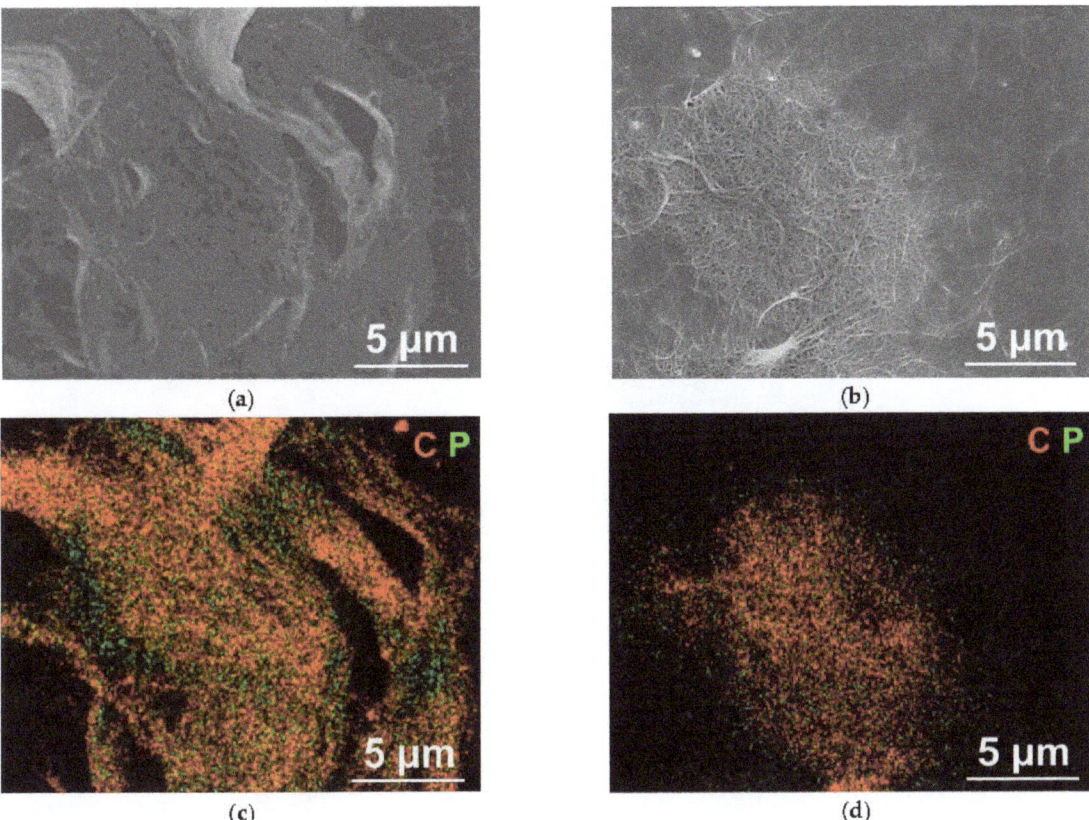

Figure 2. (a,b) SEM images and (c,d) corresponding EDS combined maps of carbon and phosphorus elements for (a,c) P@SWCNTs/P and (b,d) P@SWCNTs.

The change in the structure of the initial o-SWCNTs sample after phosphorus condensation on the outer and inner surfaces of the nanotube is supported by XRD analysis (Figure S1). The o-SWCNTs pattern contains reflections from bundled nanotubes [49]. These reflections are present in the P@SWCNTs/P and P@SWCNTs patterns and overlap with broad reflections at approximately 2θ of 18°, 24° and 32°, corresponding to fibrous red phosphorus [50].

To confirm the filling of nanotubes, we performed a study of P@SWCNTs using HAADF/STEM and EDS analysis. Figure 3a shows a HAADF image of a nanotube bundle. Bright strips along the axis of the bundle correspond to phosphorus since this element is heavier than carbon. The combined elemental map of phosphorus and carbon is depicted in Figure 3b. The phosphorus signal (green) appears evenly along the length of the bundle, while in the transverse direction it alternates with the carbon signal (red). This distribution of phosphorus indicates its location in the internal cavities of nanotubes or between neighboring nanotubes.

EDS analysis perpendicular to the bundle axis (inset in Figure 3c) found a low fraction of phosphorus (Figure 3c). The phosphorus content determined from the EDS data (Figure S2) is ~6 at% (15 wt%). The fact that the carbon profile is wider (Figure 3c) confirms the absence of phosphorus on the surface of the SWCNT bundle. The phosphorus profile is periodic due to the restrictive effect of the carbon walls. The lower intensity in the phosphorus signal is accompanied by a peak in the carbon signal. Therefore, phosphorus has a preferable location along the nanotubes. The HRTEM examination revealed the encapsulation of phosphorus in SWCNTs (Figure 3d). Phosphorus forms long ordered chains similar to the fibrous structure of red phosphorus [42].

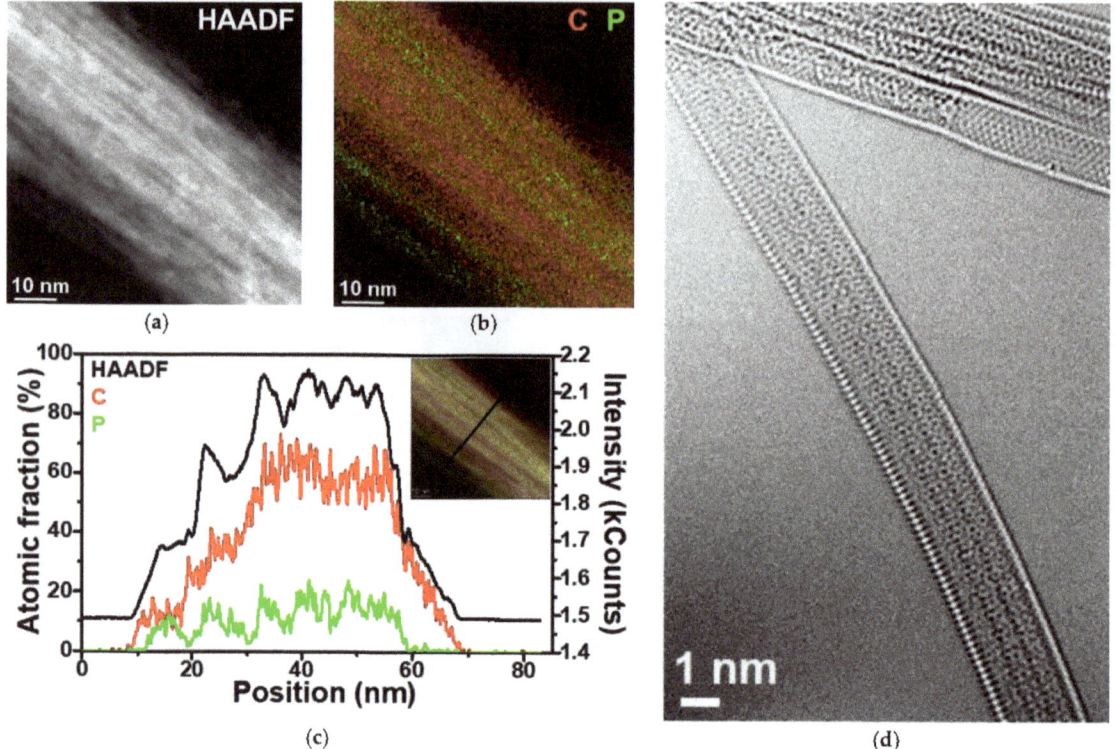

Figure 3. (**a**) HAADF image of P@SWCNTs; (**b**) corresponding EDS elemental map of carbon and phosphorus and (**c**) distribution of elements perpendicularly to the nanotube bundle (along the line shown in the inset); (**d**) HRTEM image of phosphorus-filled SWCNTs.

The phosphorus content determined from the analysis of the XPS survey spectra is ca. 16 at% (32 wt%) in P@SWCNTs/P, ca. 11 at% (23 wt%) in SWCNTs/P and ca. 9 at% (18 wt%) in P@SWCNTs (Figure S3a). The latter value agrees well with the phosphorus content obtained using EDS (Figure S2). The same ratio of reagents was used in all synthesis and the lower content of phosphorus in SWCNTs/P, where the nanotube caps are closed, as compared to P@SWCNTs/P, where the nanotubes are open, indicates that a surface is required for phosphorus condensation. During the synthesis of SWCNTs/P, only the outer surface of the nanotubes is accessible and the residual phosphorus is deposited on the walls of the ampoule.

The XPS P 2p spectra of the composite samples were fitted by three spin-orbit doublets (Figure 4a). The low-energy doublet with P $2p_{3/2}$ binding energy of 130.1 eV is assigned to elementary phosphorus, i.e., P-P species. This component dominates in the P@SWCNTs spectrum and has the lowest relative intensity in the SWCNTs/P spectrum. The P $2p_{3/2}$ line at 133.8 eV is associated with the P^{+3} state in oxidized species [46,51]. For a good fitting of the P@SWCNTs spectrum, an additional weak doublet was added between the P-P and P^{+3} components. The P $2p_{3/2}$ binding energy of 132.0 eV corresponds to the oxygen-terminated ends of phosphorus polymers [52]. We attribute this state to the ends of the encapsulated phosphorus chains, which attached to oxygen when the sample came into contact with laboratory air.

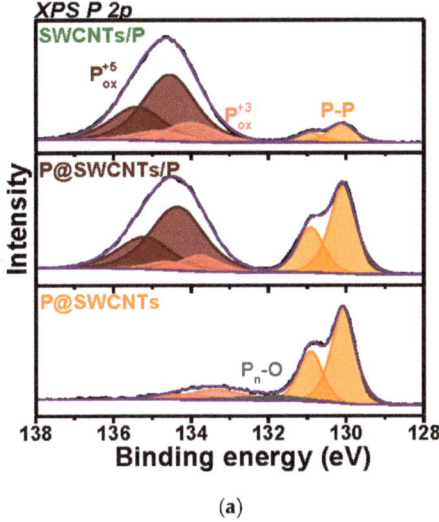
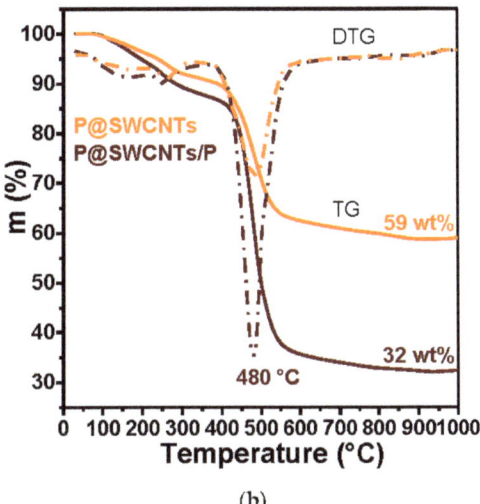

(a) (b)

Figure 4. (a) XPS P 2p spectra of P@SWCNTs (encapsulated phosphorus), P@SWCNTs/P (encapsulated and external phosphorus) and SWCNTs/P (external phosphorus); (b) TG and DTG curves measured for P@SWCNTs/P and P@SWCNTs in helium.

The P 2p spectra of P@SWCNTs/P and SWCNTs/P samples contain an intense doublet with the P $2p_{3/2}$ binding energy of 134.4 eV from the P^{+5} state in oxidized forms [51,53]. Doubles of the oxidized P^{+3} and P^{+5} states form a peak with a maximum at 134.5 eV and this peak is considerably higher in the SWCNTs/P spectrum (Figure S3b). The ratio of the area of this peak to the area under the P-P doublet is 8.8 for SWCNTs/P and 3.5 for P@SWCNTs/P. The outer surface of the SWCNTs can be covered by white phosphorus (this form is clearly visible in the ampoule after synthesis) and red phosphorus nanoparticles, observed by SEM/EDS analysis (Figure 2a,c). White phosphorus and the surface of phosphorus nanoparticles are oxidized when interacting with oxygen and water molecules from the air, while the nanotube walls protect the encapsulated phosphorus. The absence of high-energy doublet in the P 2p spectrum of P@SWCNTs (Figure 4a) confirms the cleaning of the outer surface of the nanotubes in this sample.

The decomposition of phosphorus-filled samples P@SWCNTs/P and P@SWCNTs was studied in helium by the STA method (Figure 4b). The broad peak observed in the differential TG (DTG) curves between 40–200 °C corresponds to removal of adsorbed water molecules. Next peak at ~250 °C is accompanied by the release of P, PH_2 and PH_3 species from the samples (Figure S4). This process causes a loss of ~6 wt% for P@SWCNTs/P and ~5 wt% for P@SWCNTs. The main weight loss occurs at ~480 °C due to the evaporation of P, PO and P_2. As a result of this process P@SWCNTs/P and P@SWCNTs samples lose ~51 wt% and ~28 wt%, respectively. The SWCNTs/P sample shows a similar thermal behavior (Figure S5) and has a mass loss of ca. 27 wt% over the range from 380 to 600 °C. The loss of samples stops after ~850 °C and the mass of the residue corresponds to the weight of SWCNTs in the composite. The almost two times lower SWCNT mass in the P@SWCNTs/P sample as compares to the P@SWCNTs sample (Figure 4b) is due to the fact that the former sample contains external phosphorus.

Raman spectroscopy is used to compare the structure of SWCNTs and phosphorus in reference and composite samples. The first-order region of the o-SWCNTs spectrum exhibits a weak D-band at 1357 cm^{-1} (Figure 5a) due to the low density of defects in the nanotube walls. Since the intensity of this band increases insignificantly in the spectra of composite nanomaterials, we conclude that carbon atoms do not form covalent bonds with phosphorus. The G-band at 1591 cm^{-1} with a shoulder at 1574 cm^{-1} observed in the Raman spectrum of o-SWCNTs corresponds to longitudinal optical and transverse optical phonons in semiconducting nanotubes [54]. The shoulder becomes less pronounced in SWCNTs/P and overlaps with the main G-band peak in P@SWCNTs and P@SWCNTs/P. Thus, the nanotube surface is modified in composite nanomaterials as compared to the initial o-SWCNTs. Interestingly the G-band is upshifted by ~7 cm^{-1} in the spectra of P@SWCNTs and P@SWCNTs/P. This p-doping of SWCNTs is not observed for the SWCNTs/P sample, where phosphorus located on the outer surface of the nanotubes. Therefore, only the encapsulated phosphorus accepts electron density from SWCNTs. According to density functional calculations, this occurs when phosphorus has a chain-like structure [46].

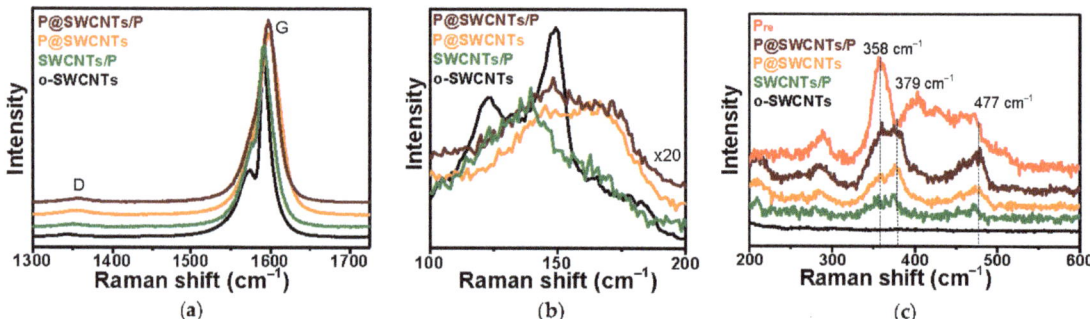

Figure 5. Raman spectra of o-SWCNTs sample and composite nanomaterials: (**a**) the first order region and (**b**) the RBM-mode region, presented with a zoom of 20; (**c**) Raman scattering from 200 to 600 cm^{-1} for reference o-SWCNTs and recrystallized phosphorus (P$_{re}$) samples and composite nanomaterials.

The peaks that appeared in the Raman spectrum of o-SWCNTs at 123 and 148 cm^{-1} (Figure 5b) correspond to the radial breathing modes (RBM) [55]. This region changes strongly in the spectra of composite nanomaterials. The radial vibrations of the SWCNTs are limited by the phosphorus coating in the SWCNTs/P sample. In the case of P@SWCNTs, upshifts of the peaks can be associated with p-doping, which changes the conditions for resonant excitation of SWCNTs [56]. Both the above effects are valid for P@SWCNTs/P, whose spectrum is a superposition of the spectra of SWCNTs/P and P@SWCNTs.

Raman scattering in phosphorous structures gives peaks in the region of 200–600 cm^{-1} [57], where the spectrum of o-SWCNTs has no features (Figure 5c). The Raman spectrum of phosphorus condensed in an ampoule without SWCNTs (sample P$_{re}$) exhibits a set of peaks characteristics of amorphous red phosphorus [58]. The SWCNT substrate significantly modifies the structure of phosphorus. The spectra of composite nanomaterials show high intensity at ~379 cm^{-1}, where the P$_{re}$ spectrum has a gap in intensity and suppressed scattering in the range of 390–446 cm^{-1}. The Raman spectrum of P@SWCNTs is very close to the spectra of red phosphorus with the Hittorf's structure [59] and fibrous structure [60]. Both of these phosphorus crystals consist of polymeric tubes with a pentagonal cross section of P$_8$ and P$_9$ cages connected by P$_2$ bridges [61]. The P@SWCNTs/P spectrum looks like a combination of the P$_{re}$ and P@SWCNTs spectra. Indeed, the sum of these spectra in a ratio of 0.3 to 1.1 gives a line that agrees well with the experimental spectrum of P@SWCNTs/P (Figure S6). The Raman spectrum of the SWCNTs/P sample has a low intensity of P-P vibrations due to the small size of elementary phosphorus cores as shown by XPS data (Figure 4a).

3.2. Electrochemical Properties

Figure 6a–e compare the first three GDC curves measured for amorphous red phosphorus, open SWCNTs and three composite materials at a current density of 0.1 A·g^{-1}. The first discharge and charge capacities of 2607 and 861 mAh·g^{-1} for red phosphorus (Figure 6a) give a very low initial Coulombic efficiency (ICE) of ~33%. A sufficient capacity drop is observed in the following second and third cycles. This is due to the low electrical conductivity of amorphous red phosphorus and the large change in its volume during lithiation. The o-SWCNTs reference sample also shows a low ICE of ~24% (Figure 6b). However, the second and third discharge-charge curves practically coincide. The large loss of capacity of o-SWCNTs in the first cycle is mainly attributed to the formation of a solid-electrolyte interphase (SEI) layer on the electrode surface, which is observed as a long plateau at ~0.9 eV during the first discharge and the irreversible incorporation of lithium into the carbon matrix [62,63].

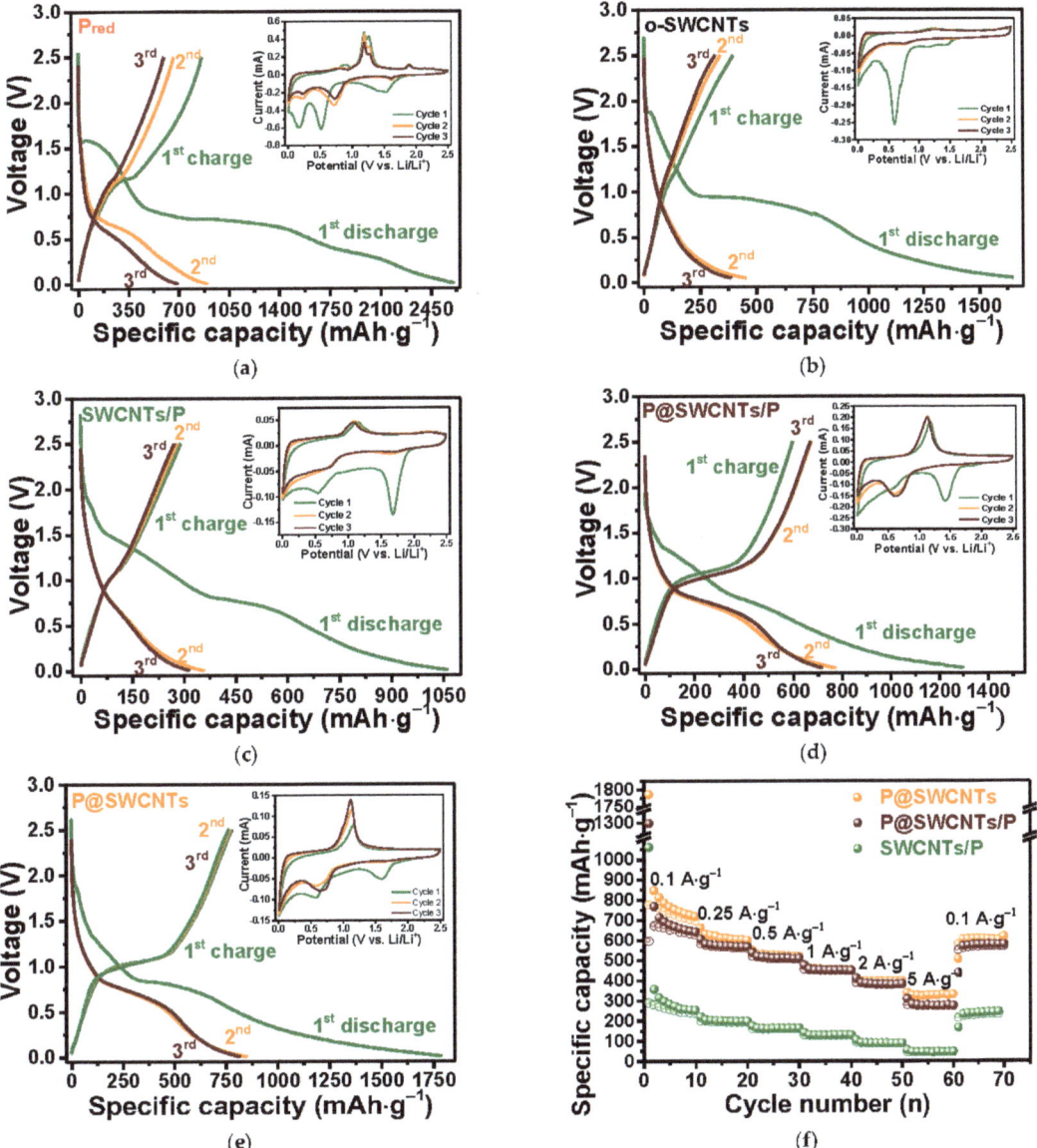

Figure 6. First three GDC curves with insets of the first three CV curves measured for (**a**) red phosphorus (P_{red}), (**b**) o-SWCNTs, (**c**) SWCNTs/P, (**d**) P@SWCNTs/P and (**e**) P@SWCNTs; (**f**) Rate capability of composite nanomaterials.

The first discharge and charge capacities are 1062 and 291 mAh·g^{-1} for SWCNTs/P (Figure 6c), 1299 and 597 mAh·g^{-1} for P@SWCNTs/P (Figure 6d) and 1787 and 779 mAh·g^{-1} for P@SWCNTs (Figure 6e). The highest value of the initial capacity of the P@SWCNTs sample can be related to two factors. First, the nanotubes in this composite are free from phosphorus coating and, therefore, lithium can accumulate in the space between them. Second, the external phosphorus in SWCNTs/P and P@SWCNTs/P is significantly oxidized as shown by the XPS P 2p spectra (Figure 4a). This oxidized phosphorus does not participate in electrochemical reactions and, in addition, can prevent the penetration of lithium ions to the elementary phosphorus core.

The ICE is 27% for SWCNTs/P, 46% for P@SWCNTs/P and 44% for P@SWCNTs. As follows from the GDC curves measured for reference samples (Figure 6a,b), both components of composite nanomaterials are responsible for the high irreversible capacity of the first cycle. It is very important to maintain good electrical contact between the phosphorus and the carbon matrix in order to maintain charge transfer [10]. Phosphorus deposition only on the outer surface of SWCNTs (SWCNTs/P sample) does not solve the problem of high irreversible capacity. However, the encapsulation of phosphorus into the internal channels of nanotubes increases the ICE value and leads to an insignificant change in the GDC curves in subsequent cycles.

CV measurements were used to determine the potentials of electrochemical processes occurring with electrode materials. The first cathodic curves for all electrodes show a set of peaks in the potential range of 0.17–1.7 V. A wide peak at ca. 1.5 V and two peaks between 0.5 and 0.17 V observed for red phosphorus (inset in Figure 6a) refer to the process of activation of the incorporation of lithium ions into phosphorus [2]. The first cathodic curve of o-SWCNTs (inset in Figure 6b) exhibits a peak at about 0.6 V due to the formation of the SEI layer [64]. This peak is also observed for all composite nanomaterials, but with a lower intensity, which indicates that the modification of the SWCNT surface as a result of the synthesis of composite nanomaterials. Irreversible and broadened cathodic peaks at 1.4–1.7 V (insets in Figure 6c–e) can be associated with both the activation of the phosphorus component and the irreversible loss of phosphorus in the SEI layer. Remarkably, the intensity of this peak decreases as the external oxidized phosphorus decreases from SWCNTs/P to P@SWCNTs/P and then to the P@SWCNTs sample.

The second and third CV curves of composite nanomaterials show two reversible redox peaks. The cathodic peak at ~0.62–0.69 V and the corresponding anodic peak at ~1.08–1.13 V (insets in Figure 6c–e) correspond to the reversible reaction P + xLi$^+$ + xe$^-$ ↔ Li$_x$P [11] with stepwise formation of LiP, Li$_2$P and Li$_3$P [3,11]. On the second and third GDC curves, these processes form plateaus, which are more repetitive during P@SWCNTs cycling (Figure 6e). The redox pair at 0.01/0.15 V vs. Li/Li$^+$ corresponds to the insertion/extraction of lithium ions into/from the SWCNT component [65].

Figure 6f compares the rate performance of composite nanomaterials for ten cycles at each applied current density. The specific capacity is calculated based on the weight of the electrode material. The SWCNTs/P sample showed the lowest values, in particular, 256, 200, 164, 130, 89 and 47 mAh·g^{-1} at current densities of 0.1, 0.25, 0.5, 1, 2 and 5 A·g^{-1}, respectively. When the current density returned to 0.1 A·g^{-1}, the SWCNTs/P electrode was still able to maintain a delithiation capacity of 245 mAh·g^{-1}. Notably, these values exceed the capacity values of red phosphorus by 20–55% (Figure S7) at current densities above 0.5 A·g^{-1}. At high cycling rates, the rate of charge transfer becomes important. We assume that in our case this effect is achieved due to the conducting property of SWCNTs. The P@SWCNTs sample showed the best performance with capacities of 734, 605, 520, 452, 398 and 328 mAh·g^{-1} at current densities of 0.1, 0.25, 0.5, 1, 2 and 5 A·g^{-1}, respectively (Figure 6f). There was no difference between delithiation and lithiation values, indicating good electrode stability. After sixty test cycles, the electrode delivered 609 mAh·g^{-1} at 0.1 A·g^{-1}. The P@SWCNTs/P electrode showed a lower capacity (by 40–90 mAh·g^{-1}) than the P@SWCNTs electrode at 0.1, 0.25 A·g^{-1} and 5 A·g^{-1} and comparable capacity at a current density of 0.5 to 2 A·g^{-1}. The capacity of this electrode was 579 mAh·g^{-1}

at following cycles at 0.1 A·g^{-1}. At all applied current densities, the specific capacity of P@SWCNTs and P@SWCNTs/P exceeds the sum of the capacities of the individual components (Figure S7).

Long-term cycling tests of composite nanomaterials were carried out at a current density of 5 A·g^{-1}. The specific capacity of SWCNTs/P continuously decreased (Figure 7a). However, after one thousand cycles, the electrode loses only ~12% of its initial capacity. The P@SWCNTs/P electrode maintained a capacity of ~296 mAh·g^{-1} for ~250 cycles, after which the capacity gradually decreased to 248 mAh·g^{-1} at 1000th cycle. The capacity loss for this electrode was ~16%. The P@SWCNTs electrode showed a slight decrease in capacity from 340 to 317 mAh·g^{-1} during cycling, which corresponds to a capacity loss of ~7%.

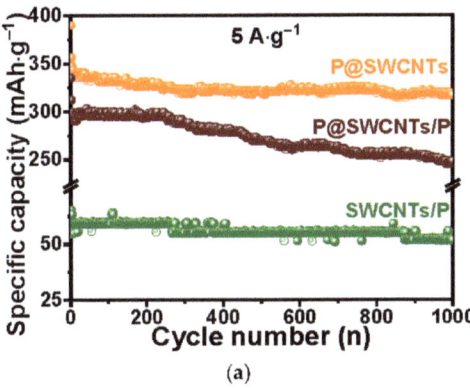
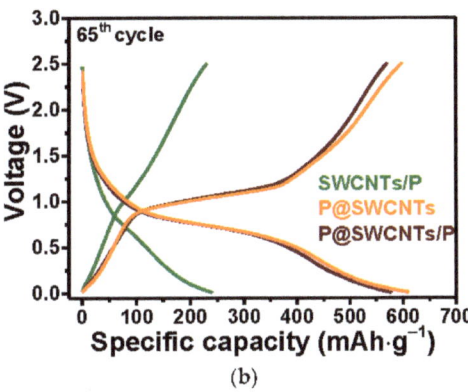

Figure 7. (a) Long-term cycling of composite nanomaterials at 5 A·g^{-1} and (b) GDC curves measured at 65th cycle at a current density of 0.1 A·g^{-1}.

The processes affecting the capacity of composite nanomaterials can be determined from the GDC curves measured after long-term cycling (Figure 7b). The P@SWCNTs/P and P@SWCNTs curves exhibit extended plateaus corresponding to reversible reactions between phosphorus and lithium ions. The contribution of the SWCNT component to the capacity is greater for P@SWCNTs. In this case, the outer surface of the SWCNTs is not contaminated with phosphorus and lithium ions can intercalate between the nanotubes. Sufficiently stable operation of all composite nanomaterials in the electrochemical cells is most likely associated with the formation of a conducting network by intertwined nanotubes.

To understand the higher rate performance of P@SWCNTs as compared to P@SWCNTs/P (Figure 7a), a kinetic analyses of the electrochemical processes was carried out. Figure 8 compares data obtained at scan rates of 0.1, 0.3, 0.5, 0.7 and 1.0 mV·s^{-1}. The CV curves show two reduction peaks and two oxidation peaks for the P@SWCNTs/P electrode (Figure 8a), while four reduction peaks and three oxidation peaks are detected for the P@SWCNTs electrode (Figure 8d). The peak R1 corresponds to the insertion of lithium ions into the SWCNT matrix and is more noticeable in the case of P@SWCNTs purified from external phosphorus. The remaining redox peaks are associated with the lithiation/delithiation of phosphorus and the difference in the number of these peaks for P@SWCNTs/P and P@SWCNTs is due to the difference in the structure of phosphorus nanoparticles in these samples [66]. The encapsulated phosphorus has a chain structure and most likely forms certain lithiated alloys. External phosphorus nanoparticles can vary significantly in size and shape and the redox peaks of their reactions with lithium ions overlap, giving broad signals in CV curves (Figure 8a).

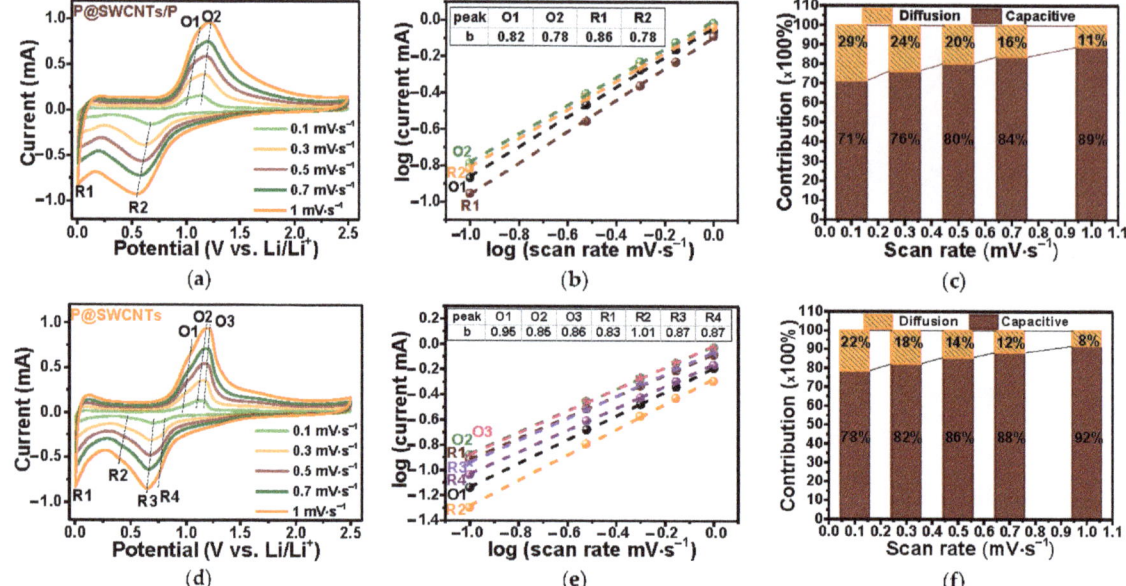

Figure 8. CV curves of (**a**) P@SWCNTs/P and (**d**) P@SWCNTs nanomaterials at different scan rates; log(i) vs. log(v) plots for selected peaks of (**b**) P@SWCNTs/P and (**e**) P@SWCNTs; the normalized ratio of capacitive and diffusion-controlled contributions at different scan rates of (**c**) P@SWCNTs/P and (**f**) P@SWCNTs.

The peak current (i) is a power-law function of the scan rate (v): $i = av^b$ [67], where a and b are adjustable parameters. At a value of b equal to 0.5, the process has a diffusion character. A b value close to 1 indicates that the surface (capacitive) process predominates [8,68–74]. The value of b strongly depends on a number of factors, such as potential, scan rate and charge storage mechanisms [75]. The dependences of log(i) on log(v) are straight lines (Figure 8b,e) and the slop of these lines relative to the x-axis gives the value of b according to the formula log(i) = log(a) + $b$log(v). The resulting values for the selected peaks are collected in the inserted tables. All values are greater than 0.5. Therefore, the electrochemical processes occurring in the P@SWCNTs/P and P@SWCNTs electrodes are mainly surface-controlled. The b values for the peaks corresponding to the interaction of lithium ions with phosphorus are slightly higher for P@SWCNTs (peaks R2, R3, R4 and O1, O2, O3) as compared to P@SWCNTs/P (peaks R2 and O1, O2). This indicates that the reactions in the P@SWCNTs electrode proceed faster [76].

The contributions of diffusion and capacitive processes to the peak current are separated by the equation: $i = k_1 v + k_2 v^{1/2}$, where the first term corresponds to surface reactions and the second term corresponds to ion intercalation [77,78]. The proportions of these two contributions at different scan rates are displayed in Figure 8c for the P@SWCNTs/P electrode and in Figure 8f for the P@SWCNTs electrode. It can be seen that the capacitive contribution is larger in the case of P@SWCNTs.

4. Discussion

To reveal the contribution of SWCNTs to the electrochemical performance of their composites with phosphorus, three samples with different distribution of components were synthesized by means of evaporation and condensation of amorphous red phosphorus.

The SWCNTs/P sample was obtained using closed nanotubes and phosphorus deposited only on the outer surface of the nanotubes. The deposit is significantly oxidized upon contact with laboratory air, as evidenced by the XPS (Figure 4a) and Raman scattering

(Figure 5c) data, in which only a small fraction of P-P bonds are detected. These bonds form red phosphorus cores coated with oxidized phosphorus. Interestingly, the SWCNTs/P electrode can operate at a high current density of 5 A g^{-1} for a thousand cycles (Figure 7a). A GDC study of the SWCNTs/P electrode after long-term cycling reveals an insignificant contribution to the capacity from the interaction of lithium ions with SWCNTs (Figure 7b). This is due to the fact that ions cannot penetrate inside capped nanotubes and into the inter-tube space filled with phosphorus. However, SWCNTs ensure the conductivity of the composite electrode and owing to this, the capacity loss is only ~12% at the 1000th cycle.

The P@SWCNTs/P sample was synthesized using pre-opened nanotubes and phosphorus condensed inside the nanotubes and on their surface. The external phosphorus deposit was removed with NaOH solution to obtain P@SWCNTs. The HAADF/STEM analysis showed that the encapsulated phosphorus was retained after the treatment used (Figure 3a–c). According to the HRTEM study, this phosphorus forms long chains inside the nanotubes (Figure 3d).

Surprisingly, the P@SWCNTs/P and P@SWCNTs samples showed very similar specific capacity values (Figure 6f), while the phosphorus content in them differed almost two times according to the XPS and TGA data. We recalculated the obtained values of capacity (C) to the mass of phosphorus determined by the TGA method. The following formula was used for calculation: C(P) = (C(composite) − C(SWCNTs in composite)/(Weight percentage of P), where the C(SWCNTs in composite) was defined as: C(SWCNTs) × Weight percentage of SWCNTs [13,79]. The C(P) values at current densities of 0.1 and 5 A·g^{-1} are 900 and 486 mAh·g^{-1} for P@SWCNTs/P and 1545 and 1006 mAh·g^{-1} for P@SWCNTs. The lower values for P@SWCNTs/P confirm that oxidized external phosphorus, which makes up about 80% of the total phosphorus content, does not contribute to the capacity.

The reported data on the electrochemical performance of P-CNT composites in LIBs are compared with the present results in Table S1. High specific capacities of composite nanomaterials were obtained using multi-walled CNTs [20,29,79,80], which make a significant contribution to the capacity due to the intercalation of lithium ions between nanotube layers. For example, the pristine multi-walled CNTs used to fabricate composite in [44] had a specific capacity of ~500 mAh·g^{-1} at a current density of 0.1 A·g^{-1}. This value is double the corresponding value of our o-SWCNTs reference sample. There are only two works on the study of P-SWCNT composites in LIBs [40,81]. Both showed much poorer electrochemical performance as compared to our P@SWCNTs electrode (Table S1). The specific capacity is sometimes given by the mass of the phosphorus component in the composite. The maximum recorded values are 1081 mAh·g^{-1} at 3.9 A·g^{-1} for the ball-milled P-CNT composite [17] and 1264 mAh·g^{-1} at 0.2 A·g^{-1} for the composite obtained from solution [79]. These values do not exceed the corresponding values for our P@SWCNTs sample.

An additional advantage of the P@SWCNTs sample is that it retains a capacity of 93% at 5 A·g^{-1} on the 1000th cycle. The higher rate capability and greater rate retention of P@SWCNTs as compared to P@SWCNTs/P can be explain by two factors. First, the oxidized layer on the surface of external phosphorus particles can slow down the diffusion of lithium ions in the P@SWCNTs/P electrode. Second, lithium ions can reach the encapsulated phosphorus in P@SWCNTs through defects present in the nanotube walls. Indeed, the second and third CV curves of the o-SWCNTs electrode demonstrate a clear redox peak at ~0.8/1.2 V (inset in Figure 6b). Such peaks were previously observed for holey graphene materials with in-plane vacancies 2–5 nm in size [82,83]. These vacancies are likely formed at the stage of SWCNT opening and purification and they are blocked by phosphorus deposits in P@SWCNTs/P. The large accumulation of lithium in P@SWCNTs at high rates (Figure 6f) can only be explained by the presence of channels additional to the open ends of the nanotube. The most probable are vacancy defects of suitable sizes in the sidewalls.

5. Conclusions

In summary, we synthesized composite nanomaterials by the vaporization-condensation process using commercial red phosphorus and SWCNTs. Under the synthesis conditions used, phosphorus chains are formed inside open SWCNTs and the SWCNT surface is covered with red phosphorus, which is oxidized upon contact with air. The external phosphorus deposit is effectively removed with an aqueous solution of sodium hydroxide. The content of phosphorus in samples with only external phosphorus SWCNTs/P, only internal phosphorus P@SWCNTs and both types of phosphorus P@SWCNTs/P is 11, 8 and 16 at%, respectively, according to XPS data. To reveal the effect of various combinations of SWCNTs and phosphorus in the composite on the electrochemical interaction with lithium ions, nanomaterials were tested as anodes in coin-cell batteries with lithium sheets as counter electrodes. The SWCNTs/P sample was found to perform better than the reference red phosphorus due to the presence of a conducting network of SWCNTs. Nanotubes practically do not contribute to the SWCNTs/P capacity because their surface is blocked by phosphorus. The creation of composite nanomaterials with internal phosphorus almost doubles the initial Coulombic efficiency as compared to SWCNTs/P and dramatically improves the specific capacity and cycling stability of the electrode. The presence of both external and internal phosphorus leads to a gradual slight degradation of the P@SWCNTs/P electrode over time, associated with slower diffusion processes into the internal volume of the material. The excellent cycling stability of P@SWCNTs over 1000 cycles at a high current density of 5 $A \cdot g^{-1}$ is associated with the synergistic effect of highly capacitive phosphorus and conductive SWCNTs, the absence of inactive oxidized phosphorus deposit and the presence of multiple channels for lithium ion diffusion to encapsulated phosphorus.

Supplementary Materials: The following supporting information can be downloaded at: https://www.mdpi.com/article/10.3390/nano13010153/s1, Figure S1: XRD patterns of P@SWCNTs/P, P@SWCNTs and o-SWCNTs; Figure S2: EDS analysis of P@SWCNTs; Figure S3: XPS survey XPS and P 2p spectra of composite nanomaterials; Figure S4: Ion current curves of evolved gases during decomposition of P@SWCNTs; Figure S5: TG and DTG curves for SWCNTs/P; Figure S6: Experimental and simulated Raman spectra of P@SWCNTs/P; Figure S7: Rate performance of o-SWCNTs and commercial red phosphorus; Table S1: Electrochemical performance of P-CNT composites in lithium-ion batteries, reported in the literature [2,17,20,29,39–41,44,79–81].

Author Contributions: Conceptualization, L.G.B. and A.V.O.; methodology, S.G.S., A.V.O. and A.L.C.; validation, S.G.S. and L.G.B.; investigation, A.A.V., I.P.A., E.V.S., P.E.P., E.A.M., E.Y.G. and A.L.C.; resources, A.L.C.; writing—original draft preparation, A.A.V.; writing—review and editing, S.G.S. and L.G.B.; supervision, L.G.B. and A.V.O.; funding acquisition, A.V.O. All authors have read and agreed to the published version of the manuscript.

Funding: This research was funded by the Russian Science Foundation (grant 22-13-00219). I.P.A., E.V.S., P.E.P., E.A.M. and E.Y.G. acknowledge the Ministry of Science and Higher Education of the Russian Federation.

Institutional Review Board Statement: Not applicable.

Informed Consent Statement: Not applicable.

Data Availability Statement: Not applicable.

Acknowledgments: We are grateful to Yu. V. Shubin for the XRD data.

Conflicts of Interest: The authors declare no conflict of interest.

References

1. Sun, Y.; Zeng, F.; Zhu, Y.; Lu, P.; Yang, D. A Review on Nanoconfinement Engineering of Red Phosphorus for Enhanced Li/Na/K-Ion Storage Performances. *J. Energy Chem.* **2021**, *61*, 531–552. [CrossRef]
2. Yuan, D.; Cheng, J.; Qu, G.; Li, X.; Ni, W.; Wang, B.; Liu, H. Amorphous Red Phosphorous Embedded in Carbon Nanotubes Scaffold as Promising Anode Materials for Lithium-Ion Batteries. *J. Power Sources* **2016**, *301*, 131–137. [CrossRef]
3. Marino, C.; Boulet, L.; Gaveau, P.; Fraisse, B.; Monconduit, L. Nanoconfined Phosphorus in Mesoporous Carbon as an Electrode for Li-Ion Batteries: Performance and Mechanism. *J. Mater. Chem.* **2012**, *22*, 22713–22720. [CrossRef]
4. Wang, L.; He, X.; Li, J.; Sun, W.; Gao, J.; Guo, J.; Jiang, C. Nano-Structured Phosphorus Composite as High-Capacity Anode Materials for Lithium Batteries. *Angew. Chem.-Int. Ed.* **2012**, *124*, 9168–9171. [CrossRef]
5. Zhu, Y.; Wen, Y.; Fan, X.; Gao, T.; Han, F.; Luo, C.; Liou, S.C.; Wang, C. Red Phosphorus-Single-Walled Carbon Nanotube Composite as a Superior Anode for Sodium Ion Batteries. *ACS Nano* **2015**, *9*, 3254–3264. [CrossRef]
6. Yu, Z.; Song, J.; Gordin, M.L.; Yi, R.; Tang, D.; Wang, D. Phosphorus-Graphene Nanosheet Hybrids as Lithium-Ion Anode with Exceptional High-Temperature. *Adv. Sci.* **2015**, *2*, 1400020. [CrossRef]
7. Zhao, D.; Zhang, L.; Fu, C.; Zhang, J.; Niu, C. The Lithium and Sodium Storage Performances of Phosphorus and Its Hierarchical Structure. *Nano Res.* **2019**, *12*, 1–17. [CrossRef]
8. Liu, S.; Feng, J.; Bian, X.; Liu, J.; Xu, H.; An, Y. A Controlled Red Phosphorus@Ni-P Core@shell Nanostructure as an Ultralong Cycle-Life and Superior High-Rate Anode for Sodium-Ion Batteries. *Energy Environ. Sci.* **2017**, *10*, 1222–1233. [CrossRef]
9. Sun, J.; Lee, H.W.; Pasta, M.; Yuan, H.; Zheng, G.; Sun, Y.; Li, Y.; Cui, Y. A Phosphorene-Graphene Hybrid Material as a High-Capacity Anode for Sodium-Ion Batteries. *Nat. Nanotechnol.* **2015**, *10*, 980–985. [CrossRef]
10. Sun, J.; Zheng, G.; Lee, H.-W.; Liu, N.; Wang, H.; Yao, H.; Yang, W.; Cui, Y. Formation of Stable Phosphorus–Carbon Bond for Enhanced Performance in Black Phosphorus Nanoparticle–Graphite Composite Battery Anodes. *Nano Lett.* **2014**, *14*, 4573–4580. [CrossRef]
11. Marino, C.; Debenedetti, A.; Fraisse, B.; Favier, F.; Monconduit, L. Activated-Phosphorus as New Electrode Material for Li-Ion Batteries. *Electrochem. Commun.* **2011**, *13*, 346–349. [CrossRef]
12. Chen, X.; Qiu, J.; Wang, Y.; Huang, F.; Peng, J.; Liu, Y.; Li, J.; Zhai, M. A Stable Polypyridinopyridine-Red Phosphorus Composite as a Superior Anode Material for Long-Cycle Lifetime Lithium-Ion Batteries. *New J. Chem.* **2019**, *43*, 6197–6204. [CrossRef]
13. Li, W.; Yang, Z.; Li, M.; Jiang, Y.; Wei, X.; Zhong, X.; Gu, L.; Yu, Y. Amorphous Red Phosphorus Embedded in Highly Ordered Mesoporous Carbon with Superior Lithium and Sodium Storage Capacity. *Nano Lett.* **2016**, *16*, 1546–1553. [CrossRef] [PubMed]
14. Bai, A.; Wang, L.; Li, J.; He, X.; Wang, J.; Wang, J. Composite of Graphite/Phosphorus as Anode for Lithium-Ion Batteries. *J. Power Sources* **2015**, *289*, 100–104. [CrossRef]
15. Xue, J.; Wang, D.; Xia, X.; Chen, Y.; Liu, H. Confined Red Phosphorus in N-Doped Hierarchically Porous Carbon for Lithium Ion Batteries with Enhanced Rate Capability and Cycle Stability. *Microporous Mesoporous Mater.* **2020**, *305*, 110365. [CrossRef]
16. Liu, Y.; Liu, Q.; Jian, C.; Cui, D.; Chen, M.; Li, Z.; Li, T.; Nilges, T.; He, K.; Jia, Z.; et al. Red-Phosphorus-Impregnated Carbon Nanofibers for Sodium-Ion Batteries and Liquefaction of Red Phosphorus. *Nat. Commun.* **2020**, *11*, 2520. [CrossRef]
17. Jiao, X.; Liu, Y.; Li, B.; Zhang, W.; He, C.; Zhang, C.; Yu, Z.; Gao, T.; Song, J. Amorphous Phosphorus-Carbon Nanotube Hybrid Anode with Ultralong Cycle Life and High-Rate Capability for Lithium-Ion Battery. *Carbon* **2019**, *148*, 518–524. [CrossRef]
18. Zhang, Z.J.; Li, W.J.; Chou, S.L.; Han, C.; Liu, H.K.; Dou, S.X. Effects of Carbon on Electrochemical Performance of Red Phosphorus (P) and Carbon Composite as Anode for Sodium Ion Batteries. *J. Mater. Sci. Technol.* **2021**, *68*, 140–146. [CrossRef]
19. Qian, J.; Wu, X.; Cao, Y.; Ai, X.; Yang, H. High Capacity and Rate Capability of Amorphous Phosphorus for Sodium Ion Batteries. *Angew. Chem.* **2013**, *125*, 4731–4734. [CrossRef]
20. Xu, Z.; Zeng, Y.; Wang, L.; Li, N.; Chen, C.; Li, C.; Li, J.; Lv, H.; Kuang, L.; Tian, X. Nanoconfined Phosphorus Film Coating on Interconnected Carbon Nanotubes as Ultrastable Anodes for Lithium Ion Batteries. *J. Power Sources* **2017**, *356*, 18–26. [CrossRef]
21. Ramireddy, T.; Xing, T.; Rahman, M.M.; Chen, Y.; Dutercq, Q.; Gunzelmann, D.; Glushenkov, A.M. Phosphorus-Carbon Nanocomposite Anodes for Lithium-Ion and Sodium-Ion Batteries. *J. Mater. Chem. A* **2015**, *3*, 5572–5584. [CrossRef]
22. Qian, J.; Qiao, D.; Ai, X.; Cao, Y.; Yang, H. Reversible 3-Li Storage Reactions of Amorphous Phosphorus as High Capacity and Cycling-Stable Anodes for Li-Ion Batteries. *Chem. Commun.* **2012**, *48*, 8931–8933. [CrossRef]
23. Li, X.; Chen, G.; Le, Z.; Li, X.; Nie, P.; Liu, X.; Xu, P.; Wu, H.B.; Liu, Z.; Lu, Y. Well-Dispersed Phosphorus Nanocrystals within Carbon via High-Energy Mechanical Milling for High Performance Lithium Storage. *Nano Energy* **2019**, *59*, 464–471. [CrossRef]
24. Li, W.J.; Chou, S.L.; Wang, J.Z.; Liu, H.K.; Dou, S.X. Simply Mixed Commercial Red Phosphorus and Carbon Nanotube Composite with Exceptionally Reversible Sodium-Ion Storage. *Nano Lett.* **2013**, *13*, 5480–5484. [CrossRef] [PubMed]
25. Yu, Z.; Song, J.; Wang, D.; Wang, D. Advanced Anode for Sodium-Ion Battery with Promising Long Cycling Stability Achieved by Tuning Phosphorus-Carbon Nanostructures. *Nano Energy* **2017**, *40*, 550–558. [CrossRef]
26. Subramaniyam, C.M.; Tai, Z.; Mahmood, N.; Zhang, D.; Liu, H.K.; Goodenough, J.B.; Dou, S.X. Unlocking the Potential of Amorphous Red Phosphorus Films as a Long-Term Stable Negative Electrode for Lithium Batteries. *J. Mater. Chem. A* **2017**, *5*, 1925–1929. [CrossRef]
27. Zhang, Y.; Wang, H.; Luo, Z.; Tan, H.T.; Li, B.; Sun, S.; Li, Z.; Zong, Y.; Xu, Z.J.; Yang, Y.; et al. An Air-Stable Densely Packed Phosphorene Graphene Composite Toward Advanced. *Adv. Energy Mater.* **2016**, *6*, 1600453. [CrossRef]
28. Yu, Y.; Li, W.; Yang, Z.; Jiang, Y.; Yu, Z.; Gu, L. Crystalline Red Phosphorus Incorporated with Porous Carbon Nanofibers as Flexible Electrode for High Performance Lithium-Ion Batteries. *Carbon* **2014**, *78*, 455–462. [CrossRef]

29. Yuan, T.; Ruan, J.; Peng, C.; Sun, H.; Pang, Y.; Yang, J.; Ma, Z.F.; Zheng, S. 3D Red Phosphorus/Sheared CNT Sponge for High Performance Lithium-Ion Battery Anodes. *Energy Storage Mater.* **2018**, *13*, 267–273. [CrossRef]
30. Li, W.; Hu, S.; Luo, X.; Li, Z.; Sun, X.; Li, M.; Liu, F.; Yu, Y. Confined Amorphous Red Phosphorus in MOF-Derived N-Doped Microporous Carbon as a Superior. *Adv. Mater.* **2017**, *29*, 1605820. [CrossRef]
31. Khlobystov, A.N. Carbon Nanotubes: From Nano Test Tube to Nano-Reactor. *ACS Nano* **2011**, *5*, 9306–9312. [CrossRef] [PubMed]
32. Thess, A.; Lee, R.; Nikolaev, P.; Dai, H.; Petit, P.; Robert, J.; Xu, C.; Lee, Y.H.; Kim, S.G.; Rinzler, A.G.; et al. Crystalline Ropes of Metallic Carbon Nanotubes. *Science* **1996**, *273*, 483–487. [CrossRef] [PubMed]
33. Liu, P.; Tan, Y.F.; Hu, D.C.M.; Jewell, D.; Duong, H.M. Multi-Property Enhancement of Aligned Carbon Nanotube Thin Films from Floating Catalyst Method. *Mater. Des.* **2016**, *108*, 754–760. [CrossRef]
34. Li, C.; Chou, T.W. A Structural Mechanics Approach for the Analysis of Carbon Nanotubes. *Int. J. Solids Struct.* **2003**, *40*, 2487–2499. [CrossRef]
35. Yu, M.F.; Files, B.S.; Arepalli, S.; Ruoff, R.S. Tensile Loading of Ropes of Single Wall Carbon Nanotubes and Their Mechanical Properties. *Phys. Rev. Lett.* **2000**, *84*, 5552–5555. [CrossRef]
36. Yu, M.F.; Lourie, O.; Dyer, M.J.; Moloni, K.; Kelly, T.F.; Ruoff, R.S. Strength and Breaking Mechanism of Multiwalled Carbon Nanotubes under Tensile Load. *Science* **2000**, *287*, 637–640. [CrossRef] [PubMed]
37. Iijima, S.; Brabec, C.; Maiti, A.; Bernholc, J. Structural Flexibility of Carbon Nanotubes. *J. Chem. Phys.* **1996**, *104*, 2089–2092. [CrossRef]
38. Falvo, M.R.; Clary, G.J.; Taylor, R.M.; Chi, V.; Brooks, F.P.; Washburn, S.; Superfine, R. Bending and Buckling of Carbon Nanotubes under Large Strain. *Nature* **1997**, *389*, 582–584. [CrossRef]
39. Li, J.; Jin, H.; Yuan, Y.; Lu, H.; Su, C.; Fan, D.; Li, Y.; Wang, J.; Lu, J.; Wang, S. Encapsulating Phosphorus inside Carbon Nanotubes via a Solution Approach for Advanced Lithium Ion Host. *Nano Energy* **2019**, *58*, 23–29. [CrossRef]
40. Smajic, J.; Alazmi, A.; Patole, S.P.; Costa, P.M.F.J. Single-Walled Carbon Nanotubes as Stabilizing Agents in Red Phosphorus Li-Ion Battery Anodes. *RSC Adv.* **2017**, *7*, 39997–40004. [CrossRef]
41. Tojo, T.; Yamaguchi, S.; Furukawa, Y.; Aoyanagi, K.; Umezaki, K.; Inada, R.; Sakurai, Y. Electrochemical Performance of Lithium Ion Battery Anode Using Phosphorus Encapsulated into Nanoporous Carbon Nanotubes. *J. Electrochem. Soc.* **2018**, *165*, A1231–A1237. [CrossRef]
42. Rybkovskiy, D.V.; Koroteev, V.O.; Impellizzeri, A.; Vorfolomeeva, A.A.; Gerasimov, E.Y.; Okotrub, A.V.; Chuvilin, A.; Bulusheva, L.G.; Ewels, C.P. "Missing" One-Dimensional Red-Phosphorus Chains Encapsulated within Single-Walled Carbon Nanotubes. *ACS Nano* **2022**, *16*, 6002–6012. [CrossRef] [PubMed]
43. Hart, M.; Chen, J.; Michaelides, A.; Sella, A.; Shaffer, M.S.P.; Salzmann, C.G. One-Dimensional Pnictogen Allotropes inside Single-Wall Carbon Nanotubes. *Inorg. Chem.* **2019**, *58*, 15216–15224. [CrossRef]
44. Zhao, D.; Zhang, J.; Fu, C.; Huang, J.; Xiao, D.; Yuen, M.M.F.; Niu, C. Enhanced Cycle Stability of Ring-Shaped Phosphorus inside Multi-Walled Carbon Nanotubes as Anodes for Lithium-Ion Batteries. *J. Mater. Chem. A* **2018**, *6*, 2540–2548. [CrossRef]
45. Hart, M.; White, E.R.; Chen, J.; McGilvery, C.M.; Pickard, C.J.; Michaelides, A.; Sella, A.; Shaffer, M.S.P.; Salzmann, C.G. Encapsulation and Polymerization of White Phosphorus Inside Single-Wall Carbon Nanotubes. *Angew. Chem.-Int. Ed.* **2017**, *129*, 8256–8260. [CrossRef]
46. Vorfolomeeva, A.A.; Pushkarevsky, N.A.; Koroteev, V.O.; Surovtsev, N.V.; Chuvilin, A.L.; Shlyakhova, E.V.; Plyusnin, P.E.; Makarova, A.A.; Okotrub, A.V.; Bulusheva, L.G. Doping of Carbon Nanotubes with Encapsulated Phosphorus Chains. *Inorg. Chem.* **2022**, *61*, 9605–9614. [CrossRef]
47. Zhang, J.; Zhao, D.; Xiao, D.; Ma, C.; Du, H.; Li, X.; Zhang, L.; Huang, J.; Huang, H.; Jia, C.L.; et al. Assembly of Ring-Shaped Phosphorus within Carbon Nanotube Nanoreactors. *Angew. Chem.-Int. Ed.* **2017**, *56*, 1850–1854. [CrossRef] [PubMed]
48. Xu, J.; Guan, L. Diameter-Selective Band Structure Modification of Single-Walled Carbon Nanotubes by Encapsulated Phosphorus Chains. *J. Phys. Chem. C* **2009**, *113*, 15099–15101. [CrossRef]
49. Allaf, R.M.; Rivero, I.V.; Spearman, S.S.; Hope-Weeks, L.J. On the Preparation of As-Produced and Purified Single-Walled Carbon Nanotube Samples for Standardized X-Ray Diffraction Characterization. *Mater. Charact.* **2011**, *62*, 857–864. [CrossRef]
50. Hu, Z.; Lu, Y.; Liu, M.; Zhang, X.; Cai, J.J. Crystalline Red Phosphorus for Selective Photocatalytic Reduction of CO_2 into CO. *J. Mater. Chem. A* **2021**, *9*, 338–348. [CrossRef]
51. Hasegawa, G.; Deguchi, T.; Kanamori, K.; Kobayashi, Y.; Kageyama, H.; Abe, T.; Nakanishi, K. High-Level Doping of Nitrogen, Phosphorus, and Sulfur into Activated Carbon Monoliths and Their Electrochemical Capacitances. *Chem. Mater.* **2015**, *27*, 4703–4712. [CrossRef]
52. Kuntz, K.L.; Wells, R.A.; Hu, J.; Yang, T.; Dong, B.; Guo, H.; Woomer, A.H.; Druffel, D.L.; Alabanza, A.; Tománek, D.; et al. Control of Surface and Edge Oxidation on Phosphorene. *ACS Appl. Mater. Interfaces* **2017**, *9*, 9126–9135. [CrossRef] [PubMed]
53. Imamura, R.; Matsui, K.; Takeda, S.; Ozaki, J.; Oya, A. A New Role for Phosphorus in Graphitization of Phenolic Resin. *Carbon* **1999**, *37*, 261–267. [CrossRef]
54. Fouquet, M.; Telg, H.; Maultzsch, J.; Wu, Y.; Chandra, B.; Hone, J.; Heinz, T.F.; Thomsen, C. Longitudinal Optical Phonons in Metallic and Semiconducting Carbon Nanotubes. *Phys. Rev. Lett.* **2009**, *102*, 075501. [CrossRef]
55. Hennrich, F.; Krupke, R.; Lebedkin, S.; Arnold, K.; Fischer, R.; Resasco, D.E.; Kappes, M.M. Raman Spectroscopy of Individual Single-Walled Carbon Nanotubes from Various Sources. *J. Phys. Chem. B* **2005**, *109*, 10567–10573. [CrossRef]

56. Kharlamova, M.V.; Kramberger, C.; Domanov, O.; Mittelberger, A.; Saito, T.; Yanagi, K.; Pichler, T.; Eder, D. Comparison of Doping Levels of Single-Walled Carbon Nanotubes Synthesized by Arc-Discharge and Chemical Vapor Deposition Methods by Encapsulated Silver Chloride. *Phys. Status Solidi Basic Res.* **2018**, *255*, 1800178. [CrossRef]
57. Impellizzeri, A.; Vorfolomeeva, A.A.; Surovtsev, N.V.; Okotrub, A.V.; Ewels, C.P.; Rybkovskiy, D. V Simulated Raman Spectra of Bulk and Low-Dimensional Phosphorus Allotropes. *Phys. Chem. Chem. Phys.* **2021**, *23*, 16611–16622. [CrossRef]
58. Amaral, P.E.M.; Nieman, G.P.; Schwenk, G.R.; Jing, H.; Zhang, R.; Cerkez, E.B.; Strongin, D.; Ji, H.F. High Electron Mobility of Amorphous Red Phosphorus Thin Films. *Angew. Chem.-Int. Ed.* **2019**, *58*, 6766–6771. [CrossRef]
59. Zhang, L.; Huang, H.; Zhang, B.; Gu, M.; Zhao, D.; Zhao, X.; Li, L.; Zhou, J.; Wu, K.; Cheng, Y.; et al. Structure and Properties of Violet Phosphorus and Its Phosphorene Exfoliation. *Angew. Chem.* **2020**, *132*, 1090–1096. [CrossRef]
60. Winchester, R.A.L.; Whitby, M.; Shaffer, M.S.P. Synthesis of Pure Phosphorus Nanostructures. *Angew. Chem.-Int. Ed.* **2009**, *48*, 3616–3621. [CrossRef]
61. Pfitzner, A. Phosphorus Remains Exciting! *Angew. Chem.-Int. Ed.* **2006**, *45*, 699–700. [CrossRef] [PubMed]
62. Song, L.; Xin, S.; Xu, D.W.; Li, H.Q.; Cong, H.P.; Yu, S.H. Graphene-Wrapped Graphitic Carbon Hollow Spheres: Bioinspired Synthesis and Applications in Batteries and Supercapacitors. *ChemNanoMat* **2016**, *2*, 540–546. [CrossRef]
63. Lin, J.; Xu, Y.; Wang, J.; Zhang, B.; Wang, C.; He, S.; Wu, J. Preinserted Li Metal Porous Carbon Nanotubes with High Coulombic Efficiency for Lithium-Ion Battery Anodes. *Chem. Eng. J.* **2019**, *373*, 78–85. [CrossRef]
64. Wu, G.T.; Wang, C.S.; Zhang, X.B.; Yang, H.S.; Qi, Z.F.; He, P.M.; Li, W.Z. Structure and Lithium Insertion Properties of Carbon Nanotubes. *J. Electrochem. Soc.* **1999**, *146*, 1696–1701. [CrossRef]
65. Landi, B.J.; Ganter, M.J.; Cress, C.D.; DiLeo, R.A.; Raffaelle, R.P. Carbon Nanotubes for Lithium Ion Batteries. *Energy Environ. Sci.* **2009**, *2*, 638–654. [CrossRef]
66. Peng, C.; Chen, H.; Zhong, G.; Tang, W.; Xiang, Y.; Liu, X.; Yang, J.; Lu, C.; Yang, Y. Capacity Fading Induced by Phase Conversion Hysteresis within Alloying Phosphorus Anode. *Nano Energy* **2019**, *58*, 560–567. [CrossRef]
67. Lindström, H.; Södergren, S.; Solbrand, A.; Rensmo, H.; Hjelm, J.; Hagfeldt, A.; Lindquist, S.E. Li$^+$ Ion Insertion in TiO$_2$ (Anatase). 1. Chronoamperometry on CVD Films and Nanoporous Films. *J. Phys. Chem. B* **1997**, *101*, 7710–7716. [CrossRef]
68. Zhao, K.; Liu, F.; Niu, C.; Xu, W.; Dong, Y.; Zhang, L.; Xie, S.; Yan, M.; Wei, Q.; Zhao, D.; et al. Graphene Oxide Wrapped Amorphous Copper Vanadium Oxide with Enhanced Capacitive Behavior for High-Rate and Long-Life Lithium-Ion Battery Anodes. *Adv. Sci.* **2015**, *2*, 1500154. [CrossRef]
69. Chen, C.; Wen, Y.; Hu, X.; Ji, X.; Yan, M.; Mai, L.; Hu, P.; Shan, B.; Huang, Y. Na$^+$ Intercalation Pseudocapacitance in Graphene-Coupled Titanium Oxide Enabling Ultra-Fast Sodium Storage and Long-Term Cycling. *Nat. Commun.* **2015**, *6*, 6929. [CrossRef]
70. Lesel, B.K.; Ko, J.S.; Dunn, B.; Tolbert, S.H. Mesoporous Li$_x$Mn$_2$O$_4$ Thin Film Cathodes for Lithium-Ion Pseudocapacitors. *ACS Nano* **2016**, *10*, 7572–7581. [CrossRef]
71. Yang, L.; Li, X.; He, S.; Du, G.; Yu, X.; Liu, J.; Gao, Q.; Hu, R.; Zhu, M. Mesoporous Mo$_2$C/N-Doped Carbon Heteronanowires as High-Rate and Long-Life Anode Materials for Li-Ion Batteries. *J. Mater. Chem. A* **2016**, *4*, 10842–10849. [CrossRef]
72. Muller, G.A.; Cook, J.B.; Kim, H.S.; Tolbert, S.H.; Dunn, B. High Performance Pseudocapacitor Based on 2D Layered Metal Chalcogenide Nanocrystals. *Nano Lett.* **2015**, *15*, 1911–1917. [CrossRef] [PubMed]
73. Chao, D.; Zhu, C.; Yang, P.; Xia, X.; Liu, J.; Wang, J.; Fan, X.; Savilov, S.V.; Lin, J.; Fan, H.J.; et al. Array of Nanosheets Render Ultrafast and High-Capacity Na-Ion Storage by Tunable Pseudocapacitance. *Nat. Commun.* **2016**, *7*, 12122. [CrossRef] [PubMed]
74. Chen, J.; Fan, X.; Ji, X.; Gao, T.; Hou, S.; Zhou, X.; Wang, L.; Wang, F.; Yang, C.; Chen, L.; et al. Intercalation of Bi Nanoparticles into Graphite Results in an Ultra-Fast and Ultra-Stable Anode Material for Sodium-Ion Batteries. *Energy Environ. Sci.* **2018**, *11*, 1218–1225. [CrossRef]
75. Mathis, T.S.; Kurra, N.; Wang, X.; Pinto, D.; Simon, P.; Gogotsi, Y. Energy Storage Data Reporting in Perspective—Guidelines for Interpreting the Performance of Electrochemical Energy Storage Systems. *Adv. Energy Mater.* **2019**, *9*, 1902007. [CrossRef]
76. Qian, Y.; Jiang, S.; Li, Y.; Yi, Z.; Zhou, J.; Li, T.; Han, Y.; Wang, Y.; Tian, J.; Lin, N.; et al. In Situ Revealing the Electroactivity of P-O and P-C Bonds in Hard Carbon for High-Capacity and Long-Life Li/K-Ion Batteries. *Adv. Energy Mater.* **2019**, *9*, 1901676. [CrossRef]
77. Augustyn, V.; Come, J.; Lowe, M.A.; Kim, J.W.; Taberna, P.L.; Tolbert, S.H.; Abruña, H.D.; Simon, P.; Dunn, B. High-Rate Electrochemical Energy Storage through Li+ Intercalation Pseudocapacitance. *Nat. Mater.* **2013**, *12*, 518–522. [CrossRef]
78. Lou, P.; Cui, Z.; Jia, Z.; Sun, J.; Tan, Y.; Guo, X. Monodispersed Carbon-Coated Cubic NiP$_2$ Nanoparticles Anchored on Carbon Nanotubes as Ultra-Long-Life Anodes for Reversible Lithium Storage. *ACS Nano* **2017**, *11*, 3705–3715. [CrossRef]
79. Sun, L.; Zhang, Y.; Zhang, D.; Liu, J.; Zhang, Y. Amorphous Red Phosphorus Anchored on Carbon Nanotubes as High Performance Electrodes for Lithium Ion Batteries. *Nano Res.* **2018**, *11*, 2733–2745. [CrossRef]
80. Zhang, L.; Yu, H.; Wang, Y. Scalable Method for Preparing Multi-Walled Carbon Nanotube Supported Red Phosphorus Nanoparticles as Anode Material in Lithium-Ion Batteries. *Mater. Lett.* **2022**, *312*, 131638. [CrossRef]
81. Smajic, J.; Alazmi, A.; Alzahrani, A.; Emwas, A.H.; Costa, P.M.F.J. The Interaction of Red Phosphorus with Supporting Carbon Additives in Lithium-Ion Battery Anodes. *J. Electroanal. Chem.* **2022**, *925*, 116852. [CrossRef]

82. Stolyarova, S.G.; Okotrub, A.V.; Shubin, Y.V.; Asanov, I.P.; Galitsky, A.A.; Bulusheva, L.G. Effect of Hot Pressing on the Electrochemical Performance of Multilayer Holey Graphene Materials in Li-Ion Batteries. *Phys. Status Solidi Basic Res.* **2018**, *255*, 1800202. [CrossRef]
83. Bulusheva, L.G.; Stolyarova, S.G.; Chuvilin, A.L.; Shubin, Y.V.; Asanov, I.P.; Sorokin, A.M.; Mel'Gunov, M.S.; Zhang, S.; Dong, Y.; Chen, X.; et al. Creation of Nanosized Holes in Graphene Planes for Improvement of Rate Capability of Lithium-Ion Batteries. *Nanotechnology* **2018**, *29*, 134001. [CrossRef] [PubMed]

Disclaimer/Publisher's Note: The statements, opinions and data contained in all publications are solely those of the individual author(s) and contributor(s) and not of MDPI and/or the editor(s). MDPI and/or the editor(s) disclaim responsibility for any injury to people or property resulting from any ideas, methods, instructions or products referred to in the content.

Article

Plasmon Effect of Ag Nanoparticles on TiO₂/rGO Nanostructures for Enhanced Energy Harvesting and Environmental Remediation

Seenidurai Athithya [1], Valparai Surangani Manikandan [1], Santhana Krishnan Harish [1,2], Kuppusamy Silambarasan [1], Shanmugam Gopalakrishnan [1,3], Hiroya Ikeda [2], Mani Navaneethan [1,3,*] and Jayaram Archana [1,*]

[1] Functional Materials and Energy Devices Laboratory, Department of Physics and Nanotechnology, SRM Institute of Science and Technology, Kattankulathur, Chennai 603 203, India
[2] Research Institute of Electronics, Shizuoka University, 3-5-1 Johoku, Naka-Ku, Hamamatsu 432-8011, Japan
[3] Nanotechnology Research Center (NRC), SRM Institute of Science and Technology, Kattankulathur, Chennai 603 203, India
* Correspondence: m.navaneethan@gmail.com (M.N.); jayaram.archana@gmail.com (J.A.)

Abstract: We report Ag nanoparticles infused with mesosphere TiO₂/reduced graphene oxide (rGO) nanosheet (TiO₂/rGO/Ag) hybrid nanostructures have been successfully fabricated using a series of solution process synthesis routes and an in-situ growth method. The prepared hybrid nanostructure is utilized for the fabrication of photovoltaic cells and the photocatalytic degradation of pollutants. The photovoltaic characteristics of a dye-sensitized solar cell (DSSC) device with plasmonic hybrid nanostructure (TiO₂/rGO/Ag) photoanode achieved a highest short-circuit current density (J_{SC}) of 16.05 mA/cm², an open circuit voltage (V_{OC}) of 0.74 V and a fill factor (FF) of 62.5%. The fabricated plasmonic DSSC device exhibited a maximum power conversion efficiency (PCE) of 7.27%, which is almost 1.7 times higher than the TiO₂-based DSSC (4.10%). For the photocatalytic degradation of pollutants, the prepared TiO₂/rGO/Ag photocatalyst exhibited superior photodegradation of methylene blue (MB) dye molecules at around 93% and the mineralization of total organic compounds (TOC) by 80% in aqueous solution after 160 min under continuous irradiation with natural sunlight. Moreover, the enhanced performance of the DSSC device and the MB dye degradation exhibited by the hybrid nanostructures are more associated with their high surface area. Therefore, the proposed plasmonic hybrid nanostructure system is a further development for photovoltaics and environmental remediation applications.

Keywords: solar energy; dye degradation; surface plasmon resonance effect; TiO₂/rGO/Ag; hybrid nanostructures

1. Introduction

The significant increase in energy requirements and the depletion of fossil fuels have caused researchers to develop energy harvesting from renewable energy resources [1,2]. The use of solar-driven photovoltaic technologies and heterogeneous photocatalysis offers an appealing solution to the current global energy crisis and environmental remediation [3–5]. In terms of solar energy conversion technologies, dye-sensitized solar cells (DSSC) and heterogeneous photocatalytic dye degradation are the most attractive and promising areas of research to address energy and environmental concerns [6,7]. Semiconducting titanium dioxide (TiO₂) nanostructures are typically potential candidates as photoanode materials in DSSC devices and photocatalysts for the heterogenous photodegradation of toxic dyes due to their unique physicochemical properties [8–10]. However, the wide bandgap of the TiO₂ nanostructures (3.2–3.3 eV) has the absorption region below 4% in the entire solar spectrum compared to that of the visible region (43% solar energy) [11]. As a result, inefficient use of the visible light spectral portion and the fast recombination of electron-hole pairs (TiO₂) are significant constraints in the large-scale development of

Citation: Athithya, S.; Manikandan, V.S.; Harish, S.K.; Silambarasan, K.; Gopalakrishnan, S.; Ikeda, H.; Navaneethan, M.; Archana, J. Plasmon Effect of Ag Nanoparticles on TiO₂/rGO Nanostructures for Enhanced Energy Harvesting and Environmental Remediation. *Nanomaterials* **2023**, *13*, 65. https://doi.org/10.3390/nano13010065

Academic Editor: Muralidharan Paramsothy

Received: 22 October 2022
Revised: 10 December 2022
Accepted: 11 December 2022
Published: 23 December 2022

Copyright: © 2022 by the authors. Licensee MDPI, Basel, Switzerland. This article is an open access article distributed under the terms and conditions of the Creative Commons Attribution (CC BY) license (https://creativecommons.org/licenses/by/4.0/).

efficient photoanodes for DSSC devices and photocatalysts [12–14]. To overcome these constraints, different TiO$_2$ nanostructure composites with two-dimensional (2D) carbon materials have been employed as an effective strategy in recent decades. In nature, 2D-single molecular layered structures with sp^2 hybrid carbon atoms, i.e., graphene, possesses high surface area (ca. 2600 m^2/g), electron mobility (ca. 15,000 m^2/V.s at room temperature) and unrestricted movement of electrons in the crystal lattice [2,15,16]. During the formation of the nanocomposite, TiO$_2$ nanostructures are bonded with the graphene surface due to the presence of intermolecular forces. This increases the number of electron spots and electron bridges, which promotes electron transport with a suppressed recombination rate at the interface [17]. Moreover, reduced graphene oxide (rGO) nanosheet composites with TiO$_2$ nanostructures offer enhanced active surface area, good electrical conductivity and a lower recombination rate of photon-induced-charge carriers for photovoltaic and photocatalytic performance [18]. Manikandan et al. reported TiO$_2$ along with rGO not only enhanced the surface area but also influenced the short-circuit current (J_{SC}) in a device due to the high carrier mobility behavior of the rGO [19].

In recent decades, the localized surface plasmon resonance (LSPR) phenomenon of noble metal nanoparticles has played a dual role as potential visible sensitizers and electron sinks in the degradation of pollutant dyes and in DSSC device performance. It is a well-known strategy to trap electrons and harvest maximum light for the development of high-efficiency photovoltaic devices. Amine-functionalized TiO$_2$ composites with GO and Ag nanoparticles exhibited high current density due to the improved electron transfer at the photoanode/electrolyte interface, as reported by Kandasamy et al. [20]. The incorporation of Ag nanoparticles (NPs) into TiO$_2$-carbon nanotube (CNT) nanocomposites exhibited a photocatalytic activity of 66% for methylene blue (MB) degradation under visible light due to the presence of the CNTs and Ag NPs, as reported by Zhao et al. [21]. MB is a well-known, highly carcinogenic thiazine pollutant that has been manufactured and used in a variety of industries for various purposes. Therefore, it is strongly recommended to remove such a persistent contaminant from any given aqueous solution [22].

Moreover, the introduction of Ag NPs onto a TiO$_2$ nanocomposite improved the absorption-coefficient of the organic dye and eventually enhanced the optical absorption in the visible-light region [23,24]. It is similarly promising to achieve superior electrical conductivity with a prolonged lifetime of the photogenerated charge carriers for rGO/TiO$_2$ nanocomposites [25]. For example, Huan et al. prepared a flower-shaped nanosheet rGO/TiO$_2$ composite material that exhibited a photocatalytic efficiency of 92.3% under UV–visible light irradiation for the degradation of a rhodamine B (RhB) solution [26]. Duygu et al. reported the development of rGO-TiO$_2$-CdO-ZnO-Ag based composites that exhibited an excellent degradation rate of methylene blue (MB) dye (15 min) with 91% photocatalytic efficiency under UV light irradiation [27]. Similarly, Zohreh et al. reported the influence of surface plasmon resonance on the photovoltaic characteristics of Ag/TiO$_2$ in a photoanode-based DSSC device with a power conversion efficiency (PCE) of 6.5% under 1 sun simulated solar irradiation [28]. As a result, the combination of the LSPR influence of Ag NPs and the conductivity of rGO nanosheets with the reduced recombination rate of the TiO$_2$/rGO/Ag hybrid nanostructures is advantageous for long-term dye degradation photocatalysts and photoanodes for DSSC devices.

In the present work, we report the in situ growth synthesis of mesosphere TiO$_2$/rGO nanosheets/Ag NPs as a plasmonic hybrid nanostructures for visible-light-responsive DSSCs and photocatalytic applications. The prepared hybrid nanostructures are further examined with respect to their structural, morphological and optical properties using various advanced characterization techniques. The influence of the LSPR properties on a constructed DSSC device with the visible-light-driven photoanode and photocatalyst dye (MB) degradation is systematically investigated with an appropriate mechanism.

2. Experimental Section

2.1. Materials and Reagents

Titanium tetra-isopropoxide (TTIP), ethylene glycol ($C_3H_6O_2$), acetone (C_3H_6O), ethanol (C_2H_6O), graphite powder, sodium nitrate ($NaNO_3$), sulfuric acid (H_2SO_4), potassium permanganate ($KMnO_4$), hydrochloric acid (HCl), hydrogen peroxide (H_2O_2), isopropyl alcohol (IPA), 1,4-benzoquinone (BQ), ethylenediaminetetraacetic acid (EDTA) and silver nitrate ($AgNO_3$) were purchased from SRL Co., Mumbai, India. All the purchased chemicals were analytical grade and used in the synthesis without further purification.

2.2. Preparation of TiO_2 Mesospheres

TiO_2 mesospheres were prepared through a combined route of sol-gel and solvothermal synthesis. In the first step (sol-gel), 3 mL of TTIP was slowly added into 150 mL of ethylene glycol, and the solution was continuously stirred for 12 h at room temperature. After a 12 h stirring process, 300 mL of acetone with 2 mL of de-ionized water (DI) were added into the solution and stirring was continued for 2 h to obtain a white suspension. Subsequently, the white suspension was collected and subjected to several centrifugations with ethanol and DI water, respectively. The obtained product was dried at 60 °C for 10 h to remove the impurities and form titanium glycolate. In the second step of the procedure (solvothermal route), 1 g of titanium glycolate was dispersed into 60 mL of a mixed solvent of ethanol and DI water under stirring. The prepared white solution was then transferred into a 100 mL autoclave and kept at 180 °C for 12 h. Finally, the TiO_2 mesospheres were obtained after annealing at 350 °C for 1 h.

2.3. Preparation of Graphene Oxide (GO)

Graphene oxide (GO) was prepared under room-temperature conditions by the oxidation of natural graphite powder using a modified Hummer's method [29]. Briefly, 1 g of natural graphite powder and 0.5 g of sodium nitrate were blended in 23 mL of concentrated sulfuric acid under vigorous stirring for 30 min. A total of 3 g of well-ground potassium permanganate was then slowly added into the mixed solution and stirred for 30 min under ice-bath conditions at 7 °C. The mixed solution was then stirred at 35 °C for 30 min. To quench the vigorous oxidation process in the solution, 3 mL of hydrogen peroxide was added to 60 mL of DI water, and the solid GO powder was obtained after washing several times with 5% of HCl and DI water, then dried overnight at 60 °C.

2.4. Preparation of Mesosphere TiO_2/rGO Sample

Prepared TiO_2 mesospheres (0.2 g) were added to a mixed solution of an equal portion of DI water (15 mL) and ethanol (15 mL) (as solution A). Then, 30 mg of GO powder was dispersed in the same ratio of solvent (DI and ethanol) under stirring for 1 h (as Solution B). Solutions A and B were then mixed with continuous stirring for 1 h. The resultant mixed solution was transferred into a 100 mL autoclave and kept in hot air at 180 °C for 12 h. The final product of mesosphere TiO_2/rGO was obtained by washing the product with DI water and drying overnight at 60 °C.

2.5. Preparation of TiO_2/rGO/Ag Hybrid Nanostructure by In Situ Growth

The TiO_2/rGO/Ag hybrid nanostructures were synthesized by an in situ hydrothermal process. In this procedure, 10 mg of silver nitrate was dissolved in 60 mL of DI water (30 mL) and ethanol (30 mL), and then 0.2 g of the prepared TiO_2/rGO was added to the above solution under continuous stirring for 1 h. The net solution was transferred into an autoclave and kept at 180 °C for 4 h. The final product was washed with DI water and dried at 60 °C.

2.6. Characterization

Structural analysis was conducted using X-ray diffraction (XRD; PANalytical, Malvern, UK) with Cu Kα radiation (λ = 1.5406 Å) in the 2-theta range between 10° and 80° with a scanning rate of 0.02°/min. Raman spectra of the prepared sample were obtained using a micro-Raman spectrometer (LABRAM HR Evolution, Horiba, Longjumeau, France) with an excitation wavelength of 532 nm. The surface morphology of the prepared samples was analyzed using high-resolution scanning electron microscopy (HR-SEM; Apreo S, Thermo Fisher Scientific, Hillsboro, OR, USA) with an acceleration voltage of 15 kV. Further analysis of the surface morphology was conducted using high-resolution transmission electron microscopy (HR-TEM; JEM-2100, JEOL, Tokyo, Japan) with an acceleration voltage of 200 kV to reveal the atomic interplanar morphology and the elemental composition of the prepared samples. Ultraviolet-visible diffuse reflectance spectroscopy (UV-DRS, V-750, JASCO, Tokyo, Japan) measurements were conducted in the range of 200 nm to 800 nm. To analyze the emission properties of the prepared sample, photoluminescence (PL; FP8600, JASCO, Tokyo, Japan) spectra were measured at room temperature. The surface area and pore size distribution of the samples were characterized by the BET (Brunauer–Emmett–Teller) and BJH (Barrett–Joner–Halenda) methods (Autosorb IQ series, Quantachrome Instruments, Boynton Beach, FL, USA). X-ray photoelectron spectroscopy (XPS) was performed via a Kratos analytical instrument (ESCA 3400, Shimadzu Corporation, Kyoto, Japan). The percentage of mineralization efficiency was determined from total organic carbon (TOC) measurements (TOC-L, Shimadzu, Kyoto, Japan). Photovoltaic characterization of the fabricated devices was performed using a solar simulator (Sciencetech, Class A, Lamp: 300 W, London, ON, Canada). The I-V measurement and incident photocurrent efficiency (IPCE) of fabricated devices was measured using the same solar simulator over the wavelength range of 200 nm to 800 nm.

2.7. Photocatalytic Experiments

The photocatalysts measurement were carried out in our laboratory, SRM Institute of Science and Technology, Chennai (28°4′ N; 82°25′ E), in April 2020 (8 April to 10 May). Daylight from 9 am to 12 pm was utilized to perform the photocatalytic experiment with an average light intensity of 68.2~89.4 mW/cm^2. The photocatalytic properties of the as-synthesized samples were performed using MB dye as a model pollutant. In a typical photocatalytic reaction, 10 ppm of MB dye was added to 50 mL of DI water and stirred for 5 min. The solution was maintained in the dark for 20 min under stirring to achieve an adsorption–desorption equilibrium. At regular time intervals of the photocatalytic dye degradation reaction solution (20 min), 3 mL aliquots of the solution were sampled and UV–Vis spectra were measured.

2.8. DSSC Device Fabrication

Prepared samples, such as TiO_2, TiO_2/rGO and TiO_2/rGO/Ag (0.25 g each), were dispersed in 2 mL of stock solution (prepared by mixing an equal amount of DI water and ethanol) and ground for 15 min. To obtain a paste-like formation, 2 mL of the stock solution was mixed with 200 µL of acetic acid and 100 µL of surfactant (Triton 100-X) during the grinding process. The resultant colloidal solution was then uniformly coated on a fluorine-doped tin oxide (FTO) substrate with optimized conditions using a doctor blade technique. After drying at 100 °C for 10 min, the coated substrate was annealed at 450 °C for 30 min. The coated FTO was then immersed in dye solution for 12 h, which consisted of 0.03 M dis-tetrabutylammonium cis-bis(isothiocyanato)bis(2,2″-bipyridyl-4,4′-dicarboxylato) ruthenium (II) (N719). The prepared photoanode was subsequently clamped with a Pt-coated counter electrode to form a sandwich type of device. An electrolyte solution consisting of 0.6 M dimethylpropylimidazolium iodide, 0.1 M lithium iodide, 0.01 M iodine and 0.5 M 4-tert-butylpyridine in acetonitrile was then filled in between the layers of the sandwich-type device.

3. Results and Discussion

3.1. Structural and Compositional Analysis

Figure 1a shows the XRD patterns of prepared samples of pristine GO, mesosphere TiO_2, TiO_2/rGO and $TiO_2/rGO/Ag$. A strong and sharp diffraction peak is observed at 9.8°, which corresponds to the (0 0 1) crystal plane of GO. Using Bragg's equation, the interlayer distance of the as-prepared GO sheet was estimated to be 0.9 nm, whereas graphite powder shows an interlayer distance around 0.33 nm [30]. The XRD pattern obtained for GO indicates that the bulk graphite is successfully reduced as rGO nanosheets [31]. Furthermore, the characteristic peaks of TiO_2 are observed at 25.28°, 37.97°, 47.95°, 53.84°, 55.02°, 62.40°, 68.70° and 75.20°, which are assigned to the (1 0 1), (0 0 4), (2 0 0), (1 0 5), (2 1 1), (2 0 4), (1 1 6) and (2 1 5) planes, respectively. In the case of $TiO_2/rGO/Ag$, the peaks observed at 38.12°, 44.26°, 64.33° and 77.35° are associated with the cubic phase of the Ag (1 1 1), (2 0 0), (2 2 0) and (3 1 1) planes (JCPDS: 04-0783). Therefore, the obtained XRD pattern of the $TiO_2/rGO/Ag$ composite reveals the coexistence of rGO, TiO_2 and Ag materials and confirms the effective formation of a hybrid nanostructure.

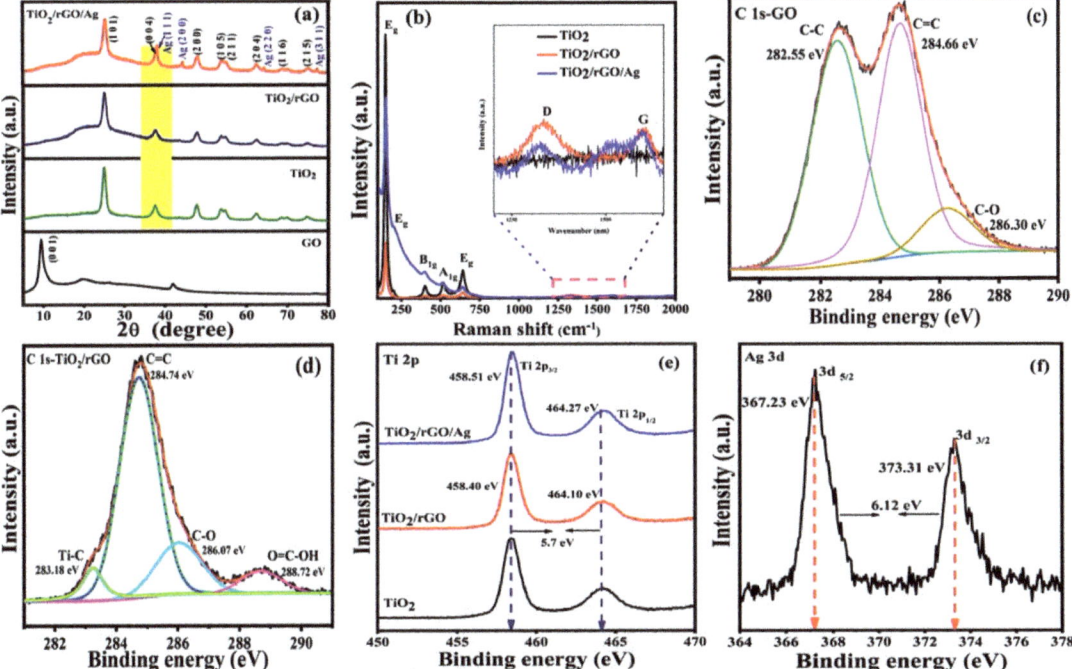

Figure 1. (a) XRD patterns, (b) Raman spectra of GO, mesosphere TiO_2, TiO_2/rGO and $TiO_2/rGO/Ag$. XPS core level spectra of the (c) C1s spectra of GO; (d) C 1s spectra of TiO_2/rGO; (e) Ti 2p spectra of mesosphere TiO_2, TiO_2/rGO and $TiO_2/rGO/Ag$; (f) Ag 3d spectra of $TiO_2/rGO/Ag$.

To explore the structural characteristics, Raman spectroscopy was performed for GO, TiO_2, TiO_2/rGO and $TiO_2/rGO/Ag$ samples, as shown in Figure 1b. The obtained Raman spectrum of pristine GO revealed the D band at 1350 cm^{-1} due to the disordered nature of the graphene structure with sp^3 defects (as given in Figure S1). In addition, the G band (at 1587 cm^{-1}) was observed in the Raman spectrum of pristine GO, which was attributed to the in-plane vibrations of C–C stretching in graphitic materials, as well as the doubly degenerate phonon mode in the Brillouin zone [32]. As shown in the inset image of Figure 1b, the G band shifted from 1587 to 1599 cm^{-1}, which is direct evidence of the effective reduction of GO to rGO with a constant D band for TiO_2/rGO [33]. In comparison with pristine GO, the I_D/I_G intensity ratio of the TiO_2/rGO and $TiO_2/rGO/Ag$ samples was slightly reduced to 0.83 from 0.87 (as shown in Supporting Information Figure S1). However, the reduction in the I_D/I_G ratio indicates a considerable reduction in the sp^2 domain size of carbon atoms, as well as the reduction of sp^3 to sp^2 [34]. In the case of $TiO_2/rGO/Ag$, the Raman vibrational modes were observed at 147, 199, 396, 515 and 639 cm^{-1}, which corresponds to the E_g, E_g, B_{1g}, A_{1g} and E_g modes of anatase TiO_2 [35]. On the other hand, the broadening of the E_g mode (147 cm^{-1}) with a considerable peak shift revealed the presence of TiO_2 mesospheres, as well as Ag NPs, on multilayer rGO. Furthermore, there were no significant changes in the observed Raman scattering modes of the plasmonic hybrid nanostructures compared with the hybrid sample (TiO_2/rGO). Nevertheless, the single Raman characteristic peak of the anatase TiO_2 phase was considerably enhanced due to the LSPR properties of Ag NPs on the $TiO_2/rGO/Ag$ hybrid surface [36]. The obtained Raman characteristic modes further confirm the successful formation of plasmonic hybrid nanostructures.

XPS measurements were performed to evaluate the chemical state and interaction of the prepared hybrid nanostructures and pristine GO samples. The surveyed spectra of TiO_2, TiO_2/rGO and $TiO_2/rGO/Ag$ indicate the presence of Ti, O, Ag and C elements without any impurities in the prepared hybrid nanostructures (as given in Supporting Information Figure S2). The deconvoluted C 1s spectrum of the as-prepared GO sample is shown in Figure 1c as follows: (i) At 282.55 eV, C bonds with sp^1 carbon atoms [37], (ii) C=C bonds denote the aromatic sp^2 structure groups present in the GO at 284.66 eV, (iii) the peak at 286.30 eV represents the carboxyl groups (C–O bonds) including epoxy and hydroxyl group [38]. In the case of the TiO_2/rGO sample, the deconvoluted XPS peak of the C 1s spectrum shows the four diverse carbon bonds with various binding energies of 283.18 eV, 284.74 eV, 286.07 eV and 288.72 eV. Thus, the peak that appears at 283.18 eV indicates the chemical bonding between the C atom and the Ti atom (Ti–C) [34]. Furthermore, the peaks with reduced intensity emerges at 284.74 eV, 286.07 eV and 288.72 eV with extensively reduced intensity purely attributed to C=C, C–O, O=C–OH bonds as oxygenated functional groups, which highlights the formation of rGO from GO [39]. Figure S3 shows the deconvoluted O 1s spectrum with two bands at 529 eV and 530 eV that correspond to the Ti–O–Ti and Ti–O–C groups. There is a slight peak shift in the TiO_2/rGO sample from 530.95 eV to 531.12 eV, which indicates the strong binding of rGO with TiO_2 mesosphere during the solvothermal process [40,41]. The chemical states of Ti species were thoroughly investigated using the XPS spectra, as shown in Figure 1e. The core level peak of Ti is split into a doublet peak observed at 458.40 eV and 464.10 eV, which corresponds to Ti $2p_{3/2}$ and Ti $2p_{1/2}$ with a splitting energy of 5.7 eV, mainly attributed to the Ti^{4+} state for the $TiO_2/rGO/Ag$ samples [42,43]. After infusing the Ag NPs into the hybrid nanostructure, the doublet peaks (Ti $2p_{3/2}$ and Ti $2p_{1/2}$) of the Ti 2p state have observed notable shifts to higher order, such as 458.40 eV to 458.51 eV (Ti $2p_{3/2}$) and 464.10 eV to 464.27 eV (Ti $2p_{1/2}$). As a result, a positive shift in the doublet peak of the Ti species directly reflects the alteration of the Fermi level, whereas the Ag species holds a lower Fermi level offset than that of TiO_2. Therefore, the resultant Fermi level alignment in the plasmonic hybrid nanostructures are more favorable for fast electron transfer processes at the $rGO/TiO_2/Ag$ interface [44]. Figure 1f shows the doublet peaks of the Ag element at binding energies of 367.23 eV and 373.31 eV, which correspond to Ag $3d_{5/2}$ and Ag $3d_{3/2}$ with a splitting energy of 6.12 eV, which confirms the presence of the Ag NPs in the prepared $TiO_2/rGO/Ag$ hybrid nanostructures [45].

3.2. Morphological Analysis

The surface morphology features of GO nanosheets, TiO_2 mesospheres, TiO_2/rGO and $TiO_2/rGO/Ag$ hybrid nanostructures were investigated using HR-SEM, TEM and HR-TEM measurements, as shown in Figure 2. The transparent ultrathin GO nanosheets are visible in the HR-SEM and HR-TEM micrographs, as shown in Figure 2(a1–a3). As-prepared TiO_2

mesosphere samples are agglomerate-free, uniformly distributed and spherical shaped, with an average size of ~570 nm (see Figure 2(b1)).

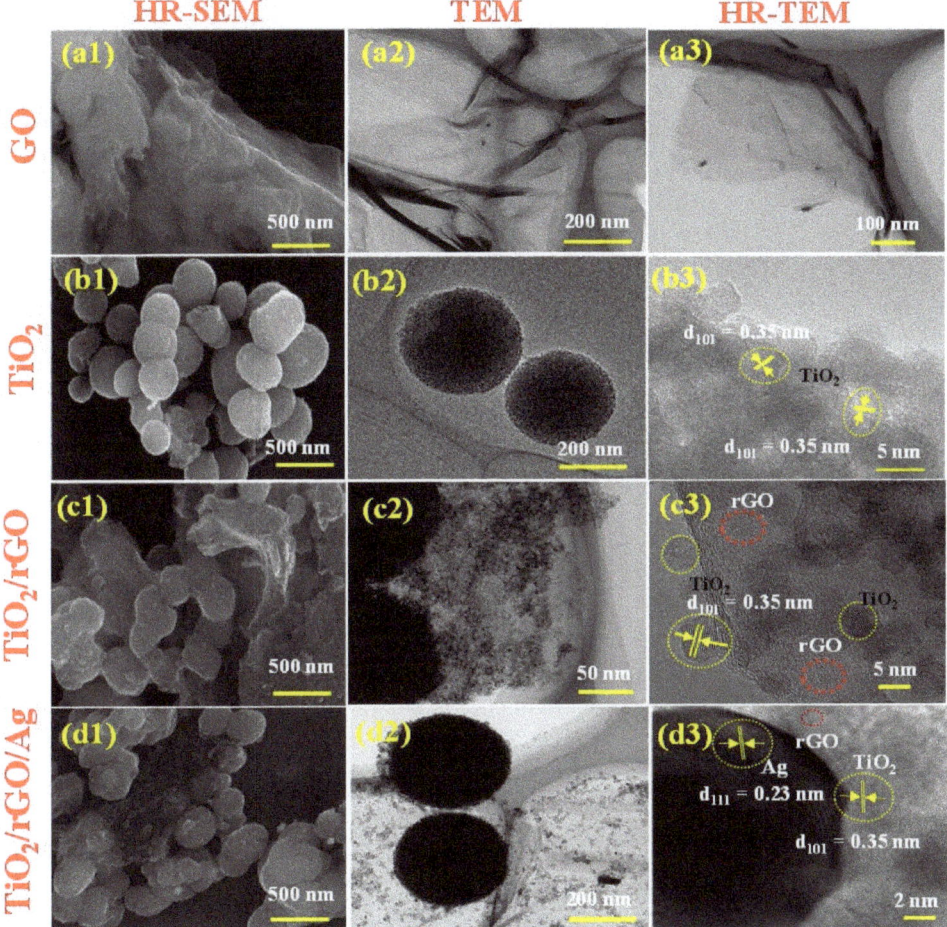

Figure 2. HR-SEM, TEM and HR-TEM images of prepared samples of GO (**a1–a3**), TiO$_2$ (**b1–b3**), TiO$_2$/rGO (**c1–c3**) and TiO$_2$/rGO/Ag (**d1–d3**).

The surface morphology and atomic interplanar distance of prepared samples were investigated using TEM and HR-TEM, as shown in Figure 2(b2,b3). Figure 2(c1–c3) shows the uniform TiO$_2$ mesospheres with an interplanar distance of 0.35 nm for the pure anatase TiO$_2$ phase over the rGO nanosheet surfaces. The presence of Ag NPs embedded in the TiO$_2$ mesosphere-coated rGO nanosheet surfaces is revealed by the TEM and HR-TEM images of the hybrid nanostructures. The interplanar distance of the prepared hybrid nanostructures was estimated from the HR-TEM image using a line profile analysis, as shown in Figure 2(d3). The calculated interplanar distance values were 0.35 nm and 0.23 nm for the anatase TiO$_2$ crystal plane of (1 0 1) and the (1 1 1) crystal plane of the Ag NPs, respectively, which is well matched with the standard JCPDS cards (Nos. 21-1272 and 04-0783) [46]. Moreover, the elemental mapping and energy dispersive X-ray spectroscopy (EDX) spectrum of the prepared hybrid nanostructures indicate the coexistence of C, Ti, O and Ag elements without the presence of any impurities in the resultant hybrid sample (as shown in Supporting Information Figure S4). Therefore, the resultant surface morphology of the prepared hybrid ternary nanostructures indicates that the plasmonic Ag NPs are uniformly embedded into the TiO_2 mesospheres anchored in the rGO nanosheet surfaces. Moreover, the thickness of the fabricated photoanode was estimated to be 14.2 μm (TiO_2) and 13.8 μm (TiO_2/rGO/Ag) and presented Figure S5a,b.

3.3. Optical and Surface Area Analysis

UV-DRS was used to investigate the optical absorption of the prepared TiO_2, TiO_2/rGO and TiO_2/rGO/Ag samples and is shown in Figure 3a. Due to the electronic transition of O 2p to Ti 3d, all of the prepared samples exhibit a typical optical absorption edge located at 392 nm of TiO_2, which corresponds to a bandgap energy of 3.3 eV for TiO_2 [47]. After the introduction of Ag NPs over the TiO_2/rGO surface, the optical absorption of the hybrid nanostructure is significantly higher in the visible region compared to the other samples, due to the LSPR of Ag NPs on the TiO_2/rGO surface [48]. Overall, the improved visible-light-region absorption observed for the prepared hybrid nanostructures indicates a greater number of photons harvested from the visible-light region, which is favorable for efficient photocatalysts and photoanodes for solar energy conversion devices.

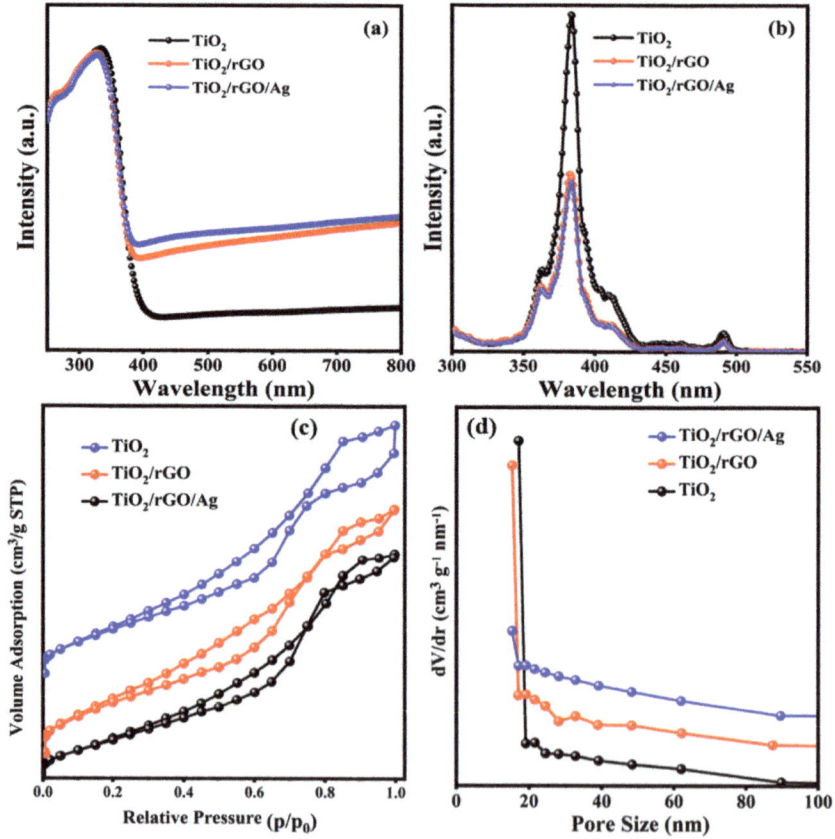

Figure 3. (a) UV-DRS absorption spectra, (b) PL spectra of mesosphere TiO_2, TiO_2/rGO and TiO_2/rGO/Ag. Nitrogen adsorption–desorption isotherms of (c) surface area and corresponding (d) pore size distribution of prepared samples TiO_2, TiO_2/rGO and TiO_2/rGO/Ag.

PL emission spectroscopy measurements are widely used to understand the desired photon-induced charge separation properties of the prepared samples. PL spectra of the prepared samples with an excitation wavelength of 300 nm and in the range from 350 to 500 nm are shown in Figure 3b. A strong band edge emission peak is observed at around 389 nm for all the prepared nanostructure samples [49]. Another emission peak centered in the visible region of 480 nm indicates the presence of oxygen vacancies on the TiO_2 surface. For the plasmonic hybrid nanostructures, the resultant emission peaks are significantly quenched, which is a direct indication of the reduction of the photon-induced electron-hole pair recombination rate that facilitates electron transfer at the TiO_2/rGO/Ag interface [50,51]. The function group of GO and prepared materials were analyzed by FTIR spectrum [52,53] (Figure S6).

The surface area and pore size distribution of the prepared samples were evaluated using the BET and BJH methods shown in Figure 3c,d. In comparison, the specific surface area of Ag NP-based hybrid nanostructures have a surface area of 297.71 m^2/g due to TiO_2 mesosphere formation with the Ag NP-coated rGO nanosheets. The estimated surface areas of the TiO_2 mesosphere and TiO_2/rGO samples were 234.16 m^2/g and 281.31 m^2/g, respectively. As shown in Figure 3d, the estimated pore sizes were 2.15 nm, 1.52 nm and 1.91 nm for the TiO_2, TiO_2/rGO and TiO_2/rGO/Ag hybrid nanostructures, respectively. The enhanced specific surface areas and pore size distributions of the hybrid nanostructures are more beneficial for the photoanodes in DSSC assemblies and photocatalysts for photocatalytic reactions.

3.4. Performance of DSSC Device

The influence of the LSPR properties of Ag NPs on the photovoltaic characteristics of a DSSC based on a TiO_2/rGO/Ag photoanode under simulated sun irradiation with an AM 1.5 G filter was investigated. The current density vs. voltage characteristics of DSSC devices assembled with various photoanodes (TiO_2, TiO_2/rGO and TiO_2/rGO/Ag) are given with error bars in Figure 4a. Compared with the TiO_2 and TiO_2/rGO photoanodes, the TiO_2/rGO/Ag photoanode show superior photovoltaic performance with a maximal J_{SC} of 16.05 mA/cm^2 and an improved open circuit voltage (Voc) of 0.74 V, with a fill factor (FF) of around 62.50%. In addition, the hybrid-nanostructure-based plasmonic DSSC device achieved a higher power conversion efficiency of 7.27%, which is 1.7 times higher than that of a pristine TiO_2 photoanode (4.01%). An overall comparison of a fabricated DSSC device's photovoltaic characteristics are given in Table 1 and SI. Figure 4b shows the IPCE of all of the fabricated DSSC devices in the range of 400 nm to 800 nm. The Ag-based DSSC device exhibited the maximum IPCE owing to the presence of N719 dye adsorption. The IPCE of the TiO_2/rGO/Ag photoanode-based plasmonic DSSC device achieved a maximum value of ca. 77.82% at an incident wavelength of 550 nm due to improved photon harvesting efficiency and a higher number of electron extraction from the N719 dye molecule than the other devices [27,28,52]. The photovoltaic performance of the fabricated TiO_2/rGO/Ag photoanode-based plasmonic DSSC cell has been compared with a recent report, as illustrated in Figure 4c [54–59]. A tentative operating mechanism of the proposed plasmonic DSSC device is illustrated to understand its enhanced photovoltaic performance. As shown in Figure 4d,e, upon 1 sun irradiation the electrons are excited from higher occupied molecular orbitals (HOMO) to lower occupied molecular orbitals (LUMO) levels of the dye molecule (N719). Under this irradiation, if the incident light coincides with the LSPR effect of Ag NPs (work function (WF) = −4.2 eV), then the electrons close to the Fermi level are excited into a higher energy state by receiving the energy from plasmon resonance, known as a hot electron, via non-radioactive process [60]. These electrons have sufficient energy to break the Schottky barrier formed between the TiO_2 mesosphere (WF = −4.4 eV) and Ag NPs and can flow through the conduction band of the TiO_2 mesosphere. Moreover, it is possible that this hot electron can be accepted and shuttled via rGO nanosheets (WF = −4.4 eV) [61–63]. The superior conductivity of rGO nanosheets could assist in accepting or transmitting electrons generated by TiO_2 or Ag NPs to the appropriate photoanode. The hybrid nanostructure attained higher efficiency than the TiO_2 mesospheres for the following reasons. (i) The large surface area provides high dye loading and efficiently scatters the incoming light within the device. (ii) The rGO nanosheets promote extraordinary charge transport where electrons come from TiO$_2$ mesospheres or Ag NPs and limit the electron-hole recombination rate. (iii) The LSPR effect of the Ag NPs extends the light absorption range from the UV to the visible region.

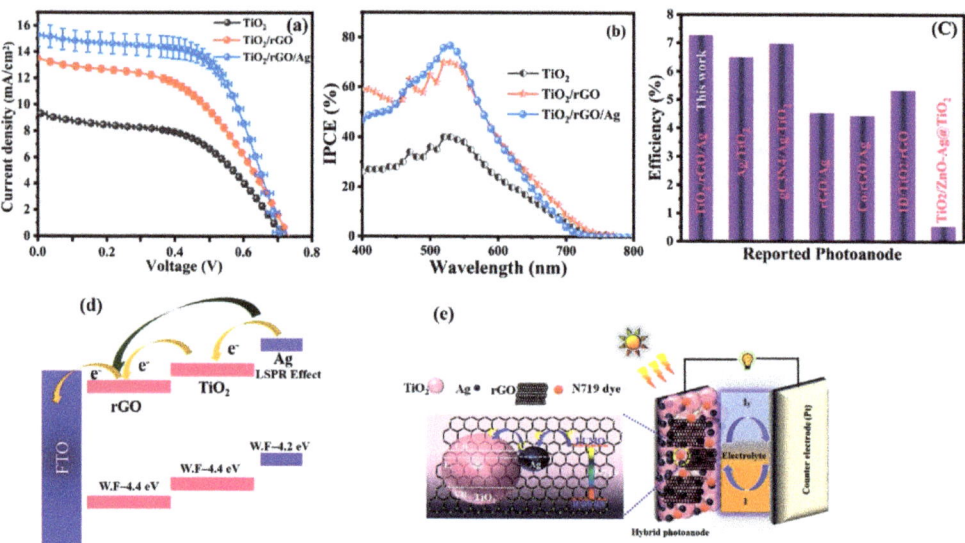

Figure 4. (a) Current density–voltage characteristics (with error bar), (b) IPCE spectra for different photoanode-based DSSC devices (mesosphere TiO$_2$, TiO$_2$/rGO and TiO$_2$/rGO/Ag) measured under one sun illumination, (c) compared efficiency of recent reported photoanode materials with TiO$_2$/rGO/Ag device performance, (d) schematic representation of the charge transfer process influenced by a plasmonic hybrid nanostructure (TiO$_2$/rGO/Ag)-based DSSC device and (e) DSSC device consists of hybrid photoanode and energy level diagram.

Table 1. Comparison of the photovoltaic properties and surface area of the DSSC device based on the mesosphere TiO$_2$, TiO$_2$/rGO and TiO$_2$/rGO/Ag photoanodes measured under AM 1.5 G one sun illumination.

Photoanode	Surface Area (m^2/g)	JSC (mA/cm^2)	VOC (V)	FF (%)	η (%)
TiO2	234.16	9.50	0.70	51.0	4.10
TiO2/rGO	281.31	13.80	0.74	55.0	5.00
TiO2/rGO/Ag	297.71	16.05	0.74	62.5	7.27

Figure S7 shows the transient photocurrent response of the TiO$_2$/rGO/Ag composite under 1 sun light irradiation with the highest photocurrent density, which is approximately one-fold greater than that of TiO$_2$/rGO. This confirms that the incorporation of Ag into the TiO$_2$/rGO/Ag nanocomposite not only enhances the light harvesting but also significantly expedites the photon-induced charge carrier separation and transport properties of Ag NP-coupled hybrid composites. The transient photocurrent response of the TiO$_2$/rGO/Ag composite was also performed under a UV filter, which confirmed the plasmonic Ag response photocurrent in the visible region. These results confirm that Ag could act as an acceptor of the photogenerated electrons by TiO$_2$/rGO and encourage fast charge transportation due to the high metallic conductivity, which effectively suppresses the charge recombination in the composites.

3.5. Photocatalytic Performance

To demonstrate the photocatalytic MB dye degradation process, TiO_2-, TiO_2/rGO- and $TiO_2/rGO/Ag$-based photocatalysts were employed under natural sunlight irradiation for three repetitions as shown in Figure S8. Figure 5a–c shows dye degradation profiles with MB as the pollutant with different irradiation times under natural sunlight for all of the prepared photocatalysts. The maximum absorption peak for the photocatalytic dye degradation process in the presence of a photocatalyst is displayed at around 664 nm due to the absorbance characteristics of MB dye. Under dark conditions, there were no significant changes in the UV–vis absorption spectra of all the prepared photocatalysts. The UV–vis absorption spectra of all the prepared photocatalysts were successfully recorded with a time interval of 20 min under continuous natural sun irradiation. The hybrid nanostructures of the $TiO_2/rGO/Ag$ photocatalyst exhibited a higher dye degradation efficiency of 93% under continuous natural sunlight irradiation for 160 min. Compared with pristine TiO_2, the enhanced dye degradation efficiency of the plasmonic hybrid photocatalyst was almost 1.3 times higher due to the LSPR properties of the Ag NPs. On the other hand, the presence of rGO/Ag on the TiO_2 surface creates a more favorable environment for fast photon-induced charge separation and transfer with an extended lifetime of charge carriers, whereas Ag NPs act as electron sinks for improved photocatalytic dye degradation [64].

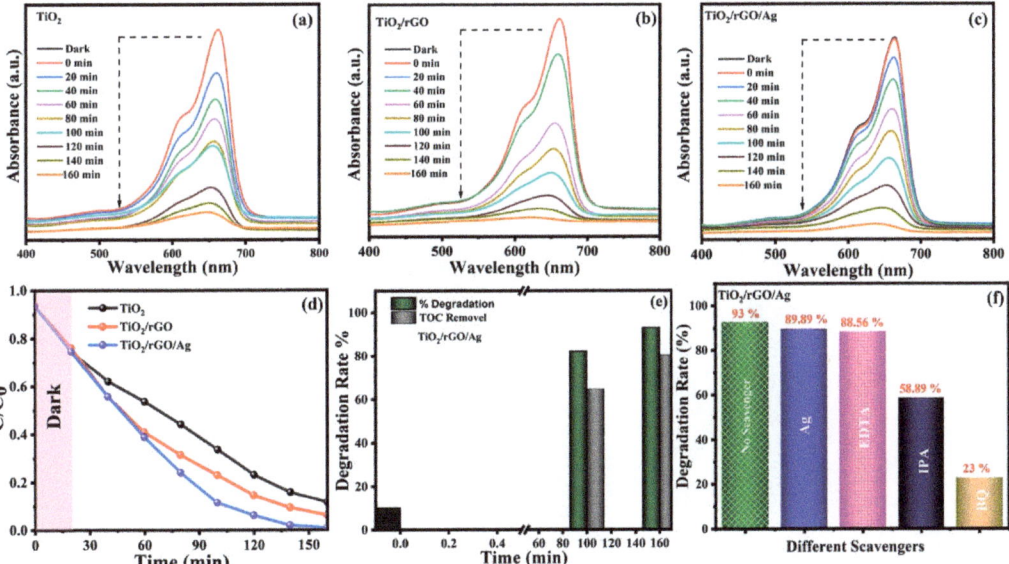

Figure 5. (**a–c**) UV absorption spectra of MB degradation with different interval time under natural sunlight, (**d**) plots of C/C_0 as a function of time (min) towards the photo degradation of MB, (**e**) TOC analysis was observed for $TiO_2/rGO/Ag$ at different intervals and the (**f**) photocatalytic degradation of MB in the presence of different scavengers of the $TiO_2/rGO/Ag$ sample under natural sunlight.

To further understand the effect of dye degradation in the form of mineralization efficiency, TOC analysis of the $TiO_2/rGO/Ag$ photocatalyst was performed with respect to various time intervals, as shown in Figure 5e. After 160 min of natural solar irradiation, the mineralization efficiency of the $TiO_2/rGO/Ag$ photocatalyst in terms of carbon content removal in MB dye was approximately 80.35% (Figure 5e). The improved TOC of the plasmonic hybrid $TiO_2/rGO/Ag$ photocatalyst suggests that it efficiently mineralized MB dye into residual organic molecules, such as CO_2 and H_2O, in the MB dye solution. Furthermore, the photocatalytic MB dye degradation activity of $TiO_2/rGO/Ag$ was evaluated by the addition of various radical scavengers (Ag, EDTA, IPA and BQ) to determine the more active species e^-, h^+, OH^- and O^{2-} in the degradation system [65,66]. Figure 5f shows various scavenger photocatalytic MB dye degradation (MB) experiments with the TiO_2/rGO/Ag system as a photocatalyst. For this study, 2 mg of various scavengers were mixed with MB dye solution in the presence of a prepared photocatalyst (15 mg) and exposed to natural sunlight for 160 min. The photocatalytic MB dye degradation efficiency of TiO_2/rGO/Ag was significantly inhibited by 58.89% and 23.00% in the presence of IPA and BQ as scavengers, respectively. This reduced the photocatalytic dye degradation efficiency, and it reveals that OH^- and O^{2-} radicals are the main active species during the degradation of MB by TiO_2/rGO/Ag under natural sunlight [67].

A possible mechanism for the photon-induced charge carrier separation and transfer in the prepared photocatalyst surface is depicted in Figure 6. The valence band (E_{VB}) and conduction band (E_{CB}) potentials of TiO_2, TiO_2/rGO and TiO_2/rGO/Ag were estimated using the Mulliken electronegativity theory [53] and the following equations:

$$E_{VB} = x - E^e + 0.5E_g \qquad (1)$$

$$E_{CB} = E_{VB} - E_g \qquad (2)$$

where x is the electronegativity (5.81 eV for TiO_2), E_{CB} is the conduction edge potential, E_{VB} is the valence band edge potential, E^e is the free energy of the electrons in the reversible hydrogen scale (4.5 eV), and E_g is the bandgap of the material. The E_{VB} and E_{CB} potentials of the TiO_2/rGO/Ag system were estimated to be -0.34 eV and 2.96 eV, respectively, while E_{VB} and E_{CB} for pure TiO_2 were estimated to be -0.32 and 2.94 eV, respectively.

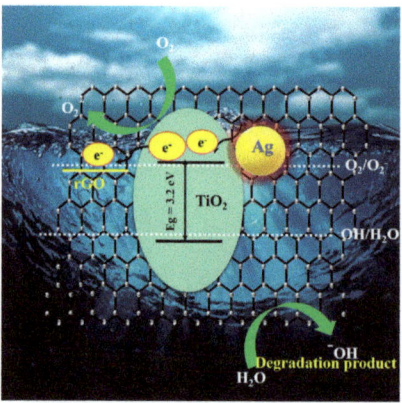

Figure 6. Schematic illustration of the possible mechanism of rGO, Ag NPs roles with mesosphere TiO_2 in MB photocatalytic degradation under natural sunlight irradiation.

Figure 6 shows the proposed photocatalytic mechanism for the natural-sunlight-responsive TiO_2-RGO-Ag nanostructure. After natural sunlight irradiation, the Ag NPs' dipolar characteristics of the SPR effect enable electrons generated from the Ag NPs to migrate to the CB of TiO_2. In contrast, electrons in the CB of TiO_2 could be transferred to the surface of the RGO nanosheets due to the favorable electron affinity of TiO_2 and the lower WF of RGO. Further, it is postulated that the photogenerated electron from Ag particles can also be transferred into the RGO sheets. TiO_2- and plasmon-excited Ag particles serve as electron transfer channels, which ensure charge separation efficiency. The oxidized Ag^+ species accept electrons from water molecules (H_2O) or OH^- adsorbed on the TiO_2 surface or from the dye molecules present in the solution and are regenerated. The reaction with H_2O or hydroxide ions produces hydroxyl radicals (OH). These radicals (O_2 and OH) are influential oxidizing agents for the degradation of MB dye molecules. Therefore, the prepared ternary nanostructures offer superior photocatalytic dye degradation (MB) due to efficient photoinduced charge carrier separation [52,53].

4. Conclusions

In this study, TiO_2/rGO/Ag hybrid nanostructures were successfully synthesized using a combination of solution processes and in situ growth and were then employed as photoanodes for DSSCs and as catalysts for photodegradation applications. The plasmon-enhanced DSSC devices demonstrate enhanced photovoltaic performance of 7.27% along with a higher short-circuit current of 16.05 mA/cm^2 and an IPCE efficiency of 77.82% at 550 nm. The results suggest that the high photovoltaic performance of the plasmon-based TiO_2/rGO/Ag device can be attributed to (i) the large specific area of TiO_2/rGO/Ag, which leads to high dye loading; (ii) TiO_2 mesospheres enhancing the light scattering effect of incoming light; and (iii) the incorporation of Ag NPs facilitating more induced photons and fast electron transport in the device. Upon natural sunlight irradiation, the prepared hybrid nanostructure shows an improved photocatalytic degradation of MB by 93% within 160 min, and the effects of different scavengers on the obtained photocatalytic activity were systematically investigated. The effects of optimum active surface area, the LSPR properties of Ag NPs and the enhanced electrical conductivity of the prepared ternary nanostructures combine to provide an enhanced visible-light-driven plasmonic DSSC device and photocatalyst for dye degradation (MB). The proposed plasmonic and hybrid-based nanostructures demonstrate an emerging strategy to establish large-scale applications of solar energy conversion technologies.

Supplementary Materials: The following supporting information can be downloaded at: https://www.mdpi.com/article/10.3390/nano13010065/s1. Figure S1: Raman spectra of prepared GO nanosheets. Figure S2: XPS survey spectrum of TiO_2, TiO_2/rGO and TiO_2/rGO/Ag. Figure S3: XPS O 1s spectra of (a) TiO_2 (b) TiO_2/rGO and (c) TiO_2/rGO/Ag. Figure S4: (i) STEM elemental mapping and (ii) EDX spectrum of TiO_2/rGO/Ag hybrid ternary nanostructures. Figure S5: The HR-SEM cross section of (a) TiO_2 (b) TiO_2/rGO/Ag photoanode. Figure S6: FTIR spectra of GO, TiO_2, TiO_2/rGO and TiO_2/rGO/Ag. Figure S7: Transient photocurrent spectra of TiO_2/rGO and TiO_2/rGO/Ag (under with and without UV filter). Figure S8: (a–c) Three different trials of UV absorption spectra of MB degradation with different intervals under natural sunlight. (d) Plots of Ln (C/C$_0$) as a function of time (min) towards the photo degradation of MB. References [52,53] are cited in supplementary materials.

Author Contributions: S.A.—designed the experiment procedure and wrote the paper, V.S.M.—data analysis, S.K.H.—draft, data analysis, conceptualization, K.S.—formal analysis, S.G.—original draft preparation, H.I.—resources, M.N.—methodology, J.A.—draft, data analysis, funding acquisition, supervision. All authors have read and agreed to the published version of the manuscript.

Funding: The financial support by "DAE-BRNS Young Scientist Research Award" No.34/20/02/2017-BRNS/34277 awarded to Dr. J. Archana to carry out the research work is gratefully acknowledged. The authors are thankful to SRMIST, Chennai, India for their SEED/START UP research grant and constant support for the research.

Data Availability Statement: The data presented in this study are available on request from the corresponding author.

Conflicts of Interest: The authors declare no conflict of interest.

References

1. Jamil, F.; Ali, H.M.; Janjua, M.M. MXene based advanced materials for thermal energy storage: A recent review. *J. Energy Storage* **2021**, *35*, 102322. [CrossRef]
2. Sözen, A.; Filiz, Ç.; Aytaç, I.; Martin, K.; Ali, H.M.; Boran, K.; Yetişken, Y. Upgrading of the performance of an air-to-air heat exchanger using graphene/water nanofluid. *Int. J. Thermophys.* **2021**, *42*, 1–15. [CrossRef]
3. Jacoby, M. Commercializing low-cost solar cells. *C&EN Glob. Enterp.* **2016**, *94*, 30–35. [CrossRef]
4. Hattori, Y.; Álvarez, S.G.; Meng, J.; Zheng, K.; Sá, J. Role of the Metal oxide electron acceptor on gold–plasmon hot-carrier dynamics and its implication to photocatalysis and photovoltaics. *ACS Appl. Nano Mater.* **2021**, *4*, 2052–2060. [CrossRef]

5. Hisatomi, T.; Kubota, J.; Domen, K. Recent advances in semiconductors for photocatalytic and photoelectrochemical water splitting. *Chem. Soc. Rev.* **2014**, *43*, 7520–7535. [CrossRef]
6. Choudhury, B.D.; Lin, C.; Shawon, S.M.A.Z.; Soliz-Martinez, J.; Huq, H.; Uddin, M.J. A photoanode with hierarchical nanoforest TiO_2 structure and silver plasmonic nanoparticles for flexible dye sensitized solar cell. *Sci. Rep.* **2021**, *11*, 7552. [CrossRef]
7. Li, Z.; Sun, Z.; Duan, Z.; Li, R.; Yang, Y.; Wang, J.; Lv, X.; Qi, W.; Wang, H. Super-hydrophobic silver-doped TiO_2 @ polycarbonate coatings created on various material substrates with visible-light photocatalysis for self-cleaning contaminant degradation. *Sci. Rep.* **2017**, *7*, 42932. [CrossRef]
8. Jing, L.; Zhou, W.; Tian, G.; Fu, H. Surface tuning for oxide-based nanomaterials as efficient photocatalysts. *Chem. Soc. Rev.* **2013**, *42*, 9509–9549. [CrossRef]
9. O'Regan, B.; Grätzel, M. A low-cost, high-efficiency solar cell based on dye-sensitized colloidal TiO_2 films. *Nature* **1991**, *353*, 737–740. [CrossRef]
10. Low, J.; Jiang, C.; Cheng, B.; Wageh, S.; Al-Ghamdi, A.A.; Yu, J. A Review of girect Z-scheme photocatalysts. *Small Methods* **2017**, *1*, 1–21. [CrossRef]
11. Graciani, J.; Álvarez, L.J.; Rodriguez, J.A.; Sanz, J.F. N Doping of rutile TiO_2 (110) surface. A theoretical DFT study. *J. Phys. Chem. C* **2008**, *112*, 2624–2631. [CrossRef]
12. Shah, S.A.S.; Zhang, K.; Park, A.R.; Kim, K.S.; Park, N.-G.; Park, J.H.; Yoo, P.J. Single-step solvothermal synthesis of mesoporous Ag–TiO_2–reduced graphene oxide ternary composites with enhanced photocatalytic activity. *Nanoscale* **2010**, *5*, 5093–5101. [CrossRef]
13. Wang, P.; Li, H.; Cao, Y.; Yu, H. Carboxyl-functionalized graphene for highly efficient H_2-evolution activity of TiO_2 photocatalyst. *Acta Phys. Chim. Sin.* **2021**, *37*, 2008047. [CrossRef]
14. Li, J.; Wu, X.; Liu, S. Fluorinated TiO_2 hollow photocatalysts for photocatalytic applications. *Acta Phys. Chim. Sin.* **2021**, *37*, 2009038. [CrossRef]
15. Schedin, F.; Geim, A.K.; Morozov, S.V.; Hill, E.W.; Blake, P.; Katsnelson, M.I.; Novoselov, K.S. Detection of individual gas molecules adsorbed on graphene. *Nat. Mater.* **2007**, *6*, 652–655. [CrossRef] [PubMed]
16. Zhang, N.; Zhang, Y.; Xu, Y.-J. Recent progress on graphene-based photocatalysts: Current status and future perspectives. *Nanoscale* **2012**, *4*, 5792–5813. [CrossRef] [PubMed]
17. Zhang, X.-Y.; Li, H.-P.; Cui, X.-L.; Lin, Y. Graphene/TiO_2 nanocomposites: Synthesis, characterization and application in hydrogen evolution from water photocatalytic splitting. *J. Mater. Chem.* **2010**, *20*, 2801–2806. [CrossRef]
18. Roy-Mayhew, J.D.; Aksay, I.A. Graphene materials and their use in dye-sensitized solar cells. *Chem. Rev.* **2014**, *114*, 6323–6348. [CrossRef] [PubMed]
19. Manikandan, V.; Palai, A.; Mohanty, S.; Nayak, S. Hydrothermally synthesized self-assembled multi-dimensional TiO_2/graphene oxide composites with efficient charge transfer kinetics fabricated as novel photoanode for dye sensitized solar cell. *J. Alloys Compd.* **2019**, *793*, 400–409. [CrossRef]
20. Kandasamy, M.; Selvaraj, M.; Kumarappan, C.; Murugesan, S. Plasmonic Ag nanoparticles anchored ethylenediamine modified TiO_2 nanowires@graphene oxide composites for dye-sensitized solar cell. *J. Alloys Compd.* **2022**, *902*, 163743. [CrossRef]
21. Zhao, D.; Yang, X.; Chen, C.; Wang, X. Enhanced photocatalytic degradation of methylene blue on multiwalled carbon nanotubes–TiO_2. *J. Colloid Interface Sci.* **2013**, *398*, 234–239. [CrossRef] [PubMed]
22. Din, M.I.; Khalid, R.; Najeeb, J.; Hussain, Z. Fundamentals and photocatalysis of methylene blue dye using various nanocatalytic assemblies- a critical review. *J. Clean. Prod.* **2021**, *298*, 126567. [CrossRef]
23. Parsa, S.M.; Yazdani, A.; Dhahad, H.; Alawee, W.H.; Hesabi, S.; Norozpour, F.; Javadi, D.; Ali, H.M.; Afrand, M. Effect of Ag, Au, TiO_2 metallic/metal oxide nanoparticles in double-slope solar stills via thermodynamic and environmental analysis. *J. Clean. Prod.* **2021**, *311*, 127689. [CrossRef]
24. Xiao, M.; Jiang, R.; Wang, F.; Fang, C.; Wang, J.; Jimmy, C.Y. Plasmon-enhanced chemical reactions. *J. Mater. Chem. A* **2013**, *1*, 5790–5805. [CrossRef]
25. Wang, P.; Wang, J.; Wang, X.; Yu, H.; Yu, J.; Lei, M.; Wang, Y. One-step synthesis of easy-recycling TiO_2-rGO nanocomposite photocatalysts with enhanced photocatalytic activity. *Appl. Catal. B Environ.* **2013**, *132–133*, 452–459. [CrossRef]
26. Xiao, H.; Wang, T. Graphene oxide (rGO)-metal oxide (TiO_2/Ag_2O) based nanocomposites for the removal of rhodamine B at UV–visible light. *J. Phys. Chem. Solids* **2021**, *154*, 110100. [CrossRef]
27. Akyüz, D. rGO-TiO_2-CdO-ZnO-Ag photocatalyst for enhancing photocatalytic degradation of methylene blue. *Opt. Mater.* **2021**, *116*, 111090. [CrossRef]
28. Mahmoudabadi, Z.D.; Eslami, E.; Narimisa, M. Synthesis of Ag/TiO_2 nanocomposite via plasma liquid interactions: Improved performance as photoanode in dye-sensitized solar cell. *J. Colloid Interface Sci.* **2018**, *529*, 538–546. [CrossRef]
29. Hummers, W.S., Jr.; Offeman, R.E. Preparation of graphitic oxide. *J. Am. Chem. Soc.* **1957**, *80*, 1339. [CrossRef]
30. Wang, T.; Tang, T.; Gao, Y.; Chen, Q.; Zhang, Z.; Bian, H. Hydrothermal preparation of Ag-TiO_2-reduced graphene oxide ternary microspheres structure composite for enhancing photocatalytic activity. *Phys. E Low-Dimens. Syst. Nanostructures* **2019**, *112*, 128–136. [CrossRef]
31. Abadikhah, H.; Kalali, E.N.; Khodi, S.; Xu, X.; Agathopoulos, S. Multifunctional thin-film nanofiltration membrane incorporated with reduced graphene oxide@TiO_2@Ag nanocomposites for high desalination performance, dye retention, and antibacterial properties. *ACS Appl. Mater. Interfaces* **2019**, *11*, 23535–23545. [CrossRef]

32. Kudin, K.N.; Ozbas, B.; Schniepp, H.C.; Prud'Homme, R.K.; Aksay, I.A.; Car, R. Raman spectra of graphite oxide and functionalized graphene sheets. *Nano Lett.* **2008**, *8*, 36–41. [CrossRef]
33. Zheng, Q.; Zhang, B.; Lin, X.; Shen, X.; Yousefi, N.; Huang, Z.-D.; Li, Z.; Kim, J.-K. Highly transparent and conducting ultralarge graphene oxide/single-walled carbon nanotube hybrid films produced by Langmuir–Blodgett assembly. *J. Mater. Chem.* **2012**, *22*, 25072–25082. [CrossRef]
34. How, G.T.S.; Pandikumar, A.; Ming, H.N.; Ngee, L.H. Highly exposed {001} facets of titanium dioxide modified with reduced graphene oxide for dopamine sensing. *Sci. Rep.* **2014**, *4*, 5044. [CrossRef]
35. Shah, S.A.S.; Park, A.R.; Zhang, K.; Park, J.H.; Yoo, P.J. Green synthesis of biphasic TiO_2–reduced graphene oxide nanocomposites with highly enhanced photocatalytic activity. *ACS Appl. Mater. Interfaces* **2012**, *4*, 3893–3901. [CrossRef] [PubMed]
36. Wen, Y.; Ding, H.; Shan, Y. Preparation and visible light photocatalytic activity of Ag/TiO_2/graphene nanocomposite. *Nanoscale* **2011**, *3*, 4411–4417. [CrossRef] [PubMed]
37. Sopinskyy, M.V.; Khomchenko, V.S.; Strelchuk, V.V.; Nikolenko, A.S.; Olchovyk, G.P.; Vishnyak, V.V.; Stonis, V.V. Possibility of graphene growth by close space sublimation. *Nanoscale Res. Lett.* **2014**, *9*, 182. [CrossRef]
38. Kisielewska, A.; Spilarewicz-Stanek, K.; Cichomski, M.; Kozłowski, W.; Piwoński, I. The role of graphene oxide and its reduced form in the in situ photocatalytic growth of silver nanoparticles on graphene-TiO_2 nanocomposites. *Appl. Surf. Sci.* **2021**, *576*, 151759. [CrossRef]
39. Pastrana-Martínez, L.M.; Morales-Torres, S.; Likodimos, V.; Falaras, P.; Figueiredo, J.L.; Faria, J.L.; Silva, A.M. Role of oxygen functionalities on the synthesis of photocatalytically active graphene–TiO_2 composites. *Appl. Catal. B Environ.* **2014**, *158–159*, 329–340. [CrossRef]
40. Leong, K.H.; Sim, L.C.; Bahnemann, D.; Jang, M.; Ibrahim, S.; Saravanan, P. Reduced graphene oxide and Ag wrapped TiO_2 photocatalyst for enhanced visible light photocatalysis. *APL Mater.* **2015**, *3*, 104503. [CrossRef]
41. Li, J.-J.; Cai, S.-C.; Yu, E.-Q.; Weng, B.; Chen, X.; Chen, J.; Jia, H.-P.; Xu, Y.-J. Efficient infrared light promoted degradation of volatile organic compounds over photo-thermal responsive Pt-rGO-TiO_2 composites. *Appl. Catal. B Environ.* **2018**, *233*, 260–271. [CrossRef]
42. Fang, R.; Miao, C.; Mou, H.; Xiao, W. Facile synthesis of $Si@TiO_2$@rGO composite with sandwich-like nanostructure as superior performance anodes for lithium ion batteries. *J. Alloys Compd.* **2020**, *818*, 152884. [CrossRef]
43. Ma, J.; Dai, J.; Duan, Y.; Zhang, J.; Qiang, L.; Xue, J. Fabrication of PANI-TiO_2/rGO hybrid composites for enhanced photocatalysis of pollutant removal and hydrogen production. *Renew. Energy* **2020**, *156*, 1008–1018. [CrossRef]
44. Wang, Y.; Yu, J.; Xiao, W.; Li, Q. Microwave-assisted hydrothermal synthesis of graphene based Au–TiO_2 photocatalysts for efficient visible-light hydrogen production. *J. Mater. Chem. A* **2014**, *2*, 3847–3855. [CrossRef]
45. Wu, Y.; Liu, H.; Zhang, J.; Chen, F. Enhanced photocatalytic activity of nitrogen-doped titania by deposited with gold. *J. Phys. Chem. C* **2009**, *113*, 14689–14695. [CrossRef]
46. Yang, D.; Wang, M.; Zou, B.; Zhang, G.L.; Lin, Z. An external template-free route to uniform semiconducting hollow mesospheres and their use in photocatalysis. *Nanoscale* **2015**, *7*, 12990–12997. [CrossRef]
47. Guo, T.-L.; Li, J.-G.; Sun, X.; Sakka, Y. Photocatalytic growth of Ag nanocrystals on hydrothermally synthesized multiphasic TiO_2/reduced graphene oxide (rGO) nanocomposites and their SERS performance. *Appl. Surf. Sci.* **2017**, *423*, 1–12. [CrossRef]
48. Alsharaeh, E.H.; Bora, T.; Soliman, A.; Ahmed, F.; Bharath, G.; Ghoniem, M.G.; Abu-Salah, K.M.; Dutta, J. Sol-Gel-sssisted microwave-derived synthesis of anatase Ag/TiO_2/GO nanohybrids toward efficient visible light phenol degradation. *Catalysts* **2017**, *7*, 133. [CrossRef]
49. Jie, Z.; Xiao, Y.; Huan, Y.; Youkang, H.; Zhiyao, Z. The preparation and characterization of TiO_2/r-GO/Ag nanocomposites and its photocatalytic activity in formaldehyde degradation. *Environ. Technol.* **2021**, *42*, 193–205. [CrossRef]
50. Li, L.; Zhai, L.; Liu, H.; Li, B.; Li, M.; Wang, B. A novel H_2O_2 photoelectrochemical sensor based on ternary RGO/Ag-TiO_2 nanotube arrays nanocomposite. *Electrochim. Acta* **2021**, *374*, 137851. [CrossRef]
51. Senthilkumar, N.; Sheet, S.; Sathishkumar, Y.; Lee, Y.S.; Phang, S.-M.; Periasamy, V.; Kumar, G.G. Titania/reduced graphene oxide composite nanofibers for the direct extraction of photosynthetic electrons from microalgae for biophotovoltaic cell applications. *Appl. Phys. A* **2018**, *124*, 769. [CrossRef]
52. Zhang, H.; Wang, X.; Li, N.; Xia, J.; Meng, Q.; Ding, J.; Lu, J. Synthesis and characterization of TiO_2/graphene oxide nanocomposites for photoreduction of heavy metal ions in reverse osmosis concentrate. *RSC Adv.* **2018**, *60*, 34241–34251. [CrossRef]
53. Qiu, B.; Zhou, Y.; Ma, Y.; Yang, X.; Sheng, W.; Xing, M.; Zhang, J. Facile synthesis of the Ti^{3+} self-doped TiO_2-graphene nanosheet composites with enhanced photocatalysis. *Sci. Rep.* **2015**, *5*, 8591. [CrossRef]
54. Li, M.; Yuan, N.; Tang, Y.; Pei, L.; Zhu, Y.; Liu, J.; Bai, L.; Li, M. Performance optimization of dye-sensitized solar cells by gradient-ascent architecture of $SiO_2@Au@TiO_2$ microspheres embedded with Au nanoparticles. *J. Mater. Sci. Technol.* **2018**, *35*, 604–609. [CrossRef]
55. Irannejad, N.; Rezaei, B.; Ensafi, A.A.; Momeni, M.M. Enhanced efficiency of dye-sensitized solar cell by using a novel modified photoanode with platinum C_3N_4 nanotubes incorporated Ag/TiO_2 nanoparticles. *Electrochim. Acta* **2017**, *247*, 764–770. [CrossRef]
56. Subramaniam, M.R.; Kumaresan, D.; Jothi, S.; McGettrick, J.D.; Watson, T.M. Reduced graphene oxide wrapped hierarchical TiO_2 nanorod composites for improved charge collection efficiency and carrier lifetime in dye sensitized solar cells. *Appl. Surf. Sci.* **2018**, *428*, 439–447. [CrossRef]

57. Ahmad, I.; Jafer, R.; Abbas, S.M.; Ahmad, N.; Rehman, A.U.; Iqbal, J.; Bashir, S.; Melaibari, A.A.; Khan, M.H. Improving energy harvesting efficiency of dye sensitized solar cell by using cobalt-rGO co-doped TiO_2 photoanode. *J. Alloys Compd.* **2022**, *891*, 162040. [CrossRef]
58. Cai, H.; Li, J.; Xu, X.; Tang, H.; Luo, J.; Binnemans, K.; Fransaer, J.; De Vos, D.E. Nanostructured composites of one-dimensional TiO_2 and reduced graphene oxide for efficient dye-sensitized solar cells. *J. Alloys Compd.* **2017**, *697*, 132–137. [CrossRef]
59. Hidayat, A.; Taufiq, A.; Supardi, Z.; Jayadininggar, S.; Sa'Adah, U.; Astarini, N.; Suprayogi, T.; Diantoro, M. Synthesis and characterization of TiO_2/ZnO-Ag@TiO_2 nanocomposite and their performance as photoanode of organic dye-sensitized solar cell. *Mater. Today Proc.* **2020**, *44*, 3395–3399. [CrossRef]
60. DuChene, J.S.; Sweeny, B.C.; Johnston-Peck, A.C.; Su, D.; Stach, E.A.; Wei, W.D. Prolonged hot electron dynamics in plasmonic-metal/semiconductor heterostructures with implications for solar photocatalysis. *Angew. Chem. Int. Ed.* **2014**, *53*, 7887–7891. [CrossRef]
61. He, Y.; Basnet, P.; Murph, S.E.H.; Zhao, Y. Ag nanoparticle embedded TiO_2 composite nanorod arrays fabricated by oblique angle deposition: Toward plasmonic photocatalysis. *ACS Appl. Mater. Interfaces* **2013**, *5*, 11818–11827. [CrossRef] [PubMed]
62. Mansfeldova, V.; Zlamalova, M.; Tarabkova, H.; Janda, P.; Vorokhta, M.; Piliai, L.; Kavan, L. Work Function of TiO_2 (anatase, rutile, and brookite) single crystals: Effects of the environment. *J. Phys. Chem. C* **2021**, *125*, 1902–1912. [CrossRef]
63. Garrafa-Gálvez, H.E.; Alvarado-Beltrán, C.G.; Almaral-Sánchez, J.L.; Hurtado-Macías, A.; Garzon-Fontecha, A.M.; Luque, P.A.; Castro-Beltrán, A. Graphene role in improved solar photocatalytic performance of TiO_2-RGO nanocomposite. *Chem. Phys.* **2019**, *521*, 35–43. [CrossRef]
64. Jaihindh, D.P.; Chen, C.-C.; Fu, Y.-P. Reduced graphene oxide-supported Ag-loaded Fe-doped TiO_2 for the degradation mechanism of methylene blue and its electrochemical properties. *RSC Adv.* **2018**, *8*, 6488–6501. [CrossRef] [PubMed]
65. Wang, W.; Xiao, K.; Zhu, L.; Yin, Y.; Wang, Z. Graphene oxide supported titanium dioxide & ferroferric oxide hybrid, a magnetically separable photocatalyst with enhanced photocatalytic activity for tetracycline hydrochloride degradation. *RSC Adv.* **2017**, *7*, 21287–21297. [CrossRef]
66. Li, T.; Wang, T.; Qu, G.; Liang, D.; Hu, S. Synthesis and photocatalytic performance of reduced graphene oxide–TiO_2 nanocomposites for orange II degradation under UV light irradiation. *Environ. Sci. Pollut. Res.* **2017**, *24*, 12416. [CrossRef]
67. Ganguly, P.; Mathew, S.; Clarizia, L.; Kumar, S.; Akande, A.; Hinder, S.; Breen, A.; Pillai, S.C. Theoretical and experimental investigation of visible light responsive $AgBiS_2$-TiO_2 heterojunctions for enhanced photocatalytic applications. *Appl. Catal. B Environ.* **2019**, *253*, 401–418. [CrossRef]

Disclaimer/Publisher's Note: The statements, opinions and data contained in all publications are solely those of the individual author(s) and contributor(s) and not of MDPI and/or the editor(s). MDPI and/or the editor(s) disclaim responsibility for any injury to people or property resulting from any ideas, methods, instructions or products referred to in the content.

Article

Electrosprayed CNTs on Electrospun PVDF-Co-HFP Membrane for Robust Membrane Distillation

Lijo Francis and Nidal Hilal *

NYUAD Water Research Center, New York University, Abu Dhabi Campus, Abu Dhabi P.O. Box 129188, United Arab Emirates
* Correspondence: nidal.hilal@nyu.edu

Abstract: In this investigation, the electrospraying of CNTs on an electrospun PVDF-Co-HFP membrane was carried out to fabricate robust membranes for the membrane distillation (MD) process. A CNT-modified PVDF-Co-HFP membrane was heat pressed and characterized for water contact angle, liquid entry pressure (LEP), pore size distribution, tensile strength, and surface morphology. A higher water contact angle, higher liquid entry pressure (LEP), and higher tensile strength were observed in the electrosprayed CNT-coated PVDF-Co-HFP membrane than in the pristine membrane. The MD process test was conducted at varying feed temperatures using a 3.5 wt. % simulated seawater feed solution. The CNT-modified membrane showed an enhancement in the temperature polarization coefficient (TPC) and water permeation flux up to 16% and 24.6%, respectively. Field-effect scanning electron microscopy (FESEM) images of the PVDF-Co-HFP and CNT-modified membranes were observed before and after the MD process. Energy dispersive spectroscopy (EDS) confirmed the presence of inorganic salt ions deposited on the membrane surface after the DCMD process. Permeate water quality and rejection of inorganic salt ions were quantitatively analyzed using ion chromatography (IC) and inductively coupled plasma-mass spectrometry (ICP-MS). The water permeation flux during the 24-h continuous DCMD operation remained constant with a >99.8% inorganic salt rejection.

Keywords: electrospinning; electrospraying; nanomaterials; nanostructured membrane; nanocomposites; seawater desalination; temperature polarization; flux

Citation: Francis, L.; Hilal, N. Electrosprayed CNTs on Electrospun PVDF-Co-HFP Membrane for Robust Membrane Distillation. *Nanomaterials* **2022**, *12*, 4331. https://doi.org/10.3390/nano12234331

Academic Editor: Muralidharan Paramsothy

Received: 17 November 2022
Accepted: 2 December 2022
Published: 6 December 2022

Publisher's Note: MDPI stays neutral with regard to jurisdictional claims in published maps and institutional affiliations.

Copyright: © 2022 by the authors. Licensee MDPI, Basel, Switzerland. This article is an open access article distributed under the terms and conditions of the Creative Commons Attribution (CC BY) license (https://creativecommons.org/licenses/by/4.0/).

1. Introduction

Electrospinning and electrospraying are versatile and cost-effective techniques for fabricating nanocomposite–nanofiber membranes or substrates for a variety of applications. Electrospinning is an electrohydrodynamic process in which a polymeric droplet is subjected to a high electric potential to generate a fast jet of polymeric dope by stretching and elongation to form nanofiber mats [1–4]. In an electrospraying process, homogeneous nanomaterial dispersions are generally used instead of polymeric dope solutions. The applications of electrospinning and/or electrospraying are not limited to applications in various industry sectors in energy, healthcare, water, and the environment [5–8]. Polymeric solution or nanomaterial dispersion parameters and electrospinning/spraying process parameters are important variables to produce engineered membranes for unique applications. Solution parameters include the concentration, viscosity, surface tension, and conductivity whereas process parameters include applied potential, the distance between needle and collector, flow rate, needle diameter, and humidity [9,10]. Incorporation of nanomaterials through blending, surface modification, or an electrospraying process onto electrospun membranes enhances the membrane characteristics, such as tensile strength, hydrophobicity, antimicrobial properties, etc. [11–20]. Mitigation of fouling and rejection of humic substances in membrane-based separation processes via engineered membranes or spacers has been studied and reported by many researchers [21–27]. Membrane distillation (MD) is a thermally driven membrane-based separation process in which a trans-membrane vapor pressure, created due to the temperature difference on the two sides of a membrane, drives water vapors from the hot side (feed solution) to the cold side of the membrane. The majority of the MD studies are reported either on water recovery applications using newly fabricated membranes or on process engineering to enhance the efficiency of the overall MD process [28–38]. MD membranes can be generally fabricated by phase inversion, sintering etching, electrospinning, etc. [39]. Simulation and modeling studies have significantly supported the MD process optimization as well as a macro-level understanding of the mechanism from the lab scale to the large scale [40–49].

In 2008, Feng et al. fabricated an electrospun PVDF membrane for the MD process for the first time. The highest water permeation flux observed at a ΔT of 60 °C was 11–12 LMH in an air gap membrane distillation (AGMD) process. The state of play and recent advances of electrospun membranes for the MD process were recently reported by Francis et al. and Tijing et al. Apart from PVDF, a variety of polymeric materials, such as polyazoles, polyimides, polyurethanes, polyacrylonitriles, polysulfones, etc., have been used for MD membrane fabrication via electrospinning [5,36,37,50,51]. Various nanomaterials, such as silica, alumina, titania, graphene, silver nanoparticles, etc., have been incorporated into the MD membrane to enhance the membrane characteristics and MD process performance [52–56].

Electrospinning of PVDF followed by the electrospraying of alumina nanoparticles was adopted by Attia et al. for the MD membrane fabrication [13]. Shahabadi et al. demonstrated an enhanced MD performance through the fabrication of an MD membrane by the electrospinning of PVDF-Co-HFP and electrospraying of titania nanoparticles [57]. Jia et al. demonstrated the fabrication of an anti-wetting/anti-fouling MD membrane by the fabrication of a superhydrophilic/superhydrophobic membrane by the deposition of octaphenylsilsesquioxane (POSS) nanoparticles on a PVDF substrate. POSS nanoparticles enhanced the surface roughness and hydrophobicity of the MD membrane [58]. They reported that the modified membrane was superhydrophobic and showed superior water permeation flux compared to the commercial membrane. Silica nanoparticles were deposited on PVDF-HFP electrospun membranes by Su et al., and they demonstrated that the fabricated membranes were superhydrophobic and had superior anti-scaling properties [59]. On the other hand, Hong et al. reported a pore-size tunable superhydrophobic membrane for the MD process. They used polydimethylsiloxane (PDMS)/PVDF-HFP for electrospraying on an electrospun polyurethane membrane and demonstrated superior MD process performance. Their observations were validated with the aid of simulation and modeling [60]. Gethard and co-workers demonstrated the fabrication of CNT-immobilized polypropylene hollow-fiber membranes for an enhanced MD process. The CNT incorporation led to a 1.85-times increase in flux and 15-times salt rejection than those compared to the parent membrane [61]. Song and coworkers reported an electrospun membrane fabrication using a polymeric dope solution of PVDF-Co-HFP blended with CNTs and applied for salty and dyeing wastewater treatment using the direct contact MD (DCMD) process. They reported enhanced membrane characteristics and DCMD performance for the CNT-incorporated electrospun MD membrane [62].

In the current investigation, a PVDF-Co-HFP dope solution was subjected to electrospinning followed by the electrospraying of carbon nanotubes (CNTs) for the fabrication of MD membranes and applied for the DCMD desalination process using 3.5 wt. % simulated seawater as the feed solution. The CNT-modified PVDF-Co-HFP composite membrane was subjected to heat pressing before the MD process testing. Membranes were characterized for water contact angle, liquid entry pressure (LEP), pore size distribution by porometry, tensile properties by universal testing machine, surface morphology by scanning electron microscopy (SEM), and energy dispersive spectrometry (EDS). Heat pressing plays a significant role to enhance the mechanical properties of the composite membrane. Surface-modified electrospun PVDF-Co-HFP membranes with electrosprayed CNTs showed enhanced MD membrane characteristics and an increased temperature polarization coefficient (TPC) with superior desalination performance.

2. Materials and Methods

2.1. Materials

Commercially available PVDF-Co-HFP was purchased from Arkema FLUORES, Colombes, France. Dimethyl acetamide, acetone, ethanol, functionalized multiwall carbon nanotubes (MWCNT), and 1H, 1H, 2H, 2H-perfluorooctyltriethoxysilane (POTS) were purchased from Sigma Aldrich, Massachusetts, MA, USA. Sea salt was purchased from Qingdao Sea-Salt Aquarium Technology Co., Ltd. (Qingdao, China).

2.2. Electrospinning of PVDF-Co-HFP and Electrospraying of CNT Dispersion

The 10 wt. % PVDF-Co-HFP solution was prepared using a solvent mixture of acetone and dimethyl acetamide in a ratio of 7:3, respectively. The polymer solution was homogenized by stirring overnight at room temperature and kept for another 10 h at room temperature to escape any trapped air bubbles in the dope solution. A volume of 10 mL polymer dope was taken in a syringe and fixed on a syringe pump of an electrospinning machine (MECC NANON, Fukuoka, Japan). A 21 G syringe needle was connected to the syringe using a plastic tube. The distance between the tip of the syringe and the rotating collector drum was adjusted to 15 cm. The positive terminal of a high-voltage power supply was connected to the syringe needle, and the negative terminal was connected to the grounded rotating drum. The polymer dope flow rate, speed of the rotating drum, and speed of the spinneret were adjusted to obtain a uniform nonwoven PVDF-Co-HFP electrospun nanofiber membrane. The 0.02% functionalized MWCNT and POTS were dispersed in a 50:50 ethanol-water mixture. The ratio of MWCNT and POTS was fixed at 1:2, respectively, in an ethanol-water mixture and subjected to ultrasonication for 3 h. POTS can act as a dispersant as well as a binder for CNTs. The CNT dispersion was taken in a syringe and subjected to electrospraying on the non-woven PVDF-Co-HFP electrospun membrane using the MECC NANON electrospinning setup used for the PVDF-Co-HFP electrospinning process. The dispersion taken in the 10 mL syringe was mounted on a syringe pump, which drives the dispersion through a plastic tube to the 18 G-sized blunted syringe needle kept 15 cm above the rotating drum. The variable parameters used for the electrospinning and electrospraying processes are shown in Table 1. Figure 1 shows the schematic representation of the electrospinning and electrospraying process.

Table 1. Electrospinning and electrospraying parameters.

Parameters	Electrospinning	Electrospraying
Solution/Dispersion Concentration	10 w/w % PVDF-Co-HFP in 7:3 Acetone:DMAc	0.02 w/w % MWCNT in 50:50 Ethanol:Water
Applied Potential (kV)	25	20
Working Distance (cm)	15	15
Flow Rate (mL/hour)	1	0.5
Speed of Collector Drum (rpm)	100	100
Spinneret Speed (mm/S)	5	5

2.3. Heat Pressing

Pristine PVDF-Co-HFP and nanocomposite PVDF-Co-HFP-CNT membranes were subjected to the heat pressing process using a Carver Press, Auto Series Plus Hydraulic Press purchased from Mitsubishi, Japan. The membrane sheet was cut into 20 × 20 cm size, sandwiched between e-PhotoInc Heat Press Transfer Teflon sheets, and placed on the stationary bottom plate of the Carver press. The temperature and load were kept at 80 °C and 2000 N. When both plates reached the desired temperature, the heated top plate was allowed to move downwards onto the bottom stationary plate and press on the membrane sandwich at the desired force for a dwell time of 30 min.

2.4. Water Contact Angle Measurements

A drop-shape analyzer (DSA 100 purchased from Kruss Scientific, Hamburg, Germany) was used to measure the water contact angle of the membrane samples. The average water contact angle value measured at five different locations on each membrane sample was considered the water contact angle of the respective membranes.

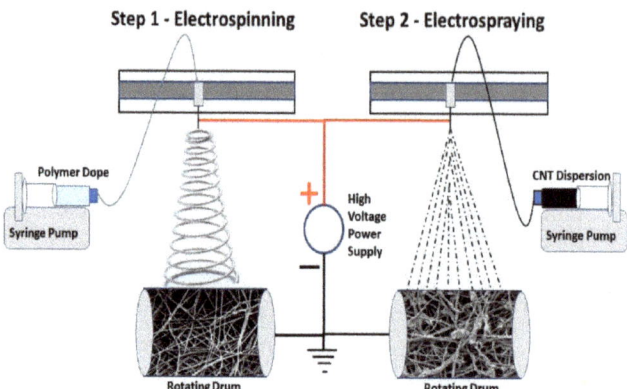

Figure 1. Schematic representation of nanocomposite membrane fabrication via electrospinning of PVDF-Co-HFP and electrospraying of CNT.

2.5. Mean Flow Pore Size, Bubble Point Measurements, and Pore Size Distribution

The average pore size, pore size distribution, and bubble point of the PVDF-Co-HFP and PVDF-Co-HFP-CNT nanocomposite membranes before and after the heat pressing process were measured using an advanced capillary flow porometer purchased from Porous Materials Inc., (New York, NY, USA) (iPore-1500A). Circular membrane samples with a diameter of 25 mm were used for the pore size analysis. The membrane samples were kept in the sample holder of the porometer, and a Galwick fluid having a very low surface tension (15.9 dynes/cm) was dripped enough to wet the hydrophobic membrane sample. The capillary flow porometric principle is used to measure the pore size distribution in which a nontoxic liquid (Galwick fluid) is allowed to spontaneously fill the membrane pores, and an essentially non-reacting gas (N_2) is passed through the sample to remove the Galwick liquid from the membrane pores. At lower pressures, the largest membrane pores will be emptied first and, as the pressure inserted by the nitrogen gas increases, the smallest pores get empty. The pore size distribution is obtained from the flow rate and pressure of the nitrogen gas. The pore size is inversely proportional to the pressure at which pores are empty.

2.6. Liquid Entry Pressure (LEP) Test

LEP is the critical pressure across a hydrophobic membrane at which the liquid starts to flow through the membrane pores [63]. In the MD process, LEP is an important parameter where the membrane has to stay non-wetted throughout the process. A bench-top LEP testing machine purchased from Convergence Minos (Convergence Industry, Drunen, The Netherlands) was used to measure the LEP of the membrane samples. Circular membrane samples with a 25 mm diameter were used for LEP measurements. Before starting the LEP test, the system automatically flushes with water to remove any air bubbles in the system. The pressure is ramped from zero in small increments (about 0.1 bar in the current investigation) until a pressure decay is observed in the system. Water starts to flow through the membrane from the point of pressure decay, and this is measured as LEP. An average value of 3 membrane samples was considered the LEP of the specific membrane.

2.7. Mechanical Characterizations

The mechanical properties of the PVDF-Co-HFP membranes and nanocomposite PVDF-Co-HFP–CNT membranes were characterized with a universal testing system (5965 model, Instron, Norwood, MA, USA). Standard dumbbell-shaped membrane samples were prepared using a Ray/Ran Hand Operated Cutter (RDM test equipment, Kemsing, UK) and subjected to the test using a 50 N load cell with a 2 mm/min strain rate.

2.8. Surface Morphology, Fiber Diameter Distribution, and EDS Images

A Thermo Fisher Field-Effect Scanning Electron Microscope (FESEM) Quanta 450, Waltham, MA, USA, was used to observe the surface morphology of the PVDF-Co-HFP and PVDF-Co-HFP-CNT membrane samples before and after the MD experiments. Gold sputter coating was employed on all membrane samples before FESEM imaging. The nanofiber diameter distribution of the electrospun membranes before and after the heat process was calculated using Image J analysis (Version 1.8). An AMETEK Octane Elect EDS detector was equipped on the FESEM machine to perform Energy-dispersive X-ray spectroscopy (EDX). Carbon sputtering was performed on the membrane samples before EDS imaging.

2.9. Temperature Polarization Coefficient (TPC)

Temperature polarization (TP) is a phenomenon that occurs in temperature-driven processes such as MD, in which the temperatures in the bulk feed (T_f) and coolant/permeate (T_p) solutions differ from the respective temperatures at the interface of the membrane and bulk solutions (feed and permeate) [64]. TP causes a reduction in the vapor pressure difference across the membrane and, thereby, the driving force of the MD process [65,66]. One of the major limitations in the MD process is the TP phenomenon. As a thermally driven separation process, heat and mass transfer are combined in the MD process, and it is very important to mitigate the TP phenomenon that occurs during the process to enhance the efficiency of the process [45,67]. The thickness of the polarization layer adjacent to the membrane surface increases as the separation process progresses, which leads to more reduction in the driving force and water vapor production. Researchers have come up with various methods to mitigate the TP phenomenon, such as employing (3D) spacers, baffles, engineered membranes, (micro) bubbling, stirring, feed flashing, employing isolation barriers, etc. [68–74]. TP phenomenon can be measured indirectly by using a term called TPC. TPC can be calculated using the following Equation (1) [44];

$$TPC = (T_{fm} - T_{pm})/(T_f - T_p) \qquad (1)$$

where T_{fm} is the feed side membrane temperature, T_{pm} is the permeate side membrane temperature, T_f is the bulk feed solution temperature, and T_p is the temperature of the bulk coolant/permeate. In an ideal condition, TPC can have a maximum value of 1.

2.10. Direct Contact Membrane Distillation (DCMD) Experiments

A fully automated MD setup purchased from Convergence Industry, The Netherlands, was used to conduct the DCMD experiments. A DCMD membrane module with an active membrane area of 60 cm^2 was used in all experiments. A 3.5 wt. % sea salt solution was prepared in deionized water and used as the feed solution, and tap water was used as the coolant. The DCMD experiments were conducted at feed solution temperatures of 35 °C, 40 °C, 45 °C, 50 °C, and 55 °C while keeping the coolant temperature constant at 15 °C. All the DCMD experiments were conducted at a flow rate of 60 L per hour on both sides of the membrane. A schematic representation of a direct contact membrane distillation (DCMD) experimental setup is shown in Figure 2.

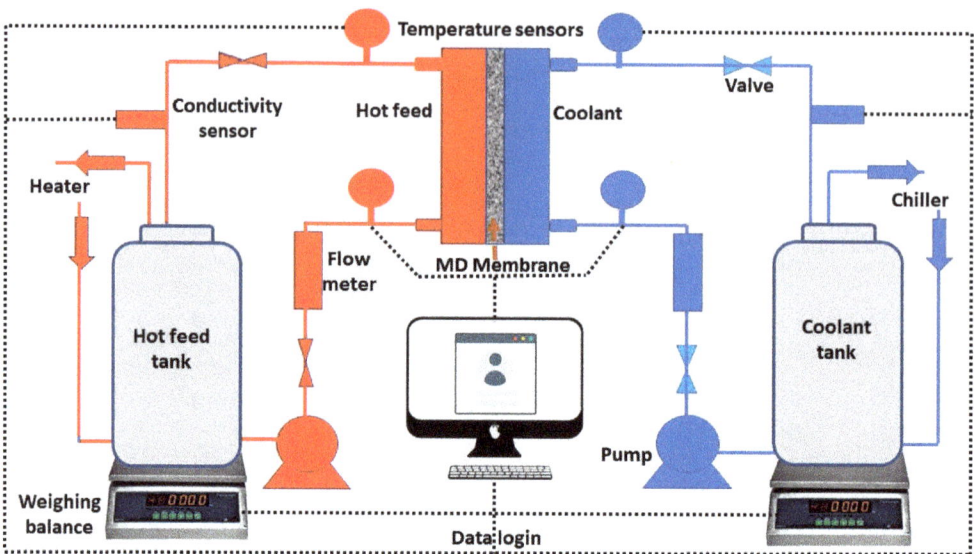

Figure 2. Schematic representation of experimental DCMD setup.

Water vapor flux was calculated using the following Equation (2):

$$J_w = (w_2 - w_1)/At \qquad (2)$$

where 'J_w' is the water vapor flux, '$(w_2 - w_1)$' is the weight of the permeate collected at specific intervals in kilograms or liters, 'A' is the effective membrane area in square meters, and 't' is the time in hours. Therefore, water vapor flux is represented in kg/m^2/h or liter/m^2/h (LMH). The temperatures of the feed and coolant at the entrance and exit of the membrane module and the conductivity of the coolant and feed solutions were measured using respective sensors periodically and recorded using a data acquisition system. Heat loss during the MD process was reduced by insulating the tubing, coolant, and feed tanks. The membrane samples were secured after 24-h continuous MD experiments for surface characterizations using FESEM and EDS. Analytical tools, such as an Inductively Coupled Plasma—Mass Spectrometer (ICP-MS Agilent 7800) purchased from Agilent Technologies, Santa Clara, CA, USA, and Ion Chromatography (IC 6000 Thermo Scientific, Waltham, MA, USA), were used for the quantitative analysis of salt ions present in the feed and permeate solutions.

3. Results and Discussion

3.1. Water Contact Angle, Heat Pressing, and Pore Size Distribution

POTS is a highly hydrophobic reagent that acts as a binder and a dispersant for CNTs in an ethanol–water mixture. Homogeneously dispersed CNTs yield an evenly distributed uniform coating on electrospun PVDF-Co-HFP membrane surfaces upon an electrohydrodynamic atomization process or electrospraying process. The observed water contact angles of the electrospun PVDF-Co-HFP and PVDF-Co-HFP-CNT membranes before and after the heat pressing process are shown in Table 2. The presence of CNTs on the PVDF membrane surface enhances the surface roughness and, thereby, the water contact angle by 3%. At the same time, the heat pressing process may cause a reduction in the surface roughness and, thereby, the average water contact angle by 3% in both the PVDF-Co-HFP membrane and CNT-modified PVDF membrane. The average flow pore size, minimum pore size, and bubble point of the electrospun PVDF-Co-HFP and composite CNT-modified PVDF-Co-HFP membranes before and after the heat pressing process are shown in Table 2.

Table 2. Water contact angle and pore size distribution of the PVDF-Co-HFP and PVDF-Co-HFP-CNT membranes.

Membrane	ESPVDF-HFP	ESPVDF-HFP-HP	ESPVDF-HFP-CNT	ESPVDF-HFP-CNT-HP
Water contact angle	136 ± 2	132 ± 3°	140 ± 3°	136 ± 2°
Minimum pore size (μm)	0.083	0.051	0.091	0.049
Mean Flow Pore Size (μm)	0.448	0.285	0.411	0.257
Pore size at the bubble point or maximum pore size (μm)	0.921	0.615	0.92	0.715

The pore size distribution profile of pristine PVDF-Co-HFP membranes and CNT-modified membranes before and after the heat pressing process is shown in Figure 3.

It is obvious from Figure 3 and Table 2 that the minimum, average, and maximum pore sizes of the CNT-modified membranes were reduced, and pore size distribution was narrowed compared to the electrospun PVDF-Co-HFP membranes after the heat pressing process. According to Shahabadi et al., narrow membrane pore size distribution is better for enhanced MD process performance [57]. Thus, the heat pressing process helps in yielding an efficient membrane with more suitable membrane characteristics for the MD process. The average pore size of the electrospun PVDF-Co-HFP and CNT-modified PVDF-Co-HFP membranes was reduced by 36.3% and 37.5%, respectively, after the heat pressing process.

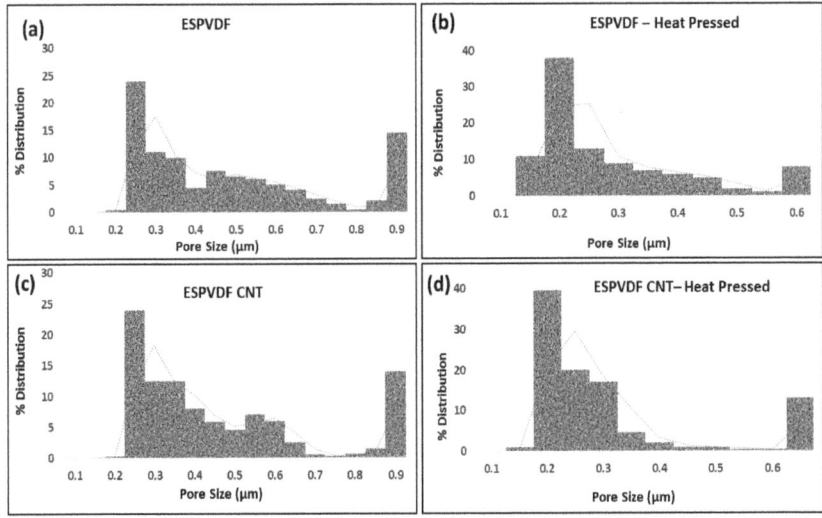

Figure 3. Percentage pore size distribution of the electrospun PVDF-Co-HFP and CNT-modified PVDF-Co-HFP membranes before and after heat pressing (a) Pristine electrospun membrane before heat pressing, (b) Pristine electrospun membrane after heat pressing, (c) CNT-coated electrospun membrane before heat pressing, and (d) CNT-coated electrospun membrane after heat pressing.

3.2. Liquid Entry Pressure and Mechanical Properties

Servi et al. conducted a scientific study of the effect of hydrophobicity on MD membrane wetting. They reported that the LEP increases with the water contact angle, which enhances the MD process performance [75]. The LEP of the electrospun PVDF-Co-HFP and PVDF-Co-HFP-CNT membranes was measured in the range of 120–125 KPa, whereas the LEP of the heat-pressed PVDF-HFP and PVDF-HFP-CNT membranes was measured in the range of 145–150 Kpa. The 20% increase in the LEP values in the heat-pressed membrane samples is attributed to the reduced mean flow pore sizes and narrow pore size distribution of the membranes, which is favorable to the MD process performance. Increased LEP values help retard the pore-wetting phenomenon, thereby, enhancing the membrane shelf life. The LEP values and tensile characteristics of pristine and modified membranes are given in Table 3.

Figure 4 shows the mechanical characteristics of the electrospun PVDF-Co-HFP membranes and CNT-modified membranes before and after heat pressing. The addition of CNTs on the electrospun PVDF-Co-HFP membrane surfaces enhanced the tensile strength from 40.5 KPa to 51 KPa. At the same time, elongation at the break of the CNT-modified PVDF-HFP membranes was reduced from 257% to 212%. The heat pressing process plays a significant role in enhancing the mechanical strength of both the PVDF-Co-HFP membranes and CNT-modified PVDF-Co-HFP membranes. After heat pressing, the tensile strength at break was increased from 40.5 Kpa to 89.5 Kpa for the PVDF-HFP membranes and 51.5 KPa to 90.6 KPa for the PVDF-HFP-CNT membranes. Thus, the heat pressing process helps to increase the tensile strength of the PVDF-Co-HFP membranes and PVDF-Co-HFP-CNT membranes to as high as 120% and 76%, respectively.

Table 3. LEP and tensile characteristics of the PVDF-HFP and PVDF-HFP-CNT membranes before and after heat pressing.

Membrane Characteristics	PVDF-HFP	PVDF-HFP-CNT	PVDF-HFP Heat Pressed	PVDF-HFP-CNT Heat Pressed
LEP (KPa)	120	125	145	150
Elongation at break (%)	257	212	221	212
Tensile Strength (KPa)	40.5	51.5	89.5	90.6

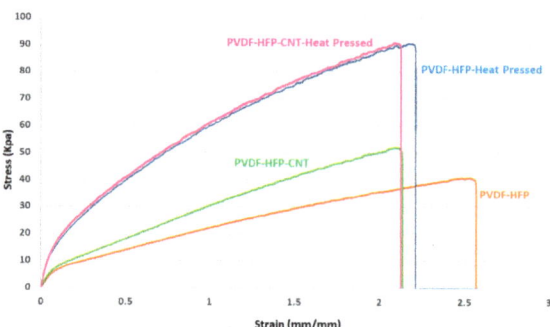

Figure 4. Mechanical characteristics of the electrospun PVDF-FHP and CNT-modified PVDF-HFP membranes before and after heat pressing.

3.3. Surface Morphology

Figure 5 shows the FESEM images of the PVDF-Co-HFP membrane and PVDF-Co-HFP-CNT membrane before and after heat pressing. Figure 5 also shows the nanofiber diameter distribution of the PVDF-Co-HFP membrane before and after heat pressing. It is evident from the high-resolution microscopic images that the nanofibers are flattened, and a reduction in the surface pore size has happened upon the heat-pressing process. Thus, a slight increase from 240 nm to 285 nm in the average fiber diameter can be observed in the heat-pressed nanofibers. It is also obvious from the plots that the fiber diameter distribution has been narrowed for heat-pressed nanofiber membranes compared to the nanofiber mats before heat pressing. These observations are strengthening the narrow pore size distribution and measured pore size values of the CNT-modified electrospun PVDF-Co-HFP nanofiber membrane. CNTs distributed on the surface of the electrospun PVDF-Co-HFP membrane are also seen from the FESEM images (Figure 5c,d). CNTs on the membrane surface could create a surface roughness, and this is the reason for the increase in the water contact angle and LEP of CNT-modified PVDF-Co-HFP membranes.

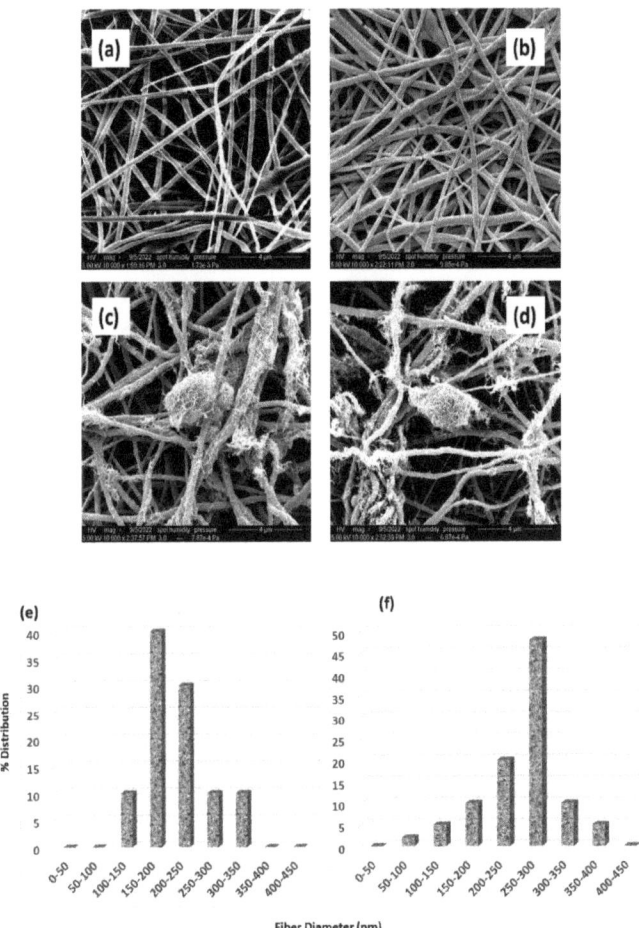

Figure 5. FESEM images of (**a**) PVDF-Co-HFP membrane before heat pressing, (**b**) PVDF-Co-HFP membrane after heat pressing, (**c**) PVDF-Co-HFP-CNT membrane before heat pressing, and (**d**) PVDF-Co-HFP-CNT membrane after heat pressing. (**e**) Nanofiber diameter distribution of PVDF-Co-HFP membrane before and (**f**) after heat pressing process.

3.4. DCMD Experiments

A series of DCMD experiments were conducted in a batch mode at various feed solution temperatures of 35 °C, 40 °C, 45 °C, 50 °C, and 55 °C while keeping the coolant temperature at 15 °C. After stabilizing the flow rate and temperatures on the feed and permeate sides, data login was initiated for the DCMD experiments. From the logged data, the water vapor flux during the DCMD test using the electrospun PVDF-Co-HFP membrane and CNT-modified electrospun-electrosprayed membrane was calculated at different feed solution temperatures. No significant flux decay was observed in each batch of the DCMD process. Salt rejection was calculated to be >99.8% in all experiments using 3.5 wt. % simulated seawater as the feed solution. Table 4 shows the percentage rejection and the results obtained from the quantitative analysis for determining the various ions present in the permeate using IC and ICP-MS analytical methods.

Table 4. Amount of inorganic salt ions present in the permeate from the DCMD process and the percentage rejection.

Salt Ions	Permeate (ppm)	Rejection (%)
Sodium	24.80	99.8
Magnesium	2.69	99.8
Potassium	1.48	99.6
Calcium	1.51	99.7
Lithium	<0.0	100
Boron	0.037	99.6
Chloride	46.38	99.8
Sulfate	5.49	99.8
Nitrate	0.99	99.8
Bromide	0.0094	99.9

The calculated water vapor flux and TPC values obtained during the DCMD process while using the electrospun PVDF-Co-HFP and PVDF-Co-HFP-CNT membranes at various feed solution temperatures are shown in Figure 6. The coolant temperature was kept constant at 15 °C in all experiments. There is an exponential relationship between the trans-membrane water vapor pressure and temperature. However, the water vapor produced during the DCMD process while using the PVDF-Co-HFP and PVDF-Co-HFP-CNT membranes reached a high of 37.5 LMH and 43.4 LMH, respectively, at a feed solution temperature of 55 °C. On the other hand, at a feed solution temperature of 35 °C, the calculated flux while using the PVDF-Co-HFP and PVDF-Co-HFP-CNT membranes was found to be 6.5 LMH and 8.1 LMH, respectively. Thus, the flux enhancement while using the CNT-modified electrospun PVDF-Co-HFP via the electrospraying technique at the feed solution temperatures of 35 °C and 55 °C was calculated to be 24.6% and 15.7%, respectively. The feed and coolant temperatures at the inlets and outlets of the membrane module were recorded using data acquisition software throughout the DCMD process. TPC can have a theoretical maximum value of 1, and a high value of TPC indicates a high efficiency of the process. Mitigation of the TP phenomenon leads to a reduction in the thickness of the polarization layer adjacent to the membrane surface [76,77]. Researchers have reported different methods for creating turbulence at the membrane surface to reduce the thickness of the polarization layer at the membrane surface to increase the TPC. In the current investigation, the TPC values were calculated, and we observed that the TPC at lower feed solution temperatures was higher than that at higher temperatures. The highest TPC values while using the PVDF-Co-HFP membrane and CNT-modified PVDF-Co-HFP membrane were calculated as 0.75 and 0.87, respectively, at a feed solution temperature of 35 °C. When the feed solution temperature increased to 55 °C, the TPC values while using the aforementioned membranes were reduced to 0.71 and 0.8, respectively. Thus, enhancement in the TPC values while using the PVDF-Co-HFP membrane and CNT-modified PVDF-Co-HFP membrane at the feed solution temperatures of 35 °C and 55 °C was calculated to be 16% and 12.6%, respectively.

CNTs distributed on the electrospun PVDF-Co-HFP membrane surface via electrospraying technique act as turbulence promoters at a micro level and may reduce the thickness of the polarization layer, which would be the reason for an enhanced TPC value and water permeation flux. Increased heat and mass transfer at higher feed solution temperatures lead to faster water vapor condensation, and this may lead to an increase in the thickness of the polarization layer and, thereby, reduced TPC values.

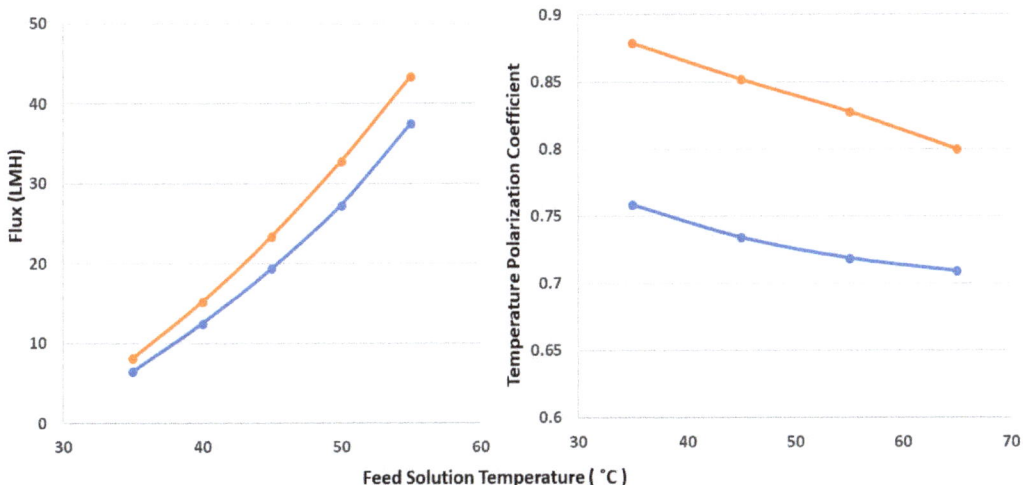

Figure 6. Water permeate flux and TPC values during a 24-h DCMD test using PVDF-Co-HFP (blue color) and CNT-coated PVDF-Co-HFP (orange color) membranes at a constant coolant temperature of 15 °C.

Figure 7 shows the FESEM images of the electrospun PVDF-Co-HFP membrane and PVDF-Co-HFP-CNT membrane after a 24-h DCMD process using 3.5 wt. % simulated seawater. CNTs distributed on the electrospun PVDF-Co-HFP membranes via the electrospraying process are visible on the membrane surface even after the 24-h DCMD operation. Different salts with their respective crystal structures are deposited on the membrane surface, which is also clearly visible in the FESEM images. As discussed in the aforementioned section, the enhanced surface roughness due to the presence of CNTs on the membrane surface act as a micro-level turbulence promoter, which leads to less salt deposition and cake layer formation on the membrane surface. Prolonged salt deposition and cake layer formation reduce the water permeation flux, which may cause a pore-wetting phenomenon that leads to the salt passage through the membrane pores to the permeate side. Therefore, surface modification of MD membranes by the electrospraying of suitable nanostructured materials on the membrane surface is advantageous to enhance the membrane characteristics/efficiency and overall MD process performance. Figure 7e,f shows the EDS images taken after the 24-h DCMD experiments. It reveals the presence of sodium, potassium, magnesium, calcium, and manganese ions on the membrane surface.

Enhancements in the hydrophobicity or water contact angle LEP and mechanical strength by the incorporation of CNTs in electrospun PVDF-Co-HFP membranes are also reported by Song et al. They have also reported an increase of 30–50% MD water flux enhancement using CNT-modified membranes [62]. In this work, the CNTs were mixed with the polymeric dope solution and subjected to electrospinning for the fabrication of the MD membranes. On the other hand, in the current study, the CNTs are coated on the surface of the PVDF-Co-HFP membrane in the second step of electrospraying. This would help the nanomaterials to spread evenly throughout the membrane surface and can be available for maximum surface roughness to impart high hydrophobicity, increased water contact angle, and high LEP values. The amount of nanomaterial loading would need to be much lower in the current method of the electrospraying process compared to the already reported studies to achieve maximum membrane efficiency in terms of water permeation flux, antifouling properties, and shelf life.

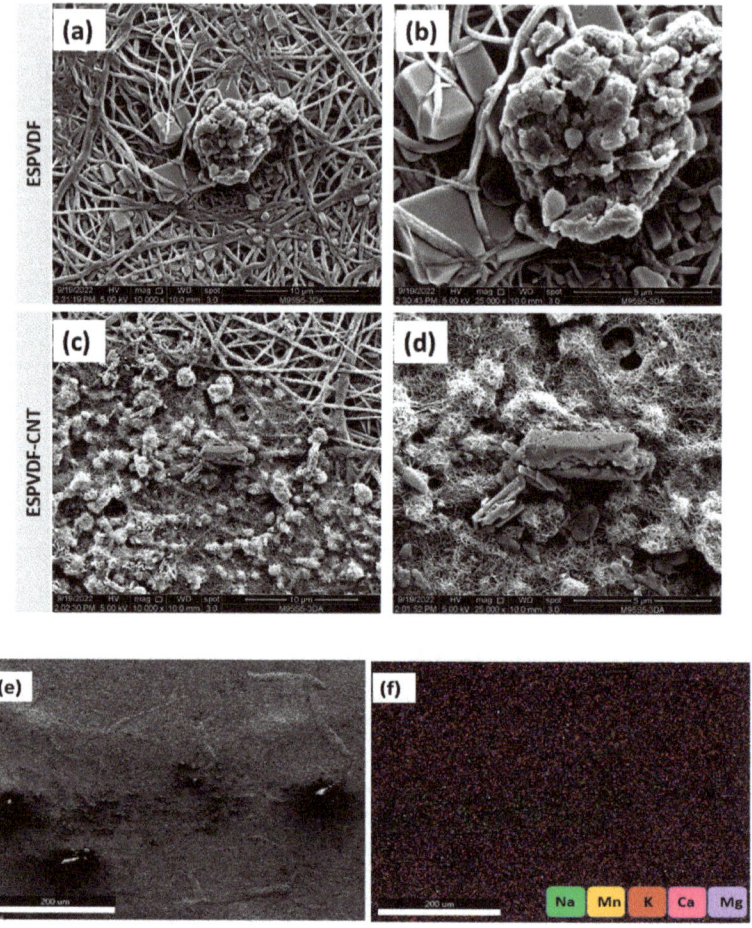

Figure 7. FESEM images of the electrospun PVDF-Co-HFP (**a**,**b**) and PVDF-Co-HFP-CNT (**c**,**d**) membranes after the 24-h DCMD operation, and the EDS images (**e**,**f**) of the CNT-modified PVDF-Co-HFP membrane after the MD process.

4. Conclusions and Future Perspectives

Membrane engineering is very important in the membrane fabrication technique to produce efficient membranes in the industry. CNTs and many other nanostructured material dispersions can be efficiently electrosprayed to get evenly distributed efficient nanomaterial coating on the membrane surface. MD membrane characteristics can be tuned by an electrospray deposition technique to get desirable MD membrane properties, such as high hydrophobicity or high water contact angle (>120°), high LEP, optimum pore size (~0.2 µm), narrow pore size distribution, etc., compared to the pristine electrospun membrane. CNT modification followed by heat pressing yields mechanically robust nanocomposite membranes with improved MD membrane characteristics. A 3% increase in the water contact angle, 20% increase in the LEP, and 42.6% reduction in the mean flow pore size towards the optimum pore size were observed in the heat pressed CNT-modified electrospun PVDF-Co-HFP membranes. The tensile strength of the heat-pressed CNT-modified membrane was significantly improved by up to 120% compared to the electrospun PVDF-Co-HFP membrane. The presence of CNTs on the membrane surface before and after the MD process are visible on the FESEM images. A water vapor flux enhancement of 15.7%,

20.6%, and 24.6% was observed at a ΔT of 20 °C, 30 °C, and 40 °C, respectively. Higher TPC values and percentage water vapor flux enhancements are observed at lower feed solution temperatures because of the higher heat loss at higher feed solution temperatures compared to the lower temperatures. A 16% and 12% enhancement in the TPC values was observed at the feed solution temperatures of 35 °C and 55 °C, respectively. A >99.8% inorganic salt rejection was observed through quantitative analytical tools (IC and ICP-MS) while conducting the DCMD process using a 3.5 wt. % simulated seawater feed solution. Thus, electrohydrodynamic atomization using appropriate nanomaterial dispersion can be recommended as an efficient tool for the surface modification of MD membranes.

CNTs have shown strong antimicrobial properties. In terms of future perspectives, the antibacterial properties of CNT-coated PVDF-Co-HFP membranes have to be explored. Simulation and modeling studies would give a clear understanding of the mechanism of heat and mass transfer during the MD process while using electrospray-deposited CNT-modified PVDF-Co-HFP membranes. Modeling tools will also give an idea of turbulence patterns at the micro level during the MD process. Atomic force microscopic (AFM) studies could reveal the nature of adhesive forces between the electrospun PVDF-Co-HFP membrane and CNTs deposited through the electrospraying process.

Author Contributions: L.F.: Conceptualization, Methodology, Experimental, Formal analysis, Data curation, Validation, Characterization, Writing the original draft. N.H.: Funding acquisition, Investigation, Supervision, Review and editing. All authors have read and agreed to the published version of the manuscript.

Funding: This work was jointly sponsored by the New York University Abu Dhabi (NYUAD) and Tamkeen under the NYUAD Research Institute Award (Project CG007). All the experiments were conducted using the research facilities at the NYUAD Water Research Center.

Data Availability Statement: Data presented in this article will be available upon request.

Conflicts of Interest: The authors declare no conflict of interest.

References

1. Xue, J.; Wu, T.; Dai, Y.; Xia, Y. Electrospinning and Electrospun Nanofibers: Methods, Materials, and Applications. *Chem. Rev.* **2019**, *119*, 5298–5415. [CrossRef] [PubMed]
2. Huang, Z.-M.; Zhang, Y.-Z.; Kotaki, M.; Ramakrishna, S. A review on polymer nanofibers by electrospinning and their applications in nanocomposites. *Compos. Sci. Technol.* **2003**, *63*, 2223–2253. [CrossRef]
3. Marsano, E.; Francis, L.; Giunco, F. Polyamide 6 nanofibrous nonwovens via electrospinning. *J. Appl. Polym. Sci.* **2010**, *117*, 1754–1765. [CrossRef]
4. Hammami, M.A.; Croissant, J.G.; Francis, L.; Alsaiari, S.K.; Anjum, D.H.; Ghaffour, N.; Khashab, N.M. Engineering Hydrophobic Organosilica Nanoparticle-Doped Nanofibers for Enhanced and Fouling Resistant Membrane Distillation. *ACS Appl. Mater. Interfaces* **2017**, *9*, 1737–1745. [CrossRef]
5. Francis, L.; Ahmed, F.E.; Hilal, N. Electrospun membranes for membrane distillation: The state of play and recent advances. *Desalination* **2022**, *526*, 115511. [CrossRef]
6. Francis, L.; Ogunbiyi, O.; Saththasivam, J.; Lawler, J.; Liu, Z. A comprehensive review of forward osmosis and niche applications. *Environ. Sci. Water Res. Technol.* **2020**, *6*, 1986–2015. [CrossRef]
7. Francis, L.; Nair, A.S.; Jose, R.; Ramakrishna, S.; Thavasi, V.; Marsano, E. Fabrication and characterization of dye-sensitized solar cells from rutile nanofibers and nanorods. *Energy* **2011**, *36*, 627–632. [CrossRef]
8. Francis, L.; Venugopal, J.; Prabhakaran, M.P.; Thavasi, V.; Marsano, E.; Ramakrishna, S. Simultaneous electrospin–electrosprayed biocomposite nanofibrous scaffolds for bone tissue regeneration. *Acta Biomater.* **2010**, *6*, 4100–4109. [CrossRef]
9. Thoppey, N.M.; Gorga, R.E.; Bochinski, J.; Clarke, L.I. Effect of Solution Parameters on Spontaneous Jet Formation and Throughput in Edge Electrospinning from a Fluid-Filled Bowl. *Macromolecules* **2012**, *45*, 6527–6537. [CrossRef]
10. Haider, A.; Haider, S.; Kang, I.-K. A comprehensive review summarizing the effect of electrospinning parameters and potential applications of nanofibers in biomedical and biotechnology. *Arab. J. Chem.* **2018**, *11*, 1165–1188. [CrossRef]
11. Vatanpour, V.; Kose-Mutlu, B.; Koyuncu, I. Electrospraying technique in fabrication of separation membranes: A review. *Desalination* **2022**, *533*, 115765. [CrossRef]
12. Lijo, F.; Marsano, E.; Vijila, C.; Barhate, R.S.; Vijay, V.K.; Ramakrishna, S.; Thavasi, V. Electrospun polyimide/titanium dioxide composite nanofibrous membrane by electrospinning and electrospraying. *J. Nanosci. Nanotechnol.* **2011**, *11*, 1154–1159. [CrossRef] [PubMed]

13. Attia, H.; Johnson, D.J.; Wright, C.J.; Hilal, N. Robust superhydrophobic electrospun membrane fabricated by combination of electrospinning and electrospraying techniques for air gap membrane distillation. *Desalination* **2018**, *446*, 70–82. [CrossRef]
14. Francis, L.; Giunco, F.; Balakrishnan, A.; Marsano, E. Synthesis, characterization and mechanical properties of nylon–silver composite nanofibers prepared by electrospinning. *Curr. Appl. Phys.* **2010**, *10*, 1005–1008. [CrossRef]
15. Francis, L.; Balakrishnan, A.; Sanosh, K.; Marsano, E. Characterization and tensile strength of HPC–PEO composite fibers produced by electrospinning. *Mater. Lett.* **2010**, *64*, 1806–1808. [CrossRef]
16. Zhang, R.; Ma, Y.; Lan, W.; Sameen, D.; Ahmed, S.; Dai, J.; Qin, W.; Li, S.; Liu, Y. Enhanced photocatalytic degradation of organic dyes by ultrasonic-assisted electrospray TiO2/graphene oxide on polyacrylonitrile/β-cyclodextrin nanofibrous membranes. *Ultrason. Sonochem.* **2020**, *70*, 105343. [CrossRef] [PubMed]
17. Hilal, N.; Kochkodan, V. Surface modified microfiltration membranes with molecularly recognising properties. *J. Membr. Sci.* **2002**, *213*, 97–113. [CrossRef]
18. Kochkodan, V.M.; Hilal, N.; Goncharuk, V.V.; Al-Khatib, L.; Levadna, T.I. Effect of the surface modification of polymer membranes on their microbiological fouling. *Colloid J.* **2006**, *68*, 267–273. [CrossRef]
19. Tawalbeh, M.; Al Mojjly, A.; Al-Othman, A.; Hilal, N. Membrane separation as a pre-treatment process for oily saline water. *Desalination* **2018**, *447*, 182–202. [CrossRef]
20. Hilal, N.; Khayet, M.; Wright, C. *Membrane Modification: Technology and Applications*, 1st ed.; 164 B/W Illustrations; CRC Press: Boca Raton, FL, USA, 2012.
21. Emadzadeh, D.; Lau, W.; Matsuura, T.; Hilal, N.; Ismail, A. The potential of thin film nanocomposite membrane in reducing organic fouling in forward osmosis process. *Desalination* **2014**, *348*, 82–88. [CrossRef]
22. Lee, W.; Ng, Z.; Hubadillah, S.; Goh, P.; Lau, W.; Othman, M.; Ismail, A.; Hilal, N. Fouling mitigation in forward osmosis and membrane distillation for desalination. *Desalination* **2020**, *480*, 114338. [CrossRef]
23. Abu Seman, M.; Khayet, M.; Hilal, N. Development of antifouling properties and performance of nanofiltration membranes modified by interfacial polymerisation. *Desalination* **2011**, *273*, 36–47. [CrossRef]
24. Giwa, A.; Hasan, S.; Yousuf, A.; Chakraborty, S.; Johnson, D.; Hilal, N. Biomimetic membranes: A critical review of recent progress. *Desalination* **2017**, *420*, 403–424. [CrossRef]
25. Khalil, A.; Ahmed, F.E.; Hashaikeh, R.; Hilal, N. 3D printed electrically conductive interdigitated spacer on ultrafiltration membrane for electrolytic cleaning and chlorination. *J. Appl. Polym. Sci.* **2022**, *139*, 52292. [CrossRef]
26. Alpatova, A.; Verbych, S.; Bryk, M.; Nigmatullin, R.; Hilal, N. Ultrafiltration of water containing natural organic matter: Heavy metal removing in the hybrid complexation–ultrafiltration process. *Sep. Purif. Technol.* **2004**, *40*, 155–162. [CrossRef]
27. Amy, G.; Ghaffour, N.; Li, Z.; Francis, L.; Linares, R.V.; Missimer, T.; Lattemann, S. Membrane-based seawater desalination: Present and future prospects. *Desalination* **2017**, *401*, 16–21. [CrossRef]
28. Gonzalez-Vogel, A.; Felis-Carrasco, F.; Rojas, O.J. 3D printed manifolds for improved flow management in electrodialysis operation for desalination. *Desalination* **2021**, *505*, 114996. [CrossRef]
29. Hammami, M.A.; Francis, L.; Croissant, J.; Ghaffour, N.; Alsaiari, S.; Khashab, N.M. Periodic Mesoporous Organosilica-Doped Nanocomposite Membranes and Systems Including Same. EP3471864B1, 2 February 2022.
30. Ghaffour, N.; Francis, L.; Li, Z.; Valladares, R.; Alsaadi, A.S.; Ghdaib, M.A.; Amy, G.L. Osmotically and Thermally Isolated forward Osmosis-Membrane Distillation (FO-MD) Integrated Module for Water Treatment Applications. US10688439B2, 23 June 2020.
31. Francis, L.; Ghaffour, N.; Alsaadi, A. Submerged Membrane Distillation for Desalination of Water. US20160310900A1, 27 October 2016.
32. Francis, L.; Ghaffour, N.; Alsaadi, A.S.; Nunes, S.P.; Amy, G.L. PVDF hollow fiber and nanofiber membranes for fresh water reclamation using membrane distillation. *J. Mater. Sci.* **2013**, *49*, 2045–2053. [CrossRef]
33. Alsaadi, A.S.; Francis, L.; Maab, H.; Amy, G.L.; Ghaffour, N. Evaluation of air gap membrane distillation process running under sub-atmospheric conditions: Experimental and simulation studies. *J. Membr. Sci.* **2015**, *489*, 73–80. [CrossRef]
34. Soukane, S.; Naceur, M.W.; Francis, L.; Alsaadi, A.; Ghaffour, N. Effect of feed flow pattern on the distribution of permeate fluxes in desalination by direct contact membrane distillation. *Desalination* **2017**, *418*, 43–59. [CrossRef]
35. Francis, L.; Maab, H.; Alsaadi, A.; Nunes, S.; Ghaffour, N.; Amy, G. Fabrication of electrospun nanofibrous membranes for membrane distillation application. *Desalination Water Treat.* **2013**, *51*, 1337–1343. [CrossRef]
36. Maab, H.; Al Saadi, A.; Francis, L.; Livazovic, S.; Ghafour, N.; Amy, G.L.; Nunes, S.P. Polyazole Hollow Fiber Membranes for Direct Contact Membrane Distillation. *Ind. Eng. Chem. Res.* **2013**, *52*, 10425–10429. [CrossRef]
37. Nunes, S.P.; Maab, H.; Francis, L. Polyazole Membrane for Water Purification. EP2626127A2, 1 June 2022.
38. Ahmed, F.E.; Hashaikeh, R.; Hilal, N. Hybrid technologies: The future of energy efficient desalination—A review. *Desalination* **2020**, *495*, 114659. [CrossRef]
39. Kebria, M.R.S.; Rahimpour, A. *Membrane Distillation: Basics, Advances, and Applications. Advances in Membrane Technologies*; Abdelrasoul, A., Ed.; IntechOpen: London, UK, 2020. [CrossRef]
40. Lee, J.-G.; Alsaadi, A.S.; Karam, A.M.; Francis, L.; Soukane, S.; Ghaffour, N. Total water production capacity inversion phenomenon in multi-stage direct contact membrane distillation: A theoretical study. *J. Membr. Sci.* **2017**, *544*, 126–134. [CrossRef]
41. Alsaadi, A.S.; Ghaffour, N.; Li, J.D.; Gray, S.; Francis, L.; Maab, H.; Amy, G.L. Modeling of air-gap membrane distillation process. In Proceedings of the AMTA/AWWA Membrane Technology Conference and Exposition, San Antonio, TX, USA, 25–28 February 2013; Volume 1, p. 33, ISBN 978-1-62748-415-2.

42. Orfi, J.; Loussif, N. Modeling of a membrane distillation unit for desalination. In *Desalination: Methods, Costs and Technology*; Nova science Publishers, Inc.: Hauppauge, NY, USA, 2010.
43. Khayet, M. Membranes and theoretical modeling of membrane distillation: A review. *Adv. Colloid Interface Sci.* **2011**, *164*, 56–88. [CrossRef]
44. Eleiwi, F.; Ghaffour, N.; Alsaadi, A.S.; Francis, L.; Laleg-Kirati, T.M. Dynamic modeling and experimental validation for direct contact membrane distillation (DCMD) process. *Desalination* **2016**, *384*, 1–11. [CrossRef]
45. Alsaadi, A.; Ghaffour, N.; Li, J.-D.; Gray, S.; Francis, L.; Maab, H.; Amy, G. Modeling of air-gap membrane distillation process: A theoretical and experimental study. *J. Membr. Sci.* **2013**, *445*, 53–65. [CrossRef]
46. Lee, J.-G.; Kim, Y.-D.; Kim, W.-S.; Francis, L.; Amy, G.; Ghaffour, N. Performance modeling of direct contact membrane distillation (DCMD) seawater desalination process using a commercial composite membrane. *J. Membr. Sci.* **2015**, *478*, 85–95. [CrossRef]
47. Zhou, Z.; Ladner, D.A. Computational Modeling of Spacers Printed Directly onto Reverse Osmosis Membranes for Enhanced Module Packing Capacity and Improved Hydrodynamics. *SSRN Electron. J.* **2022**, preprint. [CrossRef]
48. Olatunji, S.O.; Camacho, L.M. Heat and Mass Transport in Modeling Membrane Distillation Configurations: A Review. *Front. Energy Res.* **2018**, *6*, 130. [CrossRef]
49. Shirzadi, M.; Li, Z.; Yoshioka, T.; Matsuyama, H.; Fukasawa, T.; Fukui, K.; Ishigami, T. CFD Model Development and Experimental Measurements for Ammonia–Water Separation Using a Vacuum Membrane Distillation Module. *Ind. Eng. Chem. Res.* **2022**, *61*, 7381–7396. [CrossRef]
50. Maab, H.; Francis, L.; Al-Saadi, A.; Aubry, C.; Ghaffour, N.; Amy, G.; Nunes, S.P. Synthesis and fabrication of nanostructured hydrophobic polyazole membranes for low-energy water recovery. *J. Membr. Sci.* **2012**, *423–424*, 11–19. [CrossRef]
51. Tijing, L.D.; Choi, J.S.; Lee, S.; Kim, S.H.; Shon, H.K. Recent progress of membrane distillation using electrospun nanofibrous membrane. *J. Membr. Sci.* **2014**, *453*, 435–462. [CrossRef]
52. Varela-Corredor, F.; Bandini, S. Testing the applicability limits of a membrane distillation process with ceramic hydrophobized membranes: The critical wetting temperature. *Sep. Purif. Technol.* **2020**, *250*, 117205. [CrossRef]
53. Seraj, S.; Mohammadi, T.; Tofighy, M.A. Graphene-based membranes for membrane distillation applications: A review. *J. Environ. Chem. Eng.* **2022**, *10*, 107974. [CrossRef]
54. Chen, L.-H.; Chen, Y.-R.; Huang, A.; Chen, C.-H.; Su, D.-Y.; Hsu, C.-C.; Tsai, F.-Y.; Tung, K.-L. Nanostructure depositions on alumina hollow fiber membranes for enhanced wetting resistance during membrane distillation. *J. Membr. Sci.* **2018**, *564*, 227–236. [CrossRef]
55. Feng, H.; Li, H.; Li, M.; Zhang, X. Construction of omniphobic PVDF membranes for membrane distillation: Investigating the role of dimension, morphology, and coating technology of silica nanoparticles. *Desalination* **2022**, *525*, 115498. [CrossRef]
56. Francis, L.; Ghaffour, N.; Amy, G. Fabrication and Characterization of Functionally Graded Poly(vinylidine fluoride)-Silver Nanocomposite Hollow Fibers for Sustainable Water Recovery. *Sci. Adv. Mater.* **2014**, *6*, 2659–2665. [CrossRef]
57. Shahabadi, S.M.S.; Rabiee, H.; Seyedi, S.M.; Mokhtare, A.; Brant, J.A. Superhydrophobic dual layer functionalized titanium dioxide/polyvinylidene fluoride- co -hexafluoropropylene (TiO2/PH) nanofibrous membrane for high flux membrane distillation. *J. Membr. Sci.* **2017**, *537*, 140–150. [CrossRef]
58. Jia, W.; Kharraz, J.A.; Sun, J.; An, A.K. Hierarchical Janus membrane via a sequential electrospray coating method with wetting and fouling resistance for membrane distillation. *Desalination* **2021**, *520*, 115313. [CrossRef]
59. Su, C.; Horseman, T.; Cao, H.; Christie, K.S.; Li, Y.; Lin, S. Robust Superhydrophobic Membrane for Membrane Distillation with Excellent Scaling Resistance. *Environ. Sci. Technol.* **2019**, *53*, 11801–11809. [CrossRef] [PubMed]
60. Hong, S.K.; Kim, H.; Lee, H.; Lim, G.; Cho, S.J. A pore-size tunable superhydrophobic membrane for high-flux membrane distillation. *J. Membr. Sci.* **2021**, *641*, 119862. [CrossRef]
61. Gethard, K.; Sae-Khow, O.; Mitra, S. Water Desalination Using Carbon-Nanotube-Enhanced Membrane Distillation. *ACS Appl. Mater. Interfaces* **2010**, *3*, 110–114. [CrossRef] [PubMed]
62. Song, J.; Deng, Q.; Huang, M.; Kong, Z. Carbon nanotube enhanced membrane distillation for salty and dyeing wastewater treatment by electrospinning technology. *Environ. Res.* **2021**, *204*, 111892. [CrossRef]
63. Yazgan-Birgi, P.; Ali, M.I.H.; Arafat, H.A. Estimation of liquid entry pressure in hydrophobic membranes using CFD tools. *J. Membr. Sci.* **2018**, *552*, 68–76. [CrossRef]
64. Tomaszewska, M. Temperature Polarization. In *Encyclopedia of Membranes*; Springer: Berlin/Heidelberg, Germany, 2014; pp. 1–2. [CrossRef]
65. Francis, L.; Ghaffour, N.; Alsaadi, A.; Nunes, S.; Amy, G. Performance evaluation of the DCMD desalination process under bench scale and large scale module operating conditions. *J. Membr. Sci.* **2013**, *455*, 103–112. [CrossRef]
66. Alsaadi, A.S.; Francis, L.; Amy, G.L.; Ghaffour, N. Experimental and theoretical analyses of temperature polarization effect in vacuum membrane distillation. *J. Membr. Sci.* **2014**, *471*, 138–148. [CrossRef]
67. Francis, L.; Ghaffour, N.; Alsaadi, A.A.; Amy, G.L. Material gap membrane distillation: A new design for water vapor flux enhancement. *J. Membr. Sci.* **2013**, *448*, 240–247. [CrossRef]
68. Kim, Y.; Li, S.; Francis, L.; Li, Z.; Linares, R.V.; Alsaadi, A.S.; Abu-Ghdaib, M.; Son, H.S.; Amy, G.; Ghaffour, N. Osmotically and Thermally Isolated Forward Osmosis–Membrane Distillation (FO–MD) Integrated Module. *Environ. Sci. Technol.* **2019**, *53*, 3488–3498. [CrossRef]

69. Alsaadi, A.S.; Alpatova, A.; Lee, J.-G.; Francis, L.; Ghaffour, N. Flashed-feed VMD configuration as a novel method for eliminating temperature polarization effect and enhancing water vapor flux. *J. Membr. Sci.* **2018**, *563*, 175–182. [CrossRef]
70. Kim, Y.-D.; Francis, L.; Lee, J.-G.; Ham, M.-G.; Ghaffour, N. Effect of non-woven net spacer on a direct contact membrane distillation performance: Experimental and theoretical studies. *J. Membr. Sci.* **2018**, *564*, 193–203. [CrossRef]
71. Francis, L.; Ghaffour, N.; Al-Saadi, A.S.; Amy, G.L. Submerged membrane distillation for seawater desalination. *Desalination Water Treat.* **2014**, *55*, 2741–2746. [CrossRef]
72. Kim, Y.-B.; Lee, H.-S.; Francis, L.; Kim, Y.-D. Innovative swirling flow-type microbubble generator for multi-stage DCMD desalination system: Focus on the two-phase flow pattern, bubble size distribution, and its effect on MD performance. *J. Membr. Sci.* **2019**, *588*, 117197. [CrossRef]
73. Korolkov, I.V.; Gorin, Y.G.; Yeszhanov, A.B.; Kozlovskiy, A.L.; Zdorovets, M.V. Preparation of PET track-etched membranes for membrane distillation by photo-induced graft polymerization. *Mater. Chem. Phys.* **2018**, *205*, 55–63. [CrossRef]
74. Kuang, Z.; Long, R.; Liu, Z.; Liu, W. Analysis of temperature and concentration polarizations for performance improvement in direct contact membrane distillation. *Int. J. Heat Mass Transf.* **2019**, *145*, 118724. [CrossRef]
75. Servi, A.T.; Kharraz, J.; Klee, D.; Notarangelo, K.; Eyob, B.; Guillen-Burrieza, E.; Liu, A.; Arafat, H.A.; Gleason, K.K. A systematic study of the impact of hydrophobicity on the wetting of MD membranes. *J. Membr. Sci.* **2016**, *520*, 850–859. [CrossRef]
76. Alanezi, A.A.; Bassyouni, M.; Abdel-Hamid, S.M.S.; Ahmed, H.S.; Abdel-Aziz, M.H.; Zoromba, M.S.; Elhenawy, Y. Theoretical Investigation of Vapor Transport Mechanism Using Tubular Membrane Distillation Module. *Membranes* **2021**, *11*, 560. [CrossRef]
77. Tomaszewska, M. Temperature Polarization Coefficient (TPC). In *Encyclopedia of Membranes*; Springer: Berlin/Heidelberg, Germany, 2016; pp. 1880–1881. [CrossRef]

Article

Elevated Adsorption of Lead and Arsenic over Silver Nanoparticles Deposited on Poly(amidoamine) Grafted Carbon Nanotubes

Gururaj M. Neelgund [1,*], Sanjuana F. Aguilar [1], Mahaveer D. Kurkuri [2], Debora F. Rodrigues [3] and Ram L. Ray [4]

1. Department of Chemistry, Prairie View A&M University, Prairie View, TX 77446, USA
2. Centre for Research in Functional Materials (CRFM), JAIN University, Jain Global Campus, Bengaluru 562112, Karnataka, India
3. Department of Civil and Environmental Engineering, University of Houston, Houston, TX 77004, USA
4. College of Agriculture and Human Sciences, Prairie View A&M University, Prairie View, TX 77446, USA
* Correspondence: gmneelgund@pvamu.edu

Abstract: An efficient adsorbent, CNTs–PAMAM–Ag, was prepared by grafting fourth-generation aromatic poly(amidoamine) (PAMAM) to carbon nanotubes (CNTs) and successive deposition of Ag nanoparticles. The FT–IR, XRD, TEM and XPS results confirmed the successful grafting of PAMAM onto CNTs and deposition of Ag nanoparticles. The absorption efficiency of CNTs–PAMAM–Ag was evaluated by estimating the adsorption of two toxic contaminants in water, viz., Pb(II) and As(III). Using CNTs–PAMAM–Ag, about 99 and 76% of Pb(II) and As(III) adsorption, respectively, were attained within 15 min. The controlling mechanisms for Pb(II) and As(III) adsorption dynamics were revealed by applying pseudo-first and second-order kinetic models. The pseudo-second-order kinetic model followed the adsorption of Pb(II) and As(III). Therefore, the incidence of chemisorption through sharing or exchanging electrons between Pb(II) or As(III) ions and CNTs–PAMAM–Ag could be the rate-controlling step in the adsorption process. Further, the Weber–Morris intraparticle pore diffusion model was employed to find the reaction pathways and the rate-controlling step in the adsorption. It revealed that intraparticle diffusion was not a rate-controlling step in the adsorption of Pb(II) and As(III); instead, it was controlled by both intraparticle diffusion and the boundary layer effect. The adsorption equilibrium was evaluated using the Langmuir, Freundlich, and Temkin isotherm models. The kinetic data of Pb(II) and As(III) adsorption was adequately fitted to the Langmuir isotherm model compared to the Freundlich and Temkin models.

Keywords: carbon nanotubes; lead; arsenic; adsorption; poly(amidoamine); silver nanoparticles

Citation: Neelgund, G.M.; Aguilar, S.F.; Kurkuri, M.D.; Rodrigues, D.F.; Ray, R.L. Elevated Adsorption of Lead and Arsenic over Silver Nanoparticles Deposited on Poly(amidoamine) Grafted Carbon Nanotubes. *Nanomaterials* 2022, 12, 3852. https://doi.org/10.3390/nano12213852

Academic Editor: Muralidharan Paramsothy

Received: 23 September 2022
Accepted: 27 October 2022
Published: 1 November 2022

Publisher's Note: MDPI stays neutral with regard to jurisdictional claims in published maps and institutional affiliations.

Copyright: © 2022 by the authors. Licensee MDPI, Basel, Switzerland. This article is an open access article distributed under the terms and conditions of the Creative Commons Attribution (CC BY) license (https://creativecommons.org/licenses/by/4.0/).

1. Introduction

Lead and arsenic are known for their toxicity and are widely distributed in the environment, particularly in water sources [1]. These elements, under low concentration, can also be hazardous to aquatic and non-aquatic creatures and plants. Lea and arsenic are carcinogenic, mutagenic, and teratogenic [2,3]. The ingestion of these elements can cause severe adverse effects that include hypertension, neurological complications, cardiovascular disease, intestinal disorders, hematopoietic dysfunction, mental impairment, organ failure, and malfunctioning of the immune and reproductive systems [4–9]. Lead and arsenic are listed as global priority pollutants because of their high toxicity, stability, and non-biodegradability [10]. It has been reported that worldwide more than 200 million people have been affected by the consequences of these elements [4]. The primary source of exposure to these lethal elements is water through regular activities like drinking, cooking, and irrigation. Water sources are getting contaminated with arsenic by both natural processes and anthropogenic activities like weathering processes, geochemical reactions, biological activities, combustion of fossil fuels, volcanic eruptions, mining activities, leaching of artificial arsenic compounds, smelting of metals ores, desiccants, wood preservatives, agricultural pesticides, and related anthropogenic activities [11–14]. In nature, arsenic exists in organic and inorganic forms. However, the organic form of arsenic is not of importance as it undergoes biotransformation and detoxifies through methylation [15]. At the same time, the inorganic form of arsenic exists in four oxidation states viz., -3, 0, $+3$, and $+5$. Out of these, -3 and 0 oxidation states are scarce, so $+3$ and $+5$ usually exist in water. Nevertheless, the presence of either a $+3$ or $+5$ state of arsenic depends on the redox and pH conditions of water [15,16]. Accordingly, trivalent arsenite, As(III), and pentavalent arsenate, As(V), commonly occur in water. In particular, As(III) transpires in reducing conditions, and As(V) exists in oxidizing conditions of water [4]. However, both As(III) and As(V) species are highly toxic and non-biodegradable [6]. Comparatively, As(III) is more highly mobile and stable than As(V) due to its stable electronic configuration [11]. In addition, As(III) has higher cellular uptake ability and binding affinity to vicinal sulfhydryl groups that react with various proteins and inhibit their activity [6,11,17]. In specific, As(III) is about sixty times more toxic than As(V) [6].

Another toxic element, lead, contaminates the water through industrial and anthropogenic sources. The primary sources of information pollution are industries such as battery manufacturing, printing, fuels, photographic materials, ceramic, glass, and explosive manufacturing [18]. In the atmosphere, lead persists as bivalent, Pb(II). It is also likewise toxic as As(III) and widely distributed. Because of their toxicity and detrimental effects, Pb(II) and As(III) are unsafe, specifically their association with water. Therefore, eliminating these toxic components present in water is a mandatory need. For the removal of toxic Pb(II) and As(III), several methods like chemical precipitation [19], ion exchange [20], membrane technology and reverse osmosis [21], electro-dialysis [22], and adsorption [23] have been developed. Nevertheless, chemical precipitation makes it difficult to remove low-concentrated arsenic like 10 mg/L from water [24]. In addition, chemical precipitation is less effective in the removal of As(III), and it requires the pre-oxidative conversion of As(III) to As(V) during its abolition [25]. Similarly, ion exchange is less effective in removing As(III), and the development of ion exchange resins is expensive [24]. Additionally, the membrane technology-aided reverse osmosis is costly, requiring external pressure to pass the contaminated water through the membrane. Besides, the discharge of the concentrate, membrane fouling, and flux decline is inevitable in reverse osmosis [26]. In electro-dialysis, several insoluble coagulants generate and deposit over the cathode [27]. Overseeing these drawbacks, the better alternative and promising technique for removing Pb(II) and As(III) is adsorption. Adsorption is a facile, efficient, accessible, low-cost, low-energy technique [28]. Additionally, using the adsorption technique, it is possible to remove trace amounts of Pb(II) and As(III) from water. Overall, adsorption has played a significant role in eliminating various contaminants from water. Consequently, due to its exceptional benefits, adsorption is a gifted technique for efficiently removing Pb(II) and As(III).

In the process of adsorption, selecting an adsorbent is a vital step in the success of the process. For this purpose, insoluble solid materials with high specific surface area and active functional groups are elected adsorbents. In this context, several adsorbents have been developed and tested for their efficiency in removing different pollutants [6,12,29]. The dynamic adsorbents are based on activated carbon [30], activated alumina [31], inorganic minerals [32], biomass adsorbents [33–38], polymer [39–41], carbon nanotubes (CNTs) [42–44] graphene oxide [45,46], metal-organic frameworks [47], microplastics [48] and more. Among these, CNTs based adsorbents are particularly interesting because their unique physical and chemical properties are relevantly suitable for adsorption [42–44]. During the CNTs, aided adsorption process, strong interactions between CNTs and pollutants alter contaminants' mobility, bioavailability, and environmental risk [44]. The high surface area, well-defined structures, and uniform surfaces of CNTs facilitate the adsorption mechanism and process. Abundant active sites and functional groups exist over CNTs supporting the adsorption process. Furthermore, adsorption over CNTs transpires with different mechanisms, viz., ion exchange, coordination interaction, electrostatic interaction, and physical adsorption, which augment the adsorption [44]. Beyond its exceptional properties, the employment of CNTs in adsorption technology is still fragmentary, so further exploration is required. The main hindrance that restrained the application of CNTs in adsorption technology is their hydrophobic nature, which reduces the adsorption rate. However, the functionalization of CNTs introduces reactive functional groups over their surface that can significantly increase their selectivity and sensitivity toward pollutants [43]. In this fashion, CNTs can be modified to be hydrophilic for the adsorption of pollutants and more reactive by functionalizing with secondary materials. The functionalization of CNTs generates chemically active sites around defective segments such as pentagons, oriented against a tube body which generally consists of only hexagons; and it may be this that gives CNTs their remarkable ability to interact with pollutants [43,49]. Accordingly, the functionalization of CNTs is the critical step to improving their absorption efficiency. For the functionalization of CNTs, dendritic polymers are particularly interesting because of their unusual structure and properties [50]. Among different dendrimers, poly(amidoamine) (PAMAM) is predominantly interesting because of its symmetrical structure, controllable molecule chains, vast internal cavities, and abundant functional groups [51]. Based on these special structural characteristics, PAMAM persists with unique properties, such as high hydrophilicity, high dispersing ability, high bio-affinity, and ease of modification [51]. Therefore, PAMAM is especially interesting for the adsorption of heavy metal ions and plays an important role in the functionalization of nanomaterials [51–54]. A large number of amino and amide functional groups of PAMAM dendrimers can strongly chelate heavy metal ions, thus improving the enrichment efficacy [51–54]. Considering its importance, herein, CNTs were functionalized by grafting with fourth-generation aromatic PAMAM and sequential deposition of Ag nanoparticles. Thus produced, CNTs–PAMAM–Ag was explored in evaluating the adsorption efficiency of two important toxic pollutants, Pb(II) and As(III). The controlling mechanisms and dynamics of the adsorption of Pb(II) and As(III) over CNTs–PAMAM–Ag were estimated by implementing the pseudo-first and second-order kinetic models. The reaction pathways and the rate-controlling step in adsorption were evaluated using Weber–Morris intraparticle pore diffusion model. Furthermore, the adsorption equilibrium was estimated by fitting the experimental results with Langmuir, Freundlich, and Temkin isotherm models.

2. Experimental

2.1. Materials

CNTs prepared by the CVD process were received from Carbon Nanotechnology Laboratory at Rice University, Houston, TX, USA. All the chemicals were purchased from Millipore-Sigma (St. Louis, MO, USA) and used as received. The aqueous solutions were prepared using ultrapure water obtained by the Milli-Q Plus system (Millipore, Burlington, MA, USA).

2.2. Preparation of CNTs–PAMAM–Ag

The CNTs–PAMAM–Ag was prepared using the reported method [55]. In brief, 100 mg of CNTs-PAMAM, obtained by Michael's addition process, were dispersed in 15 mL of DI water by sonication, and a 10 mL aqueous solution of $AgNO_3$ (0.01 mol/L) was slowly added. The resulting suspension was allowed to stir under ambient conditions for 8 h and centrifuged. Thus, formed CNTs–PAMAM–Ag was purified by successive washings with DI water and ethanol and dried under vacuum. For control experiments, hydrophobic nature pristine CNTs were modified to hydrophilic oxidized CNTs, and PAMAM was prepared using Michael's addition procedure [55].

2.3. Adsorption Experiments

The stock solution of Pb(II) and As(III) with a 1 g/L was prepared in DI water using lead(II) nitrate and sodium arsenite, respectively. Further, the stock solution was diluted to desired concentrations using DI water. The kinetic adsorption experiments of Pb(II) and

As(III) were performed to evaluate the contact time required to attain equilibrium. In the typical experiment, 100 mg of adsorbent was dispersed into 500 mL of Pb(II) and As(III) solution with a concentration of 40 µg/L with an initial pH of 7.67 and 8.11, respectively. Then, the mixtures were allowed to stir at room temperature, and an adequate quantity of samples was collected after the required contact time. Successively, the adsorbent was separated by centrifugation, and the concentration of residual Pb(II) or As(III) in the solution was estimated by the atomic absorption spectrometer. The efficiency in adsorption of Pb(II) and As(III) as a function of time was monitored for 120 min. Then, the adsorbed quantity of Pb(II) and As(III) was evaluated using the following equation.

$$q_t = \frac{(C_0 - C_t) V}{M} \tag{1}$$

where q_t is the amount of pollutant adsorbed (mg/g) at time t; C_0 is the initial concentration of contaminant in solution (mg/L), and C_t is the concentration of pollutant in solution (mg/L) at time t; V is the volume of the solution (L), and M is the amount of adsorbent (g).

The efficiency of adsorbent in the removal of Pb(II) and As(III) was calculated by:

$$Removal\ efficiency\ (\%) = \frac{(C_0 - C_t) V}{C_0} \times 100 \tag{2}$$

For adsorption isotherms experiments, 10 mg of CNTs–PAMAM–Ag was added to 50 mL of Pb(II) or As(III) solution and allowed to stir at room temperature for 24 h to reach the equilibrium. To obtain the adsorption isotherms, the concentration of Pb(II) and As(III) solution was varied from 1 to 10 mg/L. After reaching the equilibrium, CNTs–PAMAM–Ag was separated by centrifugation, and the concentration of Pb(II) or As(III) in the solution was measured using the atomic absorption spectrometer. Then, the adsorption of Pb(II) or As(III) at equilibrium, q_e (mg/g), was determined by:

$$q_e = \frac{(C_0 - C_e) V}{M} \tag{3}$$

where q_e is the amount of Pb(II) and As(III) adsorbed (mg/g) at equilibrium.

2.4. Effect of pH on Adsorption of Pb(II)

To evaluate the effect of pH on the adsorption of Pb(II). The pH of the Pb(II) solution was varied from 4 to 12 using 0.1 M HCl and NaOH solutions. Other parameters were kept constant.

2.5. Desorption and Reuse of Adsorbent

After the adsorption experiment of Pb(II), CNTs–PAMAM–Ag was collected by centrifugation, dispersed in 0.1 M HCl solution, and stirred at room temperature for 2 h. Then, the CNTs–PAMAM–Ag was separated by centrifugation, washed with DI water, and dried under vacuum. Thus recovered CNTs–PAMAM–Ag was employed in the next cycle of adsorption of Pb(II). To estimate the reusability of CNTs–PAMAM–Ag, four adsorption–desorption cycles were performed.

2.6. Characterization

The FT-IR spectra were acquired using Thermo-Nicolet IR 2000 spectrometer (Madison, WI, USA) with KBr, and the XRD were recorded on a Scintag X-ray diffractometer (Cupertino, CA, USA), model PAD X, equipped with a Cu Kα photon source (45 kV, 40 mA) at the scanning rate of 3 °/min. Transmission electron microscopy (TEM) images were obtained with the Hitachi H-8100 microscope (Tokyo, Japan) at 200 kV. The concentration of Pb(II) and As(III) was estimated by Varian SpectrAA 220FS (Lake Forest, CA, USA) atomic absorption spectrometer.

3. Results and Discussion

The FT–IR spectrum of oxidized CNTs, presented in Figure 1a displayed the absorption bands of A2u and E1u phonon modes of CNTs at 611 and 1629 cm^{-1}, respectively [56]. The broad feature appeared in the range of 3652 and 3000 cm^{-1} corresponding to the absorption of –OH stretching of carboxyl groups. The spectrum of CNTs–PAMAM (Figure 1b) demonstrated the characteristic absorption bands of CNTs and PAMAM. The peak displayed at 3441 cm^{-1} was related to N–H stretching frequency and the band found at 1631 cm^{-1} was due to carbonyl stretching of amide (–CO–NH). The bands that appeared at 2858 and 2934 cm^{-1} corresponded to symmetric and asymmetric stretching of –CH$_2$, respectively. The bands revealed at 1515 and 837 cm^{-1} were by aromatic C–C and C–H para-aromatic out-of-plane vibrations, respectively. The bands found at 1003 and 1116 cm^{-1} was owing to aromatic –CH vibrations [55]. The bands of the aromatic ring displayed at 1515 (aromatic C–C), 1003 and 1116 (–CH), and 837 cm^{-1} (C–H para-aromatic out of plane vibration). The spectrum of CNTs–PAMAM–Ag (Figure 1c) exhibited the representative peaks of CNTs and PAMAM, however, the position of the peaks was slightly shifted, and the intensity of the bands was reduced.

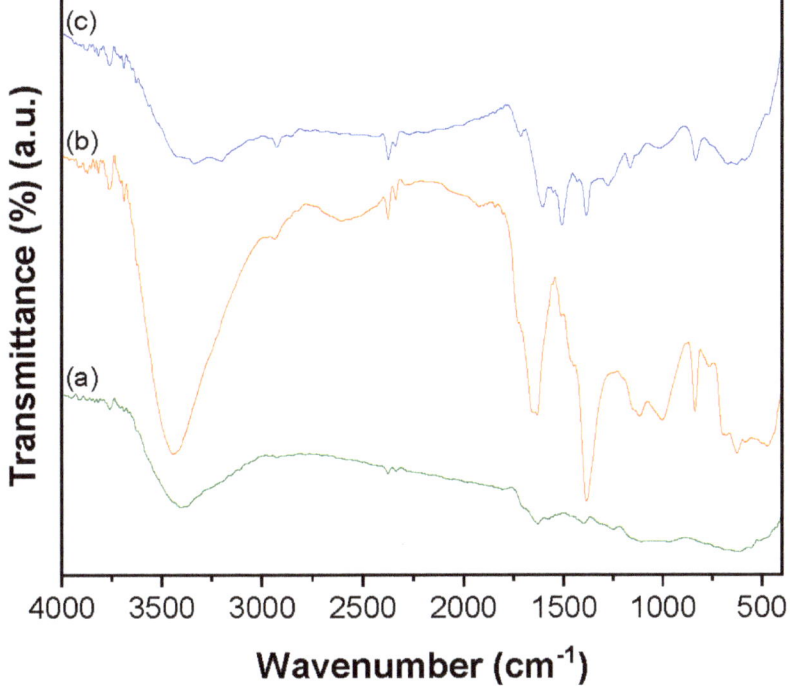

Figure 1. FT–IR spectra of (**a**) CNTs, (**b**) CNTs–PAMAM, and (**c**) CNTs–PAMAM–Ag.

The powder XRD of CNTs–PAMAM–Ag, shown in Figure 2a, depicted the intense diffractions at 38.0, 44.2, and 64.3° corresponding to (1 1 1), (2 0 0), and (2 2 0) reflections of silver with face-centered cubic (fcc) symmetry, respectively [55,57,58]. The reflections of CNTs were not observed in CNTs–PAMAM–Ag (Figure 2a); however, these were distinctly found in CNTs–PAMAM (Figure 2b). The indistinct visibility of CNTs peaks in CNTs–PAMAM–Ag could be due to the effective exfoliation of CNTs, the significant intensity of Ag reflections, and adequate coverage of the surface of CNTs by densely populated deposition Ag nanoparticles.

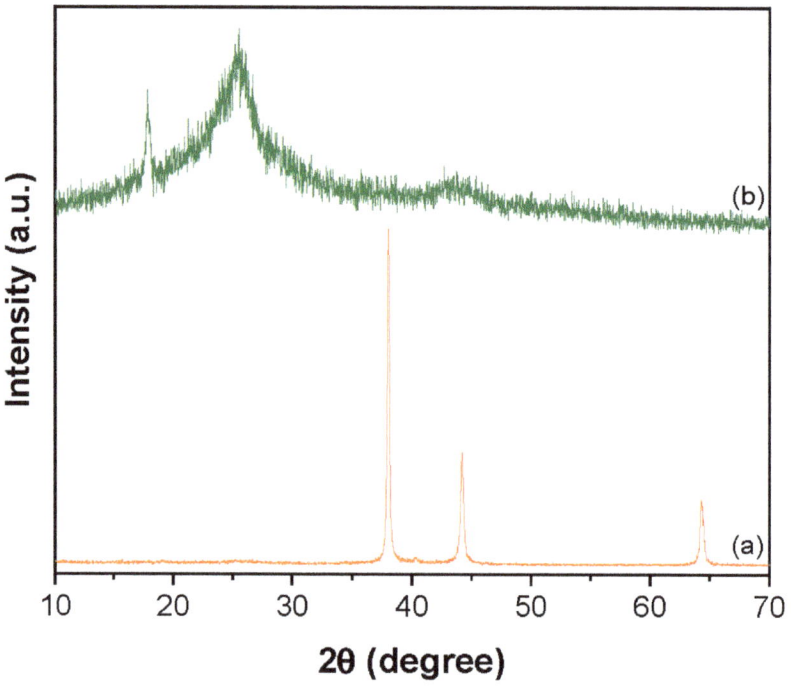

Figure 2. XRD pattern of (**a**) CNTs–PAMAM–Ag and (**b**) CNTs–PAMAM.

The TEM image of CNTs–PAMAM–Ag (Figure 3a) revealed the efficient entanglement of CNTs by successfully conjugating PAMAM chains onto their surfaces. The CNTs present in CNTs–PAMAM–Ag have a diameter of a few nanometers and a length of several micrometers. The PAMAM grafted to the surface of CNT and the deposition of spherical-shaped Ag nanoparticles are distinctly visible in Figure 3b. The average particle size of Ag nanoparticles was around 35 nm. The PAMAM is efficiently grafted over the entire surface of CNTs. However, the thickness of PAMAM was not uniform. Additionally, the homogenous distribution of Ag nanoparticles and their firm adherence is visible in Figure 3. The coarse texture was created over the surface of CNTs–PAMAM–Ag, which enables the adsorption rate of pollutants. Additionally, the hydrophilic nature of PAMAM turns the hydrophobic CNTs into hydrophilic and increases the absorption proportion. The absence of Ag nanoparticles in the void space (Figure 3) reveals the strong adherence of Ag nanoparticles to the surface PAMAM grafted CNTs. The resilient immobilization of Ag nanoparticles over PAMAM grafted CNTs caused by strong interaction ensue between them, preventing Ag nanoparticles' leaching.

The XPS survey spectrum of CNTs–PAMAM–Ag illustrated in Figure 4a confirms the presence of C, Ag, N, and O. The atomic ratio of Ag, C, and O estimated from XPS in CNTs–PAMAM–Ag was 1.17, 85.75, and 13.08%, respectively. The high-resolution spectrum of C1s (Figure 4b) exhibited a peak at 284.5 eV, divulge into three distinct peaks by Gaussian fitting. The peak at 284.9 eV was assigned to C-C bonds of sp^2 hybridized carbon atoms of CNTs [59]. The peak at 285.4 eV was due to C-C bonds occurring in structurally defective sp^3 hybridized carbon atoms and C=O bonds [59,60]. Another peak found at 290.0 eV was owing to carboxyl carbon O=C–O [61]. The high-resolution spectrum of Ag 3d (Figure 4c), disclosed into two peaks situated at 368.2 and 374.2 eV, featured the metallic state of Ag 3d5/2 and Ag 3d3/2, respectively [60]. The position of the peaks and binding energy match the value found for Ag^+ [62,63]. The difference in the binding energy between Ag 3d5/2 and Ag 3d3/2 peaks was 6 eV [60]. It confirms the presence of Ag nanoparticles in the metallic state. The N 1s spectrum shown in Figure 4d, deconvoluted into three peaks located at 394.0 eV, 399.0 eV, and 405.0 eV corresponding to –N= (quinoid imine), –NH– (benzoid amine), and positively charged nitrogen(–HN$^{·+}$– and –HN$^+$=), respectively [64]. The high-resolution spectrum of O 1s (Figure 4e) split into two peaks located at 530.1 eV by lattice O and 532.4 eV due to carbonyl (=C–O) functional groups [64].

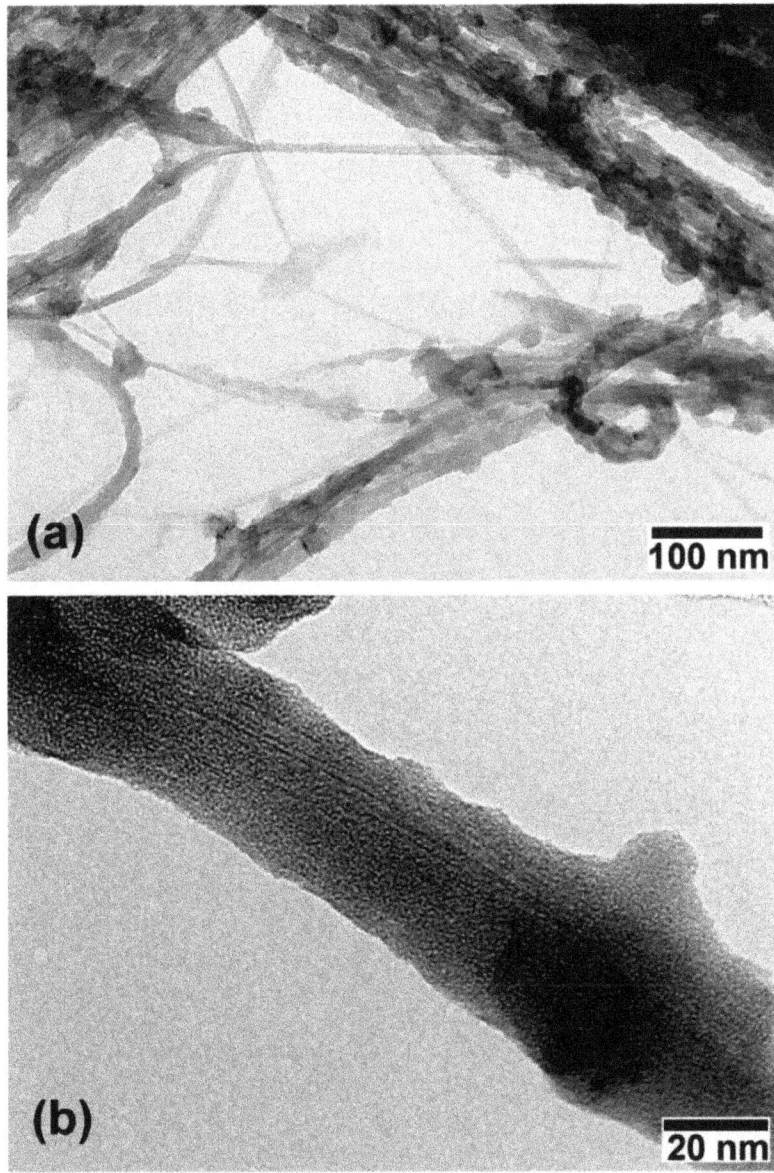

Figure 3. TEM images of CNTs–PAMAM–Ag. (**a**) revealed the efficient entanglement of CNTs by success-fully conjugating PAMAM chains onto their surfaces. (**b**) The PAMAM grafted to the surface of CNT and the deposition of spherical-shaped Ag nanoparticles.

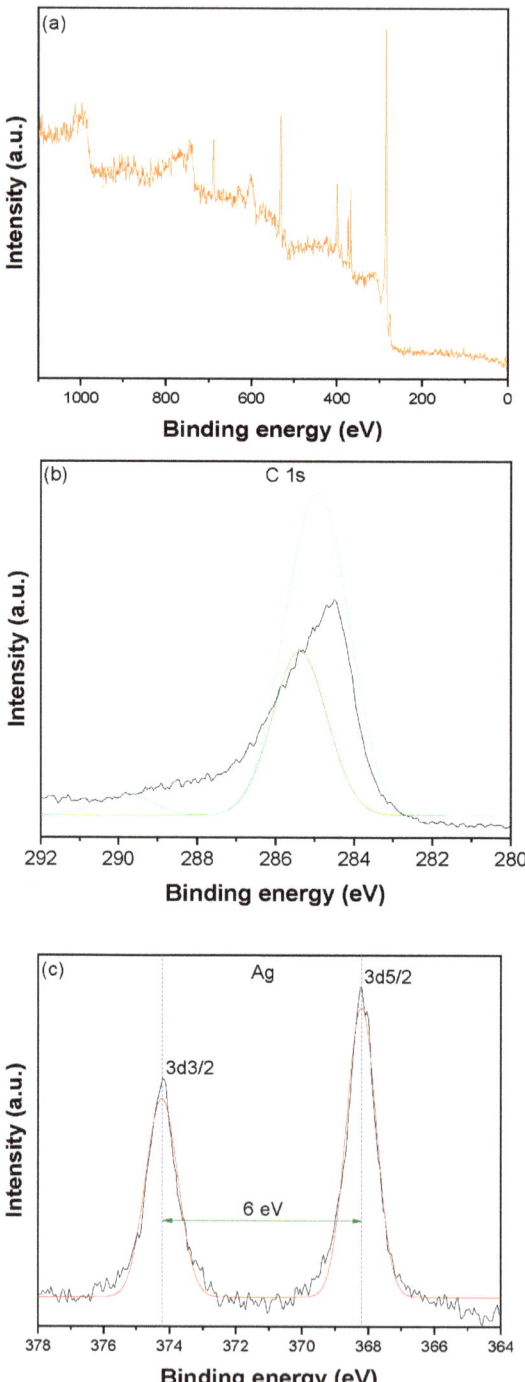

Figure 4. Cont.

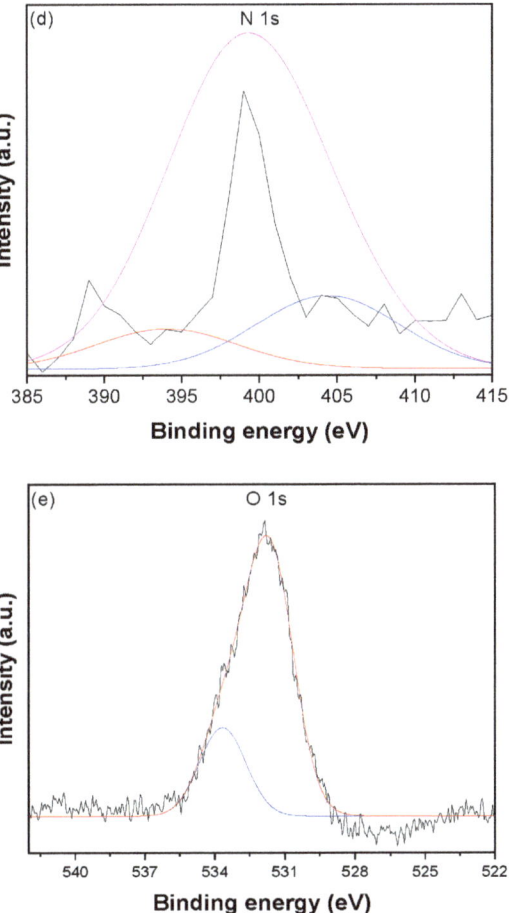

Figure 4. (**a**) XPS survey spectrum, (**b**) high-resolution spectrum of C1s, (**c**) high-resolution spectrum of Ag3d, (**d**) high-resolution spectrum of N 1s, and (**e**) high-resolution spectrum of O1s of CNTs–PAMAM–Ag.

The absorption efficiency of CNTs–PAMAM–Ag was evaluated by estimating the adsorption rate of two important toxins in water, viz., Pb(II) and As(III). Initially, the adsorption ability of oxidized CNTs and PAMAM and its improvement by their conjugation after their conjugation and deposition of Ag nanoparticles was accessed by measuring the adsorption of Pb(II) as a function of contact time. The result found for adsorption as a function of contact time for 40 µg/L concentrated Pb(II) solution is shown in Figure 5. The adsorption performance of oxidized CNTs and PAMAM was significantly improved after their conjugation and deposition of Ag nanoparticles in CNTs–PAMAM–Ag. About 99% of Pb(II) was adsorbed by CNTs–PAMAM–Ag within 15 min, while it was 58 and 38% for oxidized CNTs and PAMAM, respectively. Further, to find the controlling mechanisms of the adsorption process and its dynamics, the pseudo-first, and the second-order kinetic models were applied using the following Equations (4) and (5).

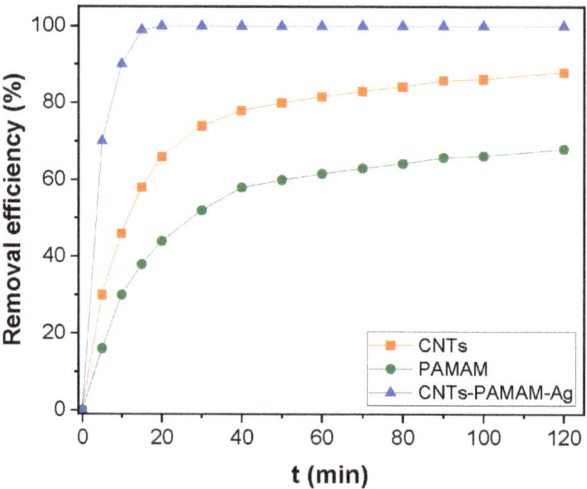

Figure 5. Kinetics of Pb(II) adsorption over oxidized CNTs, PAMAM and CNTs–PAMAM–Ag.

Pseudo-first-order model:

$$ln(q_e - q_t) = ln q_e - k_1 t \qquad (4)$$

Pseudo-second-order model:

$$\frac{t}{q_t} = \frac{1}{k_2 \, q_e^2} + \frac{t}{q_e} \qquad (5)$$

where q_e and q_t are the quantity of adsorbate (mg/g) at equilibrium and particular time t (min), respectively; k_1 (min^{-1}) and k_2 [g/(mg.min)] are the pseudo-first and second-order rate constants, respectively.

Figure 6a is the pseudo-first-order plot got for the adsorption of Pb(II) of over-oxidized CNTs, PAMAM, and CNTs–PAMAM–Ag, and Figure 6b is acquired for pseudo-second-order kinetics. The parameters estimated are illustrated in Table 1. The correlation coefficient (R^2) received for pseudo-second-order kinetics was higher than the value estimated for the pseudo-first-order kinetics. In addition, the value of q_e (exp) agreed with q_e (cal) determined from pseudo-second-order kinetics rather than the value of pseudo-first-order kinetics. Therefore, the adsorption of Pb(II) over oxidized CNTs, PAMAM, and CNTs–PAMAM–Ag takes place through pseudo-second-order kinetics. Further, the adsorption rate of Pb(II) over CNTs–PAMAM–Ag was compared with the adsorption of As(III). The kinetics profile found for the adsorption of Pb(II) and As(III) is illustrated in Figure 7 and the plot perceived for q_t versus time is given in Figure S1. It was revealed that the adsorption of Pb(II) and As(III) is time-dependent and proceeds as a function of time. Consequently, contact time is an essential factor in the adsorption of Pb(II) and As(III) and plays a significant role. The behavior in the adsorption of both Pb(II) and As(III) was identical. The adsorption rate was significant in the initial stage, and the progression in the later phase was relatively slow until the attainment of equilibrium. Within 15 min, about 99% of Pb(II) and 76% of As(III), was adsorbed. To reach the equilibrium, 20 and 70 min was needed for Pb(II) and As(III), respectively. After conquering the equilibrium, the adsorption of Pb(II) and As(III) was minute until the measured period of 120 min. The rapid adsorption in the initial stage could be due to the abundant availability of the active sites existing over the surface of CNTs–PAMAM–Ag. With the progress of time, the active sites are being saturated by the adsorption of a high number of Pb(II) and As(III) ions [65]. In addition, the repulsive forces befall the solute molecules in the solid and bulk phases [65].

The pseudo-first and pseudo-second-order plots received for adsorption of Pb(II) and As(III) over CNTs–PAMAM–Ag are demonstrated in Figure 8a,b, respectively, and the related parameters assessed are summarized in Table 1. The R^2 value for pseudo-first-order kinetics was 0.9663 and 0.9845 for Pb(II) and As(III), respectively. For pseudo-second-order kinetics, it was 0.9998 and 0.9966 for Pb(II) and As(III), respectively. The R^2 value for pseudo-second-order kinetics is higher than that of pseudo-first-order kinetics. Hence, the adsorption of Pb(II) and As(III) could be well-fitted with the pseudo-second-order kinetic model. Therefore, during adsorption, chemisorption ensues by sharing or exchanging of electrons between Pb(II) or As(III) ions and CNTs–PAMAM–Ag, which could be the rate-controlling step in the process of adsorption [5,66]. It is presumed that the pseudo-second-order adsorption process ensues through surface reactions until active sites get saturated, followed by the incidence of diffusion for complex sequential interactions [5,67].

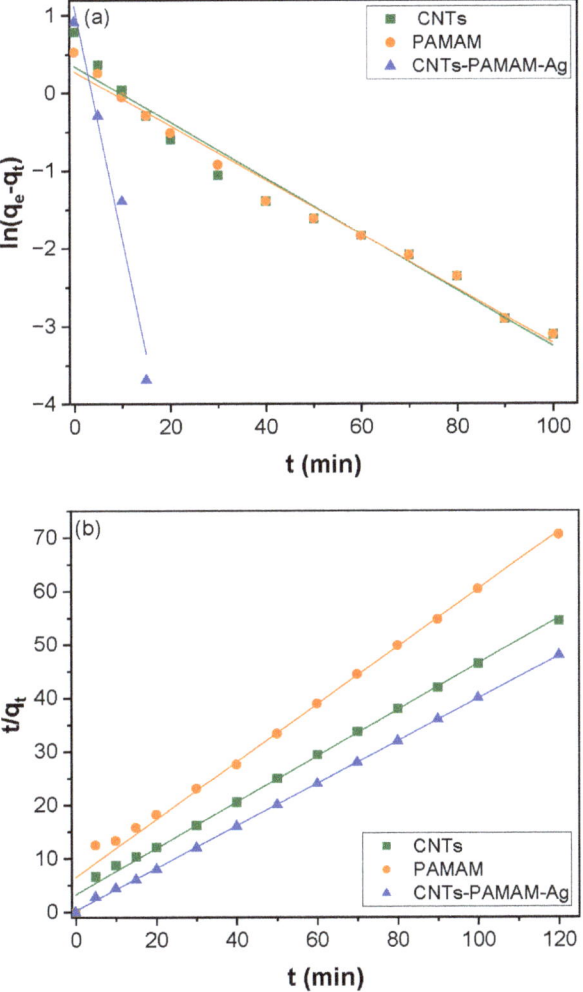

Figure 6. (**a**) Pseudo-first-order and (**b**) pseudo-second-order kinetics for Pb(II) adsorption over oxidized CNTs, PAMAM and CNTs–PAMAM–Ag.

Table 1. Parameters calculated from pseudo-first, and second-order kinetic models.

Adsorbent	Adsorbate	q_e (exp) mg/g	Pseudo-First Order Kinetic Model			Pseudo-Second-Order Kinetic Model		
			q_e (cal) mg/g	k_1 min^{-1}	R^2	q_e (cal) mg/g	k_2 g.mg^{-1}.min^{-1}	R^2
CNTs	Pb(II)	2.2	0.3483	0.0360	0.9687	2.3152	0.0573	0.9960
PAMAM	Pb(II)	1.7	0.2815	0.0349	0.9847	1.8543	0.0449	0.9896
CNTs–PAMAM–Ag	Pb(II)	2.5	1.1255	0.2983	0.9663	2.5189	0.6425	0.9993
CNTs–PAMAM–Ag	As(III)	2.5	0.7117	0.0772	0.9845	2.6309	0.0806	0.9966

To find the reaction pathways and the rate-controlling step in adsorption, Weber–Morris intraparticle pore diffusion model was used [68]. This model is based on sorbate species transport into the sorbent's pore is often the rate-controlling step in the adsorption. Thus, rate constants for intraparticle diffusion (k_{id}) were estimated using Equation (6).

$$q_t = k_{id} \, t^{0.5} + c \quad (6)$$

where q_t (mg/g) is the amount of Pb(II) and As(III) adsorbed at time t (min); c (mg/g) is the intercept that represents the boundary layer effect, and k_{id} [mg/(g.min$^{0.5}$)] is the intraparticle diffusion rate constant, which can be evaluated from the slope of the linear intraparticle diffusion plot of q_t versus $t^{0.5}$ (Figures 9 and S2). If the regression of the intraparticle diffusion plot is linear and passes through the origin, in that case, intraparticle diffusion is the sole rate-limiting step in the adsorption. However, the plot of q_t versus $t^{0.5}$, shown in Figure 9, does not pass through the origin (Figure S2). So, intraparticle diffusion is not the sole rate-limiting step in the adsorption found for Pb(II) and As(III). Moreover, if the intercept of the intraparticle diffusion plot is significant, the contribution of the surface sorption is more efficient in the rate-controlling step [69]. Contradictorily, the value of the intercept appraised for the adsorption of Pb(II) and As(III) was small (Table S1). In addition, the intraparticle diffusion plot (Figure S2) unveiled two straight lines. This specifies that intraparticle diffusion was not exclusively a rate-controlling step in the adsorption of Pb(II) and As(III); instead, it was regulated by both intraparticle diffusion and boundary layer effect [69,70]. The multilinearity found for the intraparticle diffusion plot (Figure 9) reveals that the adsorption of Pb(II) and As(III) emerge through multiple phases instead of a single [5]. Out of these, the first phase was owing to the instantaneous adsorption of Pb(II) and As(III) ions (Figure 9), and the second phase was due to the diffusion of Pb(II) and As(III) ions into the pores of CNTs–PAMAM–Ag. Finally, the third phase was from the equilibrium of adsorption that causes chemical reaction/bonding [5].

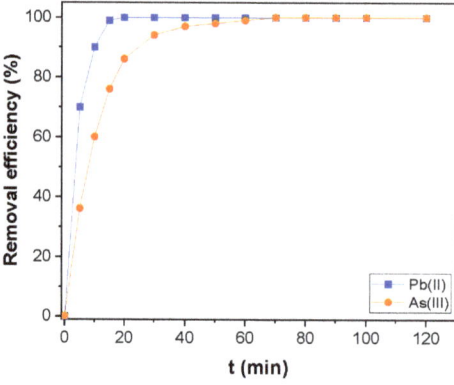

Figure 7. Kinetics of Pb(II) and As(III) adsorption over CNTs–PAMAM–Ag.

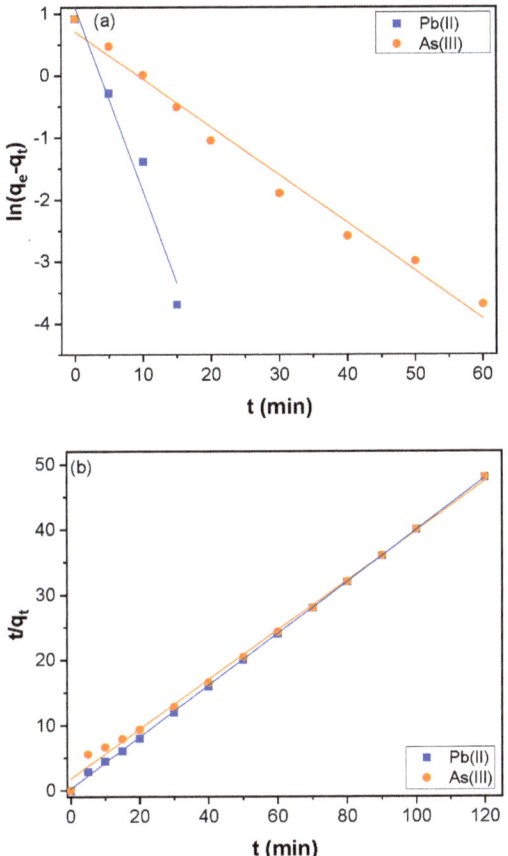

Figure 8. (**a**) Pseudo-first-order and (**b**) pseudo-second-order kinetics for Pb(II) and As(III) adsorption over CNTs–PAMAM–Ag.

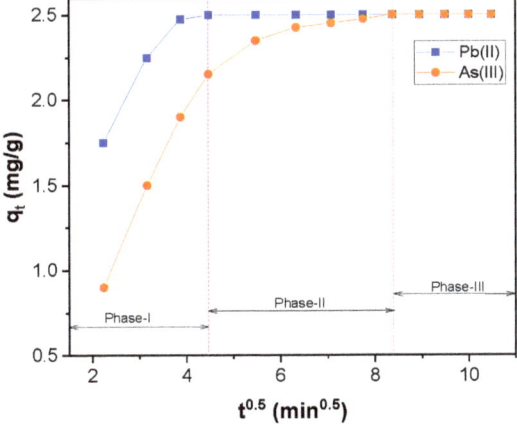

Figure 9. Weber-Morris intraparticle diffusion plot for Pb(II) and As(III) adsorption over CNTs–PAMAM–Ag.

To further explore the adsorption equilibrium, the experimental results obtained for adsorption of Pb(II) and As(III) were analyzed with three isotherm models viz., the Langmuir, the Freundlich, and the Temkin models. The Langmuir isotherm model assumes that monolayer adsorption occurs on the surface of the adsorbent [71]. Additionally, it presumes that the equivalent binding site number is specific and that adsorbate does not transmigrate [72]. Accordingly, the Langmuir model is represented by Equation (7).

$$\frac{C_e}{q_e} = \frac{C_e}{q_m} + \frac{1}{K_L\, q_m} \tag{7}$$

where q_e (mg/g) is the amount of adsorbed Pb(II) and As(III) per unit mass of CNTs–PAMAM–Ag; C_e (mg/L) is the concentration of Pb(II) and As(III) at equilibrium; q_m is the maximum amount of the Pb(II) and As(III) adsorbed per unit mass of CNTs–PAMAM–Ag to form a complete monolayer on the surface-bound at high C_e. K_L is the Langmuir adsorption constant related to the free energy of adsorption. The linear fitting for the Langmuir plot of specific adsorption (C_e/q_e) versus the equilibrium concentration (C_e) is shown in Figure 10a. The parameters calculated by the Langmuir isotherm model are tabulated in Table 2. The maximum adsorption capacity (q_m) calculated for the adsorption of Pb(II) and As(III) was 18.7 and 14.8 mg/g, respectively.

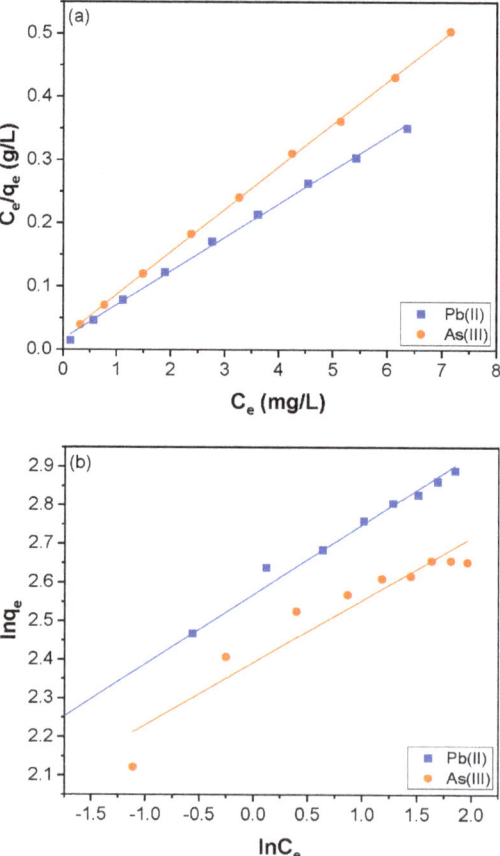

Figure 10. *Cont.*

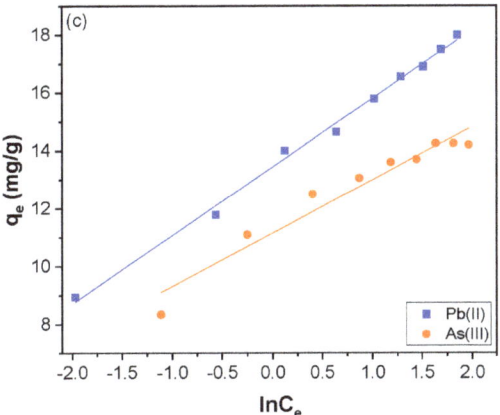

Figure 10. (**a**) Langmuir, (**b**) Freundlich, and (**c**) Temkin isotherms for Pb(II) and As(III) adsorption over CNTs–PAMAM–Ag.

Table 2. Parameters calculated from Langmuir, Freundlich, and Temkin adsorption isotherms.

Adsorbate	Langmuir Isotherm			Freundlich Isotherm			Temkin Isotherm		
	q_m mg/g	K_L L/mg	R^2_{Lan}	K_F mg/g	n	R^2_{Fre}	A	B	R^2_{Tem}
Pb(II)	18.7	3.12	0.9980	13.05	5.546	0.9916	292.3	2.4	0.9950
As(III)	14.8	3.59	0.9997	10.94	6.192	0.9100	428.9	6.1	0.9451

Furthermore, the kinetic data of adsorption of Pb(II) and As(III) was explored with the Freundlich isotherm model that defines the heterogeneous adsorption using Equation (8) [73].

$$ln\ q_e = ln\ K_F + \frac{ln\ C_e}{n} \quad (8)$$

where q_e (mg/g) is the amount of Pb(II) and As(III) adsorbed per unit mass of CNTs–PAMAM–Ag; C_e (mg/L) is the concentration of Pb(II), and As(III) at the equilibrium, K_F indicates the affinity of adsorbent of Pb(II) and As(III) and n denotes the adsorption intensity. C_e (mg/L) is the concentration of Pb(II) and As(III) at equilibrium. The Freundlich isotherm fitting plot is shown in Figure 10b, and obtained results are summarized in Table 2. The magnitude of the exponent, $1/n$, indicates the favorability of adsorption. The value of n ranging from 1 to 10 indicates the favorable conditions for adsorption [66]. The value of n calculated for Pb(II) and As(III) adsorption were 5.546 and 6.192, respectively. This represents that the adsorption of Pb(II) and As(III) over CNTs–PAMAM–Ag is favorable.

Apart, the adsorption of Pb(II) and As(III) was explored with the Temkin isotherm model [74] using Equation (9).

$$q_e = B\ ln A + B\ ln\ C_e \quad (9)$$

where $B = RT/K_T$, K_T is the Temkin constant related to the heat of adsorption (J/mol); A is the Temkin isotherm constant (L/g), R is the gas constant (8.314 J/mol K), and T the absolute temperature (K). The Temkin isotherm fitting plot of q_e versus $ln\ C_e$ is shown in Figure 10c, and the estimated parameters are listed in Table 2.

Among all the isotherm models, the Langmuir isotherm model was well suited for the adsorption of Pb(II) and As(III) (Figure 10a) with the R^2_{Lan} value of 0.9980 and 0.9997 for Pb(II) and As(III), respectively. Nonetheless, it was lower for both the Freundlich and Temkin isotherm models. It means R^2_{Lan} was high compared to R^2_{Fre} and R^2_{Tem}. Thus, the experimental kinetic data received for Pb(II) and As(III) adsorption was well fitted to the Langmuir isotherm model. So the Langmuir isotherm model satisfactorily explains the adsorption of Pb(II) and As(III) compared to the Freundlich and Temkin isotherm models. As the Langmuir equation assumes the adsorbent's surface is homogenous, a better fitting of the Langmuir isotherm model indicates the uniform distribution of active sites over the entire surface of CNTs–PAMAM–Ag and their homogeneity [66,69]. It signifies that the adsorption of Pb(II) and As(III) was dominated by the monolayer binding on the homogeneous surface of CNTs–PAMAM–Ag [66]. The maximum adsorption capacity (q_m) of CNTs–PAMAM–Ag estimated for Pb(II) and As(III) adsorption was compared to the value reported for adsorbents, which is listed in Table S2 [4,42,75–82].

Further, to investigate the effect of pH on the adsorption of Pb(II) monitored at pH of 4, 6, 8, 10, and 12. It was found that the adsorption of Pb(II) was on the pH of the solution in such a way that the maximum adsorption was perceived at around 8 (Figure S3). However, the adsorption rate was low at other measured values of pH viz., 4, 6, 10, and 12. The natural pH of the 40 µg/L Pb(II) solution used in the adsorption measurement was 8 (7.67). So the adsorption at pH 8 was recorded with an alteration of pH by adding either 0.1 M HCl or NaOH. However, the adsorption measurement at other pH values such as 4 and 6 was adjusted using 0.1 M HCl, while pH 10 and 12 were attained by adding 0.1 NaOH. Hence, the addition of HCl or NaOH could interfere with Pb(II) ions and hinder the adsorption efficiency. The recovery and repeated use of the adsorbent are essential for practical application. The efficiency of CNTs–PAMAM–Ag in the adsorption of Pb(II) was investigated for four successive cycles (Figure S4). The adsorption rate of Pb(II) was not reduced significantly in all four studied cycles. Therefore, CNTs–PAMAM–Ag is an ideal adsorbent for repeated use without losing its activity. The hierarchical architecture generated grafting of PAMAM improved the adsorption efficiency of CNTs, which was further enhanced by the deposition of Ag nanoparticles.

4. Conclusions

In conclusion, an efficient adsorbent for effective adsorption of Pb(II) and As(III) was successfully prepared by grafting fourth-generation aromatic PAMAM to CNTs and successive deposition of Ag nanoparticles. Thus produced CNTs–PAMAM–Ag was able to adsorb 99 and 76% of Pb(II) and As(III), respectively within 15 min. The kinetics data obtained for the adsorption of Pb(II) and As(III) was well fitted with the pseudo-second-order model compared to the pseudo-first-order model. It revealed the occurrence of chemisorption by sharing or exchanging electrons between Pb(II) or As(III) ions and CNTs–PAMAM–Ag. It could be the rate-controlling step in the process of adsorption. The multilinearity of the Weber–Morris plot demonstrated that intraparticle diffusion was not a rate-controlling step in the adsorption of Pb(II) and As(III); instead, it was regulated by both intraparticle diffusion and boundary layer effect. The proper fitting of kinetic data of Pb(II) and As(III) adsorption with the Langmuir isotherm model indicates the uniform distribution of active sites over the entire surface of CNTs–PAMAM–Ag and their homogeneity. In addition, it signifies that the adsorption of Pb(II) and As(III) was dominated by the monolayer binding on the homogeneous surface of CNTs–PAMAM–Ag. The adsorption ability of CNTs–PAMAM–Ag depends on the pH. The CNTs–PAMAM–Ag is an ideal adsorbent for repeated use without losing its activity. Because of its significance in Pb(II) and As(III) adsorption, CNTs–PAMAM–Ag could be an efficient adsorbent and practically applicable for the adsorption of other heavy metals and other contaminants present in water.

Supplementary Materials: The following supporting information can be downloaded at: https://www.mdpi.com/article/10.3390/nano12213852/s1. Figure S1: Plot perceived qt as a function of time for Pb(II) and As(III) adsorption over CNTs–PAMAM–Ag; Figure S2: Intraparticle diffusion model for Pb(II) and As(III) adsorption over CNTs–PAMAM–Ag; Figure S3: Effect of pH on the adsorption of Pb(II) over CNTs–PAMAM–Ag; Figure S4: Efficiency of CNTs–PAMAM–Ag in the adsorption of Pb(II) for four successive cycles. Table S1: Parameters calculated from the intra-particle diffusion plot provided in Figure S2; Table S2: Comparison of maximum adsorption capacity (qm) of CNTs–PAMAM–Ag estimated for Pb(II) and As(III) adsorption with different adsorbents.

Author Contributions: G.M.N.: data curation, writing original draft, review, and editing funding acquisition. S.F.A.: investigation, data curation. M.D.K.: data curation, writing, review, and editing. D.F.R.: writing, review, and editing. R.L.R.: writing, review, and editing. All authors have read and agreed to the published version of the manuscript.

Funding: The authors acknowledge the financial support of the National Academy of Sciences for U.S.–Egypt Science and Technology Joint Fund (2000012544) and the Welch Foundation, Texas, United States for the departmental grant (L-0002-20181021).

Data Availability Statement: Not applicable.

Conflicts of Interest: The authors declare no conflict of interest.

References

1. Rehman, K.; Fatima, F.; Waheed, I.; Akash, M.S.H. Prevalence of exposure of heavy metals and their impact on health consequences. *J. Cell. Biochem.* **2018**, *119*, 157–184. [CrossRef]
2. Zeng, Q.; Huang, Y.; Huang, L.; Hu, L.; Sun, W.; Zhong, H.; He, Z. High adsorption capacity and super selectivity for Pb(II) by a novel adsorbent: Nano humboldtine/almandine composite prepared from natural almandine. *Chemosphere* **2020**, *253*, 126650. [CrossRef]
3. Li, Z.; Gong, Y.; Zhao, D.; Deng, H.; Dang, Z.; Lin, Z. Field assessment of carboxymethyl cellulose bridged chlorapatite microparticles for immobilization of lead in soil: Effectiveness, long-term stability, and mechanism. *Sci. Total. Environ.* **2021**, *781*, 146757. [CrossRef]
4. Mudzielwana, R.; Gitari, M.W.; Ndungu, P. Enhanced As(III) and As(V) adsorption from aqueous solution by a clay based hybrid sorbent. *Front. Chem.* **2020**, *7*, 913. [CrossRef]
5. Zeng, H.; Zhai, L.; Qiao, T.; Yu, Y.; Zhang, J.; Li, D. Efficient removal of As(V) from aqueous media by magnetic nanoparticles prepared with iron-containing water treatment residuals. *Sci. Rep.* **2020**, *10*, 9335. [CrossRef]
6. Baskan, M.B.; Hadimlioglu, S. Removal of arsenate using graphene oxide-iron modified clinoptilolite-based composites: Adsorption kinetic and column study. *J. Anal. Sci. Technol.* **2021**, *12*, 22. [CrossRef]
7. Verma, L.; Siddique, M.A.; Singh, J.; Bharagava, R.N. As(III) and As(V) removal by using iron impregnated biosorbents derived from waste biomass of citrus limmeta (peel and pulp) from the aqueous solution and ground water. *J. Environ. Manag.* **2019**, *250*, 109452. [CrossRef]
8. Zhang, M.; Yin, Q.; Ji, X.; Wang, F.; Gao, X.; Zhao, M. High and fast adsorption of Cd(II) and Pb(II) ions from aqueous solutions by a waste biomass based hydrogel. *Sci. Rep.* **2020**, *10*, 3285. [CrossRef]
9. Shen, J.; Wang, N.; Wang, Y.G.; Yu, D.; Ouyang, X. Efficient adsorption of Pb(II) from aqueous solutions by metal organic framework (Zn-BDC) coated magnetic montmorillonite. *Polymers* **2018**, *10*, 1383. [CrossRef]
10. Wang, N.; Xu, X.; Li, H.; Yuan, L.; Yu, H. Enhanced selective adsorption of Pb(II) from aqueous solutions by one-pot synthesis of xanthate-modified chitosan sponge: Behaviors and mechanisms. *Ind. Eng. Chem. Res.* **2016**, *55*, 12222–12231. [CrossRef]
11. Mandal, S.; Sahu, M.K.; Patel, R.K. Adsorption studies of arsenic(III) removal from water by zirconium polyacrylamide hybrid material (ZrPACM-43). *Water Resour. Ind.* **2013**, *4*, 51–67. [CrossRef]
12. Mohan, D.; Charles, P. Arsenic removal from water/wastewater using adsorbents-a critical review. *J. Hazard. Mater.* **2007**, *142*, 1–53. [CrossRef]
13. Choong, Y.S.T.; Chuah, G.T.; Robia, H.Y.; Koay, L.F.G.; Azni, I. Arsenic toxicity, health hazards and removal techniques from water: An overview. *Desalination* **2007**, *217*, 139–166. [CrossRef]
14. Shevade, S.; Ford, R. Use of synthetic zeolites for arsenate removal from pollutant water. *Water Res.* **2004**, *38*, 3197–3204. [CrossRef]
15. Yao, S.; Liu, Z.; Shi, Z. Arsenic removal from aqueous solutions by adsorption onto iron oxide/activated carbon magnetic composite. *J. Environ. Health Sci. Eng.* **2014**, *12*, 58. [CrossRef]
16. Tallman, D.E.; Shaikh, A.U. Redox stability of inorganic arsenic (III) and arsenic (V) in aqueous solution. *Anal. Chem.* **1980**, *52*, 196–199. [CrossRef]
17. Aposhian, H.V.; Maiorino, R.M.; Dart, R.C.; Perry, D.F. Urinary excretion of meso-2, 3-dimercaptosuccinic acid in human subjects. *J. Clin. Pharm. Ther.* **1989**, *45*, 520–526. [CrossRef]

18. Jeyakumar, R.P.S.; Chandrasekaran, V. Adsorption of lead (II) ions by activated carbons prepared from marine green algae: Equilibrium and kinetics studies. *Int. J. Ind. Chem.* **2014**, *5*, 10. [CrossRef]
19. Monrad, M.; Ersboll, A.K.; Sorensen, M.; Baastrup, R.; Hansen, B.; Gammelmark, A.; Tjonneland, A.; Overvadd, K.; Raaschou-Nielsen, O. Low-level arsenic in drinking water and risk of incident myocardial infarction: A cohort study. *Environ. Res.* **2017**, *154*, 318–324. [CrossRef]
20. Lee, C.; Alvarez, P.J.J.; Nam, A.; Park, S.; Do, T.; Choi, U.; Lee, S. Arsenic (V) removal using an amine-doped acrylic ion exchange fiber: Kinetic, equilibrium, and regeneration studies. *J. Hazard. Mater.* **2017**, *325*, 223–229. [CrossRef]
21. Yoon, J.; Amy, G.; Chung, J.; Sohn, J.; Yoon, Y. Removal of toxic ions (chromate, arsenate, and perchlorate) using reverse osmosis, nanofiltration, and ultrafiltration membranes. *Chemosphere* **2009**, *77*, 228–235. [CrossRef]
22. Vasudevan, S.; Lakshmi, J. Electrochemical removal of boron from water: Adsorption and thermodynamic studies. *Can. J. Chem. Eng.* **2012**, *90*, 1017–1026. [CrossRef]
23. Ungureanu, G.; Santos, S.; Boaventura, R.; Botelho, C. Arsenic and antimony in water and wastewater: Overview of removal techniques with special reference to latest advances in adsorption. *J. Environ. Manag.* **2015**, *151*, 326–342. [CrossRef]
24. Hao, L.; Liu, M.; Wang, N.; Li, G. A critical review on arsenic removal from water using iron-based adsorbents. *RSC Adv.* **2018**, *8*, 39545–39560. [CrossRef]
25. Sun, T.; Zhao, Z.; Liang, Z.; Liu, J.; Shi, W.; Cui, F. Efficient As(III) removal by magnetic $CuO-Fe_3O_4$ nanoparticles through photo-oxidation and adsorption under light irradiation. *J. Colloid Interface Sci.* **2017**, *495*, 168–177. [CrossRef]
26. Hao, L.; Wang, N.; Wang, C.; Li, G. Arsenic removal from water and river water by the combined adsorption—UF membrane process. *Chemosphere* **2018**, *202*, 768–776. [CrossRef]
27. Song, P.; Yang, Z.; Zeng, G.; Yang, X.; Xu, H.; Wang, L.; Xu, R.; Xiong, W.; Ahmad, K. Electrocoagulation treatment of arsenic in wastewaters: A comprehensive review. *Chem. Eng. J.* **2017**, *317*, 707–725. [CrossRef]
28. Chowdhury, R. Using adsorption and sulphide precipitation as the principal removal mechanisms of arsenic from a constructed wetland—A critical review. *Chem. Ecol.* **2017**, *33*, 560–571. [CrossRef]
29. Pandey, P.K.; Choubey, S.; Verma, Y.; Pandey, M.; Chandrashekhar, K. Biosorptive removal of arsenic from drinking water. *Bioresource Technol.* **2009**, *100*, 634–637. [CrossRef]
30. Tao, H.C.; Zhang, H.R.; Li, J.B.; Ding, W.Y. Biomass based activated carbon obtained from sludge and sugarcane bagasse for removing lead ion from wastewater. *Bioresour. Technol.* **2015**, *192*, 611–617. [CrossRef]
31. Inchaurrondo, N.; Luc, C.; Mori, F.; Pintar, A.; Zerjav, G.; Valiente, M.; Palet, C. Synthesis and adsorption behavior of mesoporous alumina and Fe-doped alumina for the removal of dominant arsenic species in contaminated waters. *J. Environ. Chem. Eng.* **2019**, *7*, 102901. [CrossRef]
32. Gao, M.; Ma, Q.; Lin, Q.; Chang, J.; Ma, H. A novel approach to extract SiO_2 from fly ash and its considerable adsorption properties. *Mater. Des.* **2017**, *116*, 666–675. [CrossRef]
33. Tang, Y.; Chen, L.; Wei, X.; Yao, Q.; Li, T. Removal of lead ions from aqueous solution by the dried aquatic plant, *Lemna perpusilla* torr. *J. Hazard. Mater.* **2013**, *244*, 603–612. [CrossRef] [PubMed]
34. Wang, J.; Wei, J.; Li, J. Rice straw modified by click reaction for selective extraction of noble metal ions. *Bioresour. Technol.* **2015**, *177*, 182–187. [CrossRef] [PubMed]
35. Liu, X.; Li, G.; Chen, C.; Zhang, X.; Zhou, K.; Long, X. Banana stem and leaf biochar as an effective adsorbent for cadmium and lead in aqueous solution. *Sci. Rep.* **2022**, *12*, 1584. [CrossRef]
36. Rezaei, M.; Pourang, N.; Moradi, A.M. Removal of lead from aqueous solutions using three biosorbents of aquatic origin with the emphasis on the affective factors. *Sci. Rep.* **2022**, *12*, 751. [CrossRef]
37. Tian, X.; Xie, Q.; Chai, G.; Li, G. Simultaneous adsorption of As(III) and Cd(II) by ferrihydrite-modified biochar in aqueous solution and their mutual effects. *Sci. Rep.* **2022**, *12*, 5918. [CrossRef]
38. Chen, Y.; Lin, Q.; Wen, X.; He, J.; Luo, H.; Zhong, Q.; Wu, L.; Li, J. Simultaneous adsorption of As(III) and Pb(II) by the iron-sulfur codoped biochar composite: Competitive and synergistic effects. *J. Environ. Sci.* **2023**, *125*, 14–25. [CrossRef]
39. Lv, Q.; Hu, X.; Zhang, X.; Huang, L.; Liu, Z.; Sun, G. Highly efficient removal of trace metal ions by using poly(acrylic acid) hydrogel adsorbent. *Mater. Des.* **2019**, *181*, 107934. [CrossRef]
40. Arshad, F.; Selvaraj, M.; Zain, J.; Banat, F.; Haija, M.A. Polyethylenimine modified graphene oxide hydrogel composite as an efficient adsorbent for heavy metal ions. *Sep. Purif. Technol.* **2019**, *209*, 870–880. [CrossRef]
41. Kheyrabadi, F.B.; Zare, E.N. Antimicrobial nanocomposite adsorbent based on poly(meta-phenylenediamine) for remediation of lead (II) from water medium. *Sci. Rep.* **2022**, *12*, 4632. [CrossRef] [PubMed]
42. Wang, Z.; Xu, W.; Jie, F.; Zhao, Z.; Zhou, K.; Liu, H. The selective adsorption performance and mechanism of multiwall magnetic carbon nanotubes for heavy metals in wastewater. *Sci. Rep.* **2021**, *11*, 16878. [CrossRef]
43. Fiyadh, S.S.; AlSaadi, M.A.; Jaafar, W.Z.; Alomar, M.K.; Fayaed, S.S.; Mohd, N.S.; Hin, L.S.; El-Shafie, A. Review on heavy metal adsorption processes by carbon nanotubes. *J. Clean. Prod.* **2019**, *230*, 783–793. [CrossRef]
44. Pan, B.; Xing, B. Adsorption mechanisms of organic chemicals on carbon nanotubes. *Environ. Sci. Technol.* **2008**, *42*, 9005–9013. [CrossRef]
45. Olanipekun, O.; Oyefusi, A.; Neelgund, G.M.; Oki, A. Adsorption of lead over graphite oxide. *Spectrochim. Acta A* **2014**, *118*, 857–860. [CrossRef] [PubMed]

46. Olanipekun, O.; Oyefusi, A.; Neelgund, G.M.; Oki, A. Synthesis and characterization of reduced graphite oxide–polymer composites and their application in adsorption of lead. *Spectrochim. Acta A* **2015**, *149*, 991–996. [CrossRef] [PubMed]
47. Govarthanan, M.; Jeon, C.; Kim, W. Synthesis and characterization of lanthanum-based metal organic framework decorated polyaniline for effective adsorption of lead ions from aqueous solutions. *Environ. Pollut.* **2022**, *303*, 119049. [CrossRef] [PubMed]
48. Li, W.; Zu, B.; Yang, Q.; Huang, Y.; Li, J. Adsorption of lead and cadmium by microplastics and their desorption behavior as vectors in the gastrointestinal environment. *J. Environ. Chem. Eng.* **2022**, *10*, 107379. [CrossRef]
49. Abbasi, M. Synthesis and characterization of magnetic nanocomposite of chitosan/SiO_2/carbon nanotubes and its application for dyes removal. *J. Clean. Prod.* **2017**, *145*, 105–113. [CrossRef]
50. Sun, J.; Hong, C.; Pan, C. Surface modification of carbon nanotubes with dendrimers or hyperbranched polymers. *Polym. Chem.* **2011**, *2*, 998–1007. [CrossRef]
51. Guo, D.; Huang, S.; Zhu, Y. The adsorption of heavy metal ions by poly (amidoamine) dendrimer-functionalized nanomaterials: A review. *Nanomaterials* **2022**, *12*, 1831. [CrossRef] [PubMed]
52. Geng, X.; Qu, R.; Kong, X.; Geng, S.; Zhang, Y.; Sun, C.; Ji, C. Facile synthesis of cross-linked hyperbranched polyamidoamines dendrimers for efficient Hg(II) removal from water. *Front. Chem.* **2021**, *9*, 743429. [CrossRef] [PubMed]
53. Yuan, Y.; Wu, Y.; Wang, H.; Tong, Y.; Sheng, X.; Sun, Y.; Zhou, X.; Zhou, Q. Simultaneous enrichment and determination of cadmium and mercury ions using magnetic PAMAM dendrimers as the adsorbents for magnetic solid phase extraction coupled with high performance liquid chromatography. *J. Hazard. Mater.* **2020**, *386*, 121658. [CrossRef] [PubMed]
54. Pawlaczyk, M.; Schroeder, G. Adsorption studies of Cu(II) ions on dendrimer-grafted silica-based materials. *J. Mol. Liq.* **2019**, *281*, 176–185. [CrossRef]
55. Neelgund, G.M.; Oki, A. Deposition of silver nanoparticles on dendrimer functionalized multiwalled carbon nanotubes: Synthesis, characterization and antimicrobial activity. *J. Nanosci. Nanotechnol.* **2011**, *11*, 3621–3629. [CrossRef]
56. Pan, B.; Cui, D.; Gao, F.; He, R. Growth of multi-amine terminated poly(amidoamine) dendrimers on the surface of carbon nanotubes. *Nanotechnology* **2006**, *17*, 2483–2489. [CrossRef]
57. Neelgund, G.M.; Oki, A. Photothermal effect of Ag nanoparticles deposited over poly(amidoamine) grafted carbon nanotubes. *J. Photochem. Photobiol. A* **2018**, *364*, 309–315. [CrossRef]
58. Neelgund, G.M.; Hrehorova, E.; Joyce, M.; Bliznyuk, V. Synthesis and characterization of polyaniline derivative and silver nanoparticle composites. *Polym. Int.* **2008**, *57*, 1083–1089. [CrossRef]
59. Zhao, C.; Guo, J.; Yu, C.; Zhang, Z.; Sun, Z.; Piao, X. Fabrication of CNTs-Ag-TiO_2 ternary structure for enhancing visible light photocatalytic degradation of organic dye pollutant. *Mater. Chem. Phys.* **2020**, *248*, 122873. [CrossRef]
60. Wei, X.; Wang, C.; Ding, S.; Yang, K.; Tian, F.; Li, F. One-step synthesis of Ag nanoparticles/carbon dots/TiO_2 nanotube arrays composite photocatalyst with enhanced photocatalytic activity. *J. Environ. Chem. Eng.* **2021**, *9*, 104729. [CrossRef]
61. Sharifian, K.; Mahdikhah, V.; Sheibani, S. Ternary Ag@$SrTiO_3$@CNT plasmonic nanocomposites for the efficient photodegradation of organic dyes under the visible light irradiation. *Ceram. Int.* **2021**, *47*, 22741–22752. [CrossRef]
62. Fan, G.; Du, B.; Zhou, J.; Yu, W.; Chen, Z.; Yang, S. Stable Ag_2O/g-C_3N_4 p-n heterojunction photocatalysts for efficient inactivation of harmful algae under visible light. *Appl. Catal. B* **2020**, *265*, 118610. [CrossRef]
63. Jiao, J.; Wan, J.; Ma, Y.; Wang, Y. Enhanced photocatalytic activity of AgNPs-in-CNTs with hydrogen peroxide under visible light irradiation. *Environ. Sci. Pollut. Res.* **2019**, *26*, 26389–26396. [CrossRef]
64. Lin, Y.; Wu, S.; Yang, C.; Chen, M.; Li, X. Preparation of size-controlled silver phosphate catalysts and their enhanced photocatalysis performance via synergetic effect with MWCNTs and PANI. *Appl. Catal. B* **2019**, *245*, 71–86. [CrossRef]
65. Joshi, S.; Sharma, M.; Kumari, A.; Shrestha, S.; Shrestha, B. Arsenic removal from water by adsorption onto iron oxide/nano-porous carbon magnetic composite. *Appl. Sci.* **2019**, *9*, 3732. [CrossRef]
66. Ploychompoo, S.; Chen, J.; Luo, H.; Liang, Q. Fast and efficient aqueous arsenic removal by functionalized MIL-100(Fe)/rGO/d-MnO_2 ternary composites: Adsorption performance and mechanism. *J. Environ. Sci.* **2020**, *91*, 22–34. [CrossRef] [PubMed]
67. Xie, F.; Wu, F.; Liu, G.; Mu, Y.; Feng, C.; Wang, H.; Giesy, J.P. Removal of phosphate from eutrophic lakes through adsorption by in situ formation of magnesium hydroxide from diatomite. *Environ. Sci. Technol.* **2014**, *48*, 582–590. [CrossRef]
68. Weber, W.J., Jr.; Morris, J.C. Kinetics of adsorption on carbon from solution. *J. Sanit. Eng. Div. Proceed. Am. Soc. Civil Eng.* **1963**, *89*, 31–59. [CrossRef]
69. Hameed, B.H.; Salman, J.M.; Ahmad, A.L. Adsorption isotherm and kinetic modeling of 2,4-D pesticide on activated carbon derived from date stones. *J. Hazard. Mater.* **2009**, *163*, 121–126. [CrossRef]
70. Hamayun, M.; Mahmood, T.; Naeem, A.; Muska, M.; Din, S.U.; Waseem, M. Equilibrium and kinetics studies of arsenate adsorption by $FePO_4$. *Chemosphere* **2013**, *99*, 207–215. [CrossRef]
71. Rashid, M.; Price, N.T.; Pinilla, M.A.G.; O'Shea, K.E. Effective removal of phosphate from aqueous solution using humic acid coated magnetite nanoparticles. *Water Res.* **2017**, *123*, 353–360. [CrossRef] [PubMed]
72. Reed, B.E.; Matsumoto, M.R. Modeling cadmium adsorption by activated carbon using the Langmuir and Freundlich isotherm expressions. *Sep. Sci. Technol.* **1993**, *28*, 2179–2195. [CrossRef]
73. Freundlich, H. Über die adsorption in lösungen (Adsorption in solution). *Z. Phys. Chem.* **1906**, *57*, 384–470. [CrossRef]
74. Temkin, M.J.; Pyzhev, V. Recent modifications to Langmuir Isotherms. *Acta Physiochim. USSR* **1940**, *12*, 217–222.

75. Mittal, A.; Naushad, M.; Sharma, G.; ALothman, Z.A.; Wabaidur, S.M.; Alam, M. Fabrication of MWCNTs/ThO$_2$ nanocomposite and its adsorption behavior for the removal of Pb(II) metal from aqueous medium. *Desalin. Water Treat. Sci. Eng.* **2016**, *57*, 21863–21869. [CrossRef]
76. Li, S.; Gong, Y.; Yang, Y.; He, C.; Hu, L.; Zhu, L.; Sun, L.; Shu, D. Recyclable CNTs/Fe$_3$O$_4$ magnetic nanocomposites as adsorbents to remove bisphenol A from water and their regeneration. *Chem. Eng. J.* **2015**, *260*, 231–239. [CrossRef]
77. Kabbashi, N.A.; Atieh, M.A.; Al-Mamun, A.; Mirghami, M.E.S.; Alam, M.D.Z.; Yahya, N. Kinetic adsorption of application of carbon nanotubes for Pb(II) removal from aqueous solution. *J. Environ. Sci.* **2009**, *2*, 539–544. [CrossRef]
78. Zhang, H.; Dang, Q.; Liu, C.; Cha, D.; Yu, Z.; Zhu, W.; Fan, B.; Zhang, H. Uptake of Pb(II) and Cd(II) on chitosan microsphere surface successively grafted by methyl acrylate and diethylenetriamine. *ACS Appl. Mater. Interfaces* **2017**, *2*, 2. [CrossRef]
79. Lee, S.M.; Lalhmunsiama, T.; Tiwari, D. Porous hybrid materials in the remediation of water contaminated with As(III) and As(V). *Chem. Eng. J.* **2015**, *270*, 496–507. [CrossRef]
80. Yin, H.; Kong, M.; Gu, X.; Chen, H. Removal of arsenic from water by porous charred granulated attapulgite-supported hydrated iron oxide in bath and column modes. *J. Clean. Product.* **2017**, *166*, 88–97. [CrossRef]
81. Tiwari, D.; Lee, S.M. Novel hybrid materials in the remediation of ground waters contaminated with As(III) and As(V). *Chem. Eng. J.* **2012**, *204–206*, 23–31. [CrossRef]
82. Ren, X.; Zhang, Z.; Luo, H.; Hu, B.; Dang, Z.; Yang, C.; Li, L. Adsorption of arsenic on modified montmorillonite. *Appl. Clay Sci.* **2014**, *97–98*, 17–23. [CrossRef]

Article

Ionic Liquids as Alternative Curing Agents for Conductive Epoxy/CNT Nanocomposites with Improved Adhesive Properties

Lidia Orduna, Itziar Otaegi, Nora Aranburu and Gonzalo Guerrica-Echevarría *

POLYMAT and Department of Advanced Polymers and Materials: Physics, Chemistry and Technology, Faculty of Chemistry, University of the Basque Country (UPV/EHU), Paseo Manuel de Lardizabal 3, 20018 Donostia-San Sebastián, Spain
* Correspondence: gonzalo.gerrika@ehu.eus

Abstract: Good dispersion of carbon nanotubes (CNTs) together with effective curing were obtained in epoxy/CNT nanocomposites (NCs) using three different ionic liquids (ILs). Compared to conventional amine-cured epoxy systems, lower electrical percolation thresholds were obtained in some of the IL-based epoxy systems. For example, the percolation threshold of the trihexyltetradecylphosphonium dicyanamide (IL-P-DCA)-based system was 0.001 wt.%. The addition of CNTs was not found to have any significant effect on the thermal or low-strain mechanical properties of the nanocomposites, but it did improve their adhesive properties considerably compared to the unfilled systems. This study demonstrates that ILs can be used to successfully replace traditional amine-based curing agents for the production of electrically conductive epoxy/CNT NCs and adhesives, as a similar or better balance of properties was achieved. This represents a step towards greater sustainability given that the vapor pressure of ILs is low, and the amount needed to effectively cure epoxy resins is significantly lower than any of their counterparts.

Keywords: epoxy resin; ionic liquid; carbon nanotube; nanocomposite; curing agent; dispersing agent; mechanical properties; lap shear; conductivity

1. Introduction

Epoxy resins are widely used due to their excellent mechanical and adhesive properties. One way to further improve these properties and acquire others—such as electrical conductivity—is by adding carbon-based nanofillers such as carbon nanotubes (CNTs). CNTs are an allotropic form of carbon made up of one (SWCNT) or multiple (MWCNT) one-atom-thick graphene sheets that are rolled into hollow cylinders with a nanometric diameter. They present outstanding mechanical and electrical properties. However, the nanofiller must be properly dispersed for the nanocomposites (NCs) to obtain a good balance of properties.

Epoxy resin/CNT NCs cured with traditional curing agents, such as amines, anhydrides, polyamides and imidazoles, have been widely studied [1–5]. It is well known that the addition of CNTs improves low-strain mechanical properties [1,2,6–10]. However, high concentrations of CNTs can cause aggregates to form [1–3], thus limiting their effectiveness as reinforcing agents. Regarding their adhesive properties, while a small number of CNTs can increase lap shear strength, larger quantities can cause more and larger aggregates to form, thus conferring a negative effect on their adhesive properties [4,5,11].

In fact, the strong propensity of CNTs to form bundles and aggregates (due to π–π stacking and Van der Waals forces of attraction) is one of their main drawbacks, as it limits their potential for improving mechanical and adhesive properties. In recent years, the use of ionic liquids (ILs) has been reported to enhance the dispersion of CNTs, thereby improving the properties of NCs based on different thermoplastic [12–14] and thermosetting [15–18] polymers as a result. This is because ILs interact with CNTs through cation–π interactions [19,20] and interrupt the π–π forces of attraction [21]. Improved dispersion levels [13,22–25] have led to both increases in electrical conductivity [12,25–27] and decreases in the percolation threshold [12,25,27].

ILs are generally defined as salts that melt at temperatures below 100 °C. They have a low vapor pressure, are ionically conductive, and are thermally and chemically resistant. Due to these properties, they are used in a wide variety of applications including as solvents in synthesis [28], electrolytes in batteries [29], and as catalysts [30]. In the field of polymer science and technology, they are also used as compatibilizers for immiscible blends [31–33], as plasticizers [34,35], as curing agents for epoxy resins [36–44] and, as previously mentioned, as dispersing agents for nanofillers.

Regarding the role of ILs as curing agents in epoxy-based systems (as we reported in previous works [45,46]), imidazolium- [38–41] and phosphonium- [36,37,42–46] based ILs are effective substitutes for traditional volatile and toxic curing agents. In addition, as they act as initiators rather than as comonomers [42], less IL is required to cure the epoxy resin effectively. Several papers have been published on the effect of ILs (as dispersing and/or curing agents) on the mechanical, electrical, and/or adhesive properties of epoxy/CNT systems [6–9,15,16,18,36,37,47,48].

Santos et al. [15] investigated the properties of epoxy/MWCNT NCs with the IL tributyl(ethyl)-phosphonium diethylphosphate used as the curing and dispersing agent. They observed that the mixing procedure significantly affected the degree of dispersion of the CNTs. In their study, the percolation threshold (p_c) was reached at 0.016 vol% when 10 phr of IL was used and at 0.047 vol% when 30 phr of IL was used. Using a different phosphonium-based IL (tri(hexyl)tetradecyl phosphonium bis(2,4,4-trimethylpentyl) phosphinate) as the dispersing and curing agent in epoxy/MWCNT NCs, Soares et al. [36] and Maka et al. [37] achieved p_c-s of between 0.25 and 0.5 phr MWCNTs [36] and below 0.25 wt% [37], respectively.

Alves et al. [16] and Lopes Pereira et al. [48] studied the role of the IL 1-butyl-3-methyl-imidazolium tetrafluoroborate as a dispersing agent in epoxy NCs containing 1 phr MWCNTs and cured with a commercial hardener. They observed that the electrical conductivity of the NC containing IL was three orders of magnitude greater than the NC without IL [48]. However, the addition of IL effected a decrease in the lap shear strength (from 20.4 MPa (the epoxy/CNT) to 14.5 MPa (the epoxy/CNT/IL)) caused by its lubricating effect [16]. Hameed et al. [9] attained individually dispersed MWCNTs using the same IL (1-butyl-3-methylimidazolium tetrafluoroborate) as the dispersing and co-curing agent (with a commercial amine). The authors reported that the diameter of the CNTs in the NCs containing the IL was larger, suggesting that the CNTs were wrapped in or covered by the IL. The addition of 0.5 wt% CNTs caused a 13% increase in Young's modulus and a 23% increase in tensile strength.

Waters et al. [8,18] and Throckmorton et al. [47] obtained epoxy/SWCNT composites using an IL (1-ethyl-3-methylimidazolium dicyanamide) as the initiator and dispersing agent. The authors reported a percolation threshold of 0.01 wt% SWCNTs for the system containing the IL, which was significantly lower than that of the corresponding traditional amine-cured system (0.4 wt%) [18,47]. Indeed, when the processing conditions were further optimized, an even lower percolation threshold (0.005 wt%) was achieved [18]. Regarding the system's mechanical properties, the addition of 0.1 wt% SWCNTs to the epoxy/IL system led to a 9% improvement in Young's modulus, which the authors attributed to the good dispersion of the nanotubes [8].

Kleinschmidt et al. [6] analyzed the effect of the IL 1-n-butyl-3-methylimidazolium bis(trifluoromethanesulfonyl)imide as the dispersing agent of an epoxy-based NC containing 0.1 wt% MWCNTs cured with two amines. They observed that while the addition of the CNTs led to a decrease in the tensile strength of the neat epoxy resin, the addition of the IL led to an improvement, which the authors attributed to the improved adhesion between the CNTs and the epoxy resin. Finally, the tensile moduli of the epoxy resin and of the epoxy/CNT and epoxy/CNT/IL nanocomposites were all similar.

Gholami et al. [7] also analyzed the dispersion efficiency of different choline chloride-based ILs in epoxy systems containing 0.3 wt% MWCNTs and cured with commercial hardeners. They observed that the addition of the glycerol and choline chloride-based ILs improved the electrical conductivity of the epoxy/CNT system by three orders of magnitude (from 10^{-8} S/cm to 10^{-5} S/cm). They attributed this to the ability of the IL to arrange the electrically conductive CNTs in the matrix and form a more complete conductive network. The enhanced dispersion of the CNTs also led to a 12% improvement in the tensile strength (from 64 MPa to 72 MPa) of the epoxy/CNT NCs.

However, while the dispersive effect of ILs in epoxy/CNT systems has been discussed quite widely in the literature, the role of ILs as effective curing agents has received less attention. Moreover, to the best of our knowledge, no studies have been conducted on the effect of different ILs and CNT concentrations on the final mechanical, electrical, and adhesive properties of epoxy NCs. Therefore, in this study, three different ionic liquids were selected (based on a previous study [46]) and used as curing/dispersing agents for epoxy/CNT NCs, with the objective of optimizing the performance of epoxy NCs without using traditional, volatile curing agents. With this aim in mind, the nanostructure and the thermal, electrical, mechanical, and adhesive properties of the NCs were determined and compared.

2. Materials and Methods

2.1. Materials

The ionic liquids used in this study were as follows: (a) trihexyltetradecylphosphonium bis(2,4,4-trimethylpentyl)phosphinate (IL-P-TMPP), (b) 1-ethyl-3-methylimidazolium dicyanamide (IL-I-DCA) from Sigma Aldrich, and (c) trihexyltetradecylphosphonium dicyanamide (IL-P-DCA) from IoLiTec-Ionic Liquid Technologies GmbH. Table 1 shows the structures and properties of all three. The epoxy resin employed was a diglycidyl ether of bisphenol A (DGEBA) (Nazza, Eurotex) (epoxy equivalent: 186 g; density (at 20 °C): 1.17 g/cm^3; viscosity (at 25 °C): 11,500–13,500 mPa·s). A traditional amine-based curing agent, 2,2'-dimethyl-4,4'-methylenebis(cyclohexylamine) (Aradur) (Huntsman), was used as a reference. The carbon nanotubes used in this work were of the Nanocyl NC7000 series (L = 1.5 μm, D = 9.5 nm, 250–300 m^2/g surface area, 90% purity) supplied by Nanocyl.

Table 1. Structures, properties, and abbreviations of the three ILs used in this study.

Abbreviation	Structure	Properties
IL-P-TMPP		Molecular weight (g/mol): 773.27 Density (20 °C) (g/cm^3): 0.895
IL-P-DCA		Molecular weight (g/mol): 549.90 Density (20 °C) (g/cm^3): 0.9
IL-I-DCA		Molecular weight (g/mol): 177.21 Density (20 °C) (g/cm^3): 1.060

2.2. Preparation of Samples

The epoxy resin was previously degassed in a vacuum oven at 80 °C for one hour. For the reference unfilled epoxy/IL and epoxy/Aradur systems, 10 phr of IL (selected as the optimum content based on our previous work [46]) and the stoichiometric concentration of the amine, respectively, were added to the epoxy resin, and the resulting mixtures were mechanically mixed at 50 °C using a Heidolph RZR2000 digital rod stirrer until completely homogeneous mixtures were obtained. They were then poured into the corresponding molds or between substrates and were cured according to the curing protocols shown in Table 2.

Table 2. Curing protocol used for the epoxy/IL and the epoxy/Aradur systems.

Curing Agent	Curing Protocol
IL-P-TMPP	2 h 80 °C/2 h 120 °C/1 h 150 °C/1 h 170 °C
IL-P-DCA	2 h 120 °C/2 h 140 °C/1 h 170 °C
IL-I-DCA	2 h 110 °C/1 h 140 °C/1 h 170 °C
Aradur	2 h 80 °C/2 h 120 °C/1 h 170 °C/1 h 200 °C

For the epoxy/CNT/IL systems, after degassing the DGEBA, the CNTs were added at concentrations ranging from 0 to 0.25% and mechanically mixed at 2000 rpm for 20 min using a EUROSTAR power control-visc digital stirrer. They were then ultrasonicated in a Hielscher UP400s at an amplitude of 100% for 20 min. Next, the IL (10 phr) was added and mechanically mixed at 50 °C for 5 min using a Heidolph RZR2000 digital rod stirrer. Finally, the mixture was poured into molds or placed between lap shear test substrates and the appropriate curing protocol was applied (Table 2). Figure 1 shows the experimental flowchart.

Figure 1. Experimental flowchart illustrating the steps used in the preparation of the samples for the different systems in the study.

2.3. Characterization

2.3.1. Thermal Properties

Dynamic mechanical analysis (DMA) was used to determine the glass transition temperature and to calculate the crosslinking density of the systems. Rectangular, nominally sized specimens measuring 17.5 × 6.0 × 2.0 mm^3 were tested using a TA Q800 viscoelastometer in single-cantilever bending mode, with a frequency of 1 Hz and an amplitude of 15 μm. The heating rate was set at 4 °C/min and the temperature interval ranged from −100 °C to 250 °C. One specimen was tested per composition. The elasticity theory (Equation (1)) was used to calculate the crosslinking density (v_e):

$$v_e = \frac{E_r}{3RT_r} \qquad (1)$$

where E_r is the storage modulus in the rubbery state at a reference temperature (T_r = 245 °C) and R is the ideal gas constant (8.314 J/mol·K).

2.3.2. Nanostructure

The nanostructure and the dispersion level of the CNTs were analyzed by transmission electron microscopy (TEM). A Tecnai G2 20 Twin microscope was used at an accelerating voltage of 200 kV. The samples were cut at a 45° angle using a Leica EM UCG ultramicrotome with a diamond blade.

2.3.3. Electrical Properties

The electrical conductivity through the sample was measured using a digital Keithley 6487 picoammeter on circular samples (Ø70 mm × 2 mm thick), to which 1 V was applied for 1 min. The electrical conductivity (σ) was calculated using Equation (2):

$$\sigma = \frac{1}{\rho} = \frac{thickness\ (\text{cm}) \times I}{22.9 \times V} \quad (\text{S/cm}) \tag{2}$$

where ρ is the resistivity, V is the voltage applied, I is the intensity of the current, and 22.9 is the geometrical constant (specific area between electrodes). Three measurements were made for each reported value.

The percolation threshold was calculated using Equation (3):

$$\sigma(p) = B(p - p_c)^t \tag{3}$$

where $\sigma(p)$ is the conductivity, B is a constant, t is the critical exponent, p is the nanofiller concentration, and p_c is the percolation concentration. The experimental data were fitted by plotting $\log(\sigma)$ vs. $\log(p - p_c)$ and increasing p_c until the best linear fit was obtained.

2.3.4. Mechanical Properties

Three-point bending tests were carried out in an Instron 5569 universal testing machine. The span was set at 64 mm with a crosshead speed of 2 mm/min. At least 5 specimens were tested for each composition.

The flexural modulus (E_f), the flexural strength (σ_F), and the deformation at break (ε_F) were calculated according to the ISO 178 standard, using Equations (4), (5) and (6), respectively:

$$E_f = \frac{FL^3}{4sbh^3} \tag{4}$$

$$\sigma_F = \frac{3F_{max}L}{2bh^2} \tag{5}$$

$$\varepsilon_F\ (\%) = \frac{6sh}{L^2} \times 100 \tag{6}$$

where F is the load, F_{max} is the maximum load, L is the span, s is the degree of deflection, and b and h are the width and thickness of the specimen, respectively.

Notched Charpy impact tests were carried out using a Ceast 6548/000 pendulum with an impactor of 2 J. Notched specimens with a radius of 0.25 mm and a depth of 2.54 mm were used. At least 8 specimens were tested for each composition.

2.3.5. Adhesive Properties

The adhesive properties were studied by performing lap shear strength tests according to the ASTM D-1002 standard. The substrates—aluminum 2021-T351 alloy sheets measuring 100 mm × 25 mm × 1.6 mm—were purchased from Rocholl GmbH, and the tested adhesion area measured 12.5 mm × 25 mm. An Instron 5569 universal testing machine in tensile mode was used, employing a constant speed of 1 mm/min. The lap shear strength was calculated by dividing the maximum force by the adhesion area. For each reported value, 10 specimens were tested.

3. Results and Discussion
3.1. Thermal Properties

Figure 2 shows the tanδ and the storage modulus vs temperature curves of the samples cured with the different ILs at different CNT concentrations. The main peak of the tanδ curves, which appears at high temperatures (shown in the inset), indicates the glass transition temperature (T_g). The maximums of the peaks for the IL-based systems, along with the calculated crosslinking density values, are summarized in Table 3.

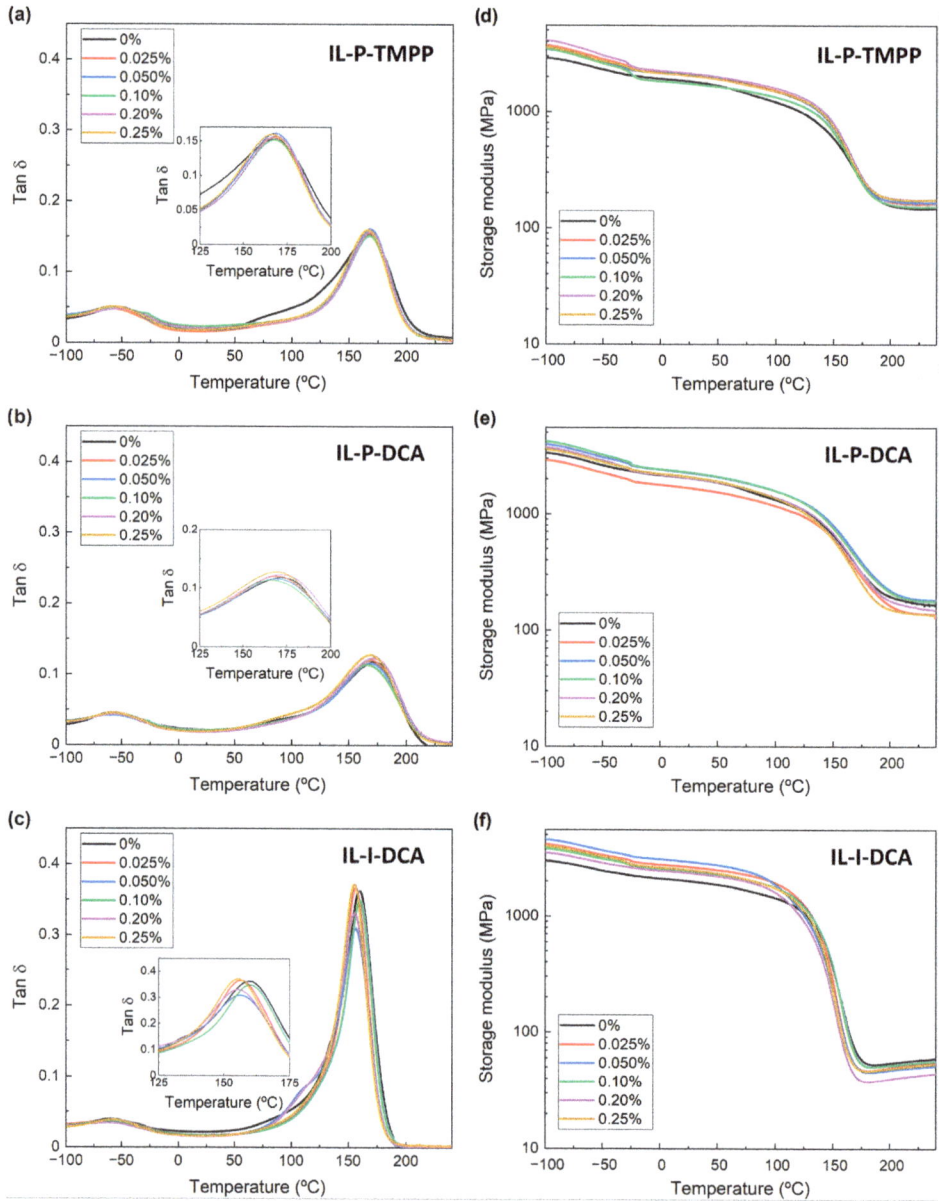

Figure 2. Tan δ (**a**–**c**) and storage modulus (**d**–**f**) obtained by DMTA for IL-P-TMPP (**a**,**d**), IL-P-DCA (**b**,**e**), and IL-I-DCA (**c**,**f**) epoxy/IL NCs at different CNT fractions.

Table 3. T_g and crosslinking density values for the epoxy/IL NCs in the study at different CNT fractions. The T_g of the reference epoxy/Aradur system was 189 °C and the crosslinking density was 2790 mol/m^3.

CNT wt.%	T_g (°C)	ν_e (mol/m^3)	CNT wt.%	T_g (°C)	ν_e (mol/m^3)	CNT wt.%	T_g (°C)	ν_e (mol/m^3)
IL-P-TMPP			IL-P-DCA			IL-I-DCA		
0	168	11509	0	172	12616	0	160	4625
0.025	168	12610	0.025	169	10,828 *	0.025	157	4310
0.05	169	12660	0.05	169	14,146 *	0.05	156	4015
0.1	168	11847	0.1	166	13,669	0.1	160	4398
0.2	169	13594	0.2	173	11,437	0.2	155	3599
0.25	167	13590	0.25	170	10,448	0.25	155	4158

* Calculated at lower T_r due to the instability of the curve at 245 °C.

As Figure 2 and Table 3 show, neither the glass transition temperature nor the crosslinking density changed significantly upon the addition of the CNTs. In fact, both higher [49–51] and lower [52–54] T_g values have been reported when CNTs were added to epoxy resins, so there is no general consensus in the literature regarding the impact of CNTs on the T_g of conventionally cured epoxy/CNT systems. Decreases in T_g are generally attributed to the steric hindrance of the CNTs towards the curing reaction [54,55], which gives rise to samples with less cross-linkage. By contrast, increases in T_g are usually linked to decreases in the mobility of the macromolecular chains caused by the presence of CNTs. This effect is more pronounced the greater the degree of dispersion or adhesion, i.e., the two parameters reported to most heavily impact the T_g [51]. Regarding amine-cured epoxy systems with ILs as dispersing agents, both increases [9] and decreases [17,48,56] in the T_g have been reported. The effect of the presence of ILs is not completely clear in these cases because they are also known to act as effective plasticizers. Both decreases [36] and increases [15,37] in T_g have also been reported for epoxy/CNT NCs where the ILs acted as both curing and dispersing agents. The behavior of the epoxy/Aradur system in this study was similar to that of the epoxy/IL systems; the addition of CNTs did not significantly affect the T_g or the crosslinking density of the unfilled system (the T_g and crosslinking density of the 0.2 wt.% CNT composition were 191 °C and 3090 mol/m^3, respectively).

3.2. Nanostructure

The nanostructures of the epoxy/IL systems were analyzed by transmission electron microscopy (TEM). Figure 3a–c show representative micrographs of the IL-P-DCA, IL-P-TMPP, and IL-I-DCA epoxy/IL systems, respectively, each containing 0.2 wt.% of CNTs. As a reference, Figure 3d shows a representative TEM micrograph of the epoxy/Aradur NC with the same concentration of CNTs. As can be seen in Figure 3a–c, good dispersion of the nanofiller was achieved in the case of all three ILs, with mostly individually dispersed CNTs and far fewer small aggregates. When the different ILs are compared, it is evident that, in Figure 3a, there are fewer and smaller CNT aggregates than those appearing in Figure 3b,c, indicating that the CNTs were best dispersed in the epoxy/CNT/IL-P-DCA system, and more poorly dispersed in the epoxy/CNT/IL-P-TMPP and epoxy/CNT/IL-I-DCA systems. A similar nanostructure was observed in the reference epoxy/CNT/Aradur system (shown here in Figure 3d). As expected, these differences significantly affected the electrical properties of the NCs and are discussed below.

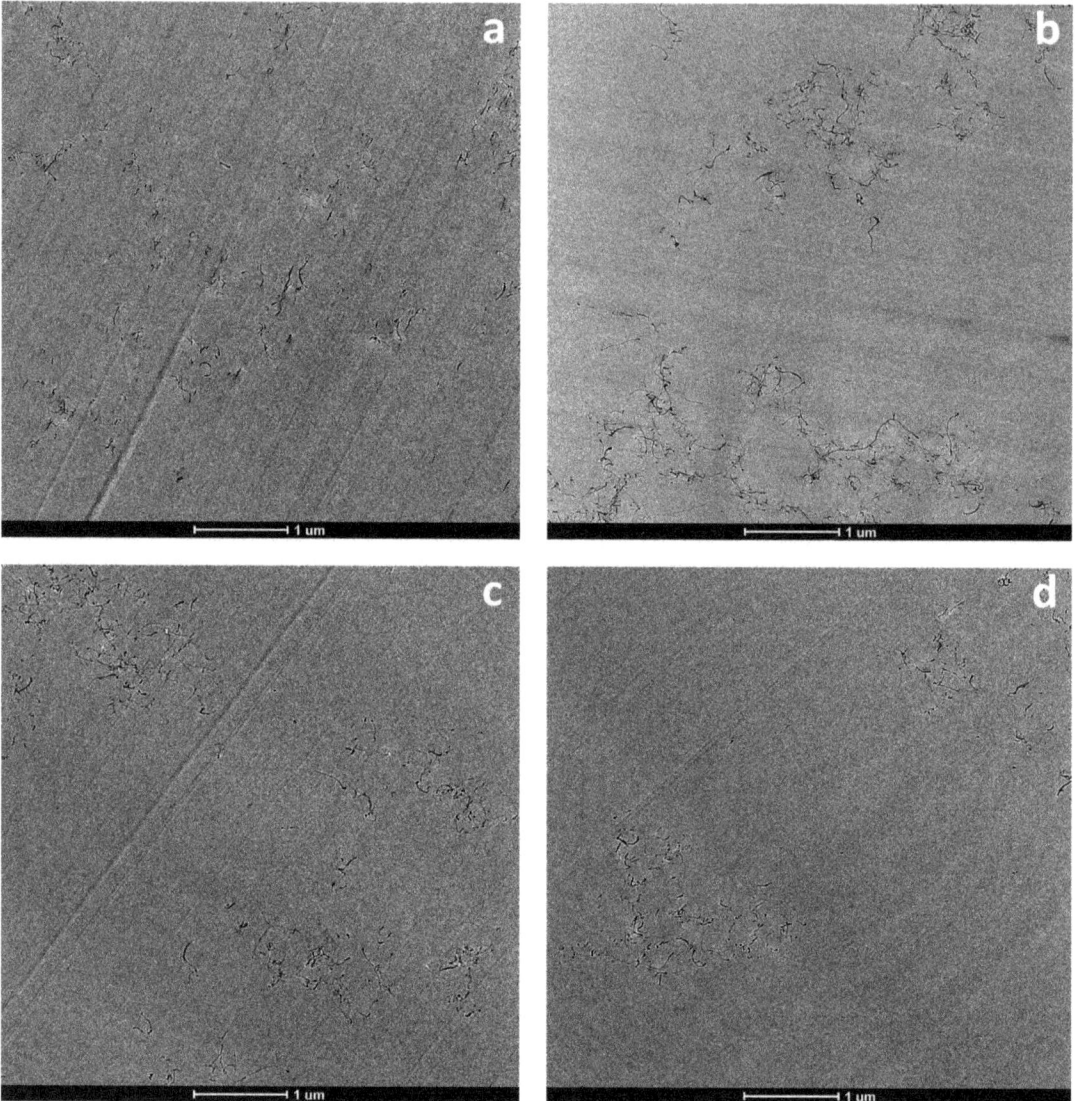

Figure 3. TEM micrographs of the epoxy/IL-P-DCA (**a**), epoxy/IL-P-TMPP (**b**), epoxy/IL-I-DCA (**c**), and the reference epoxy/Aradur systems containing 0.2 wt.% CNT (**d**).

The achievement of good dispersion of CNTs is known to be complicated due to interactions between the sp^2 orbitals perpendicular to the layers, which lead to the reaggregation of the CNTs. ILs have been reported to be effective in terms of dispersing CNTs, although how they actually execute this is a controversial topic judging from the bibliography contained herein. In the case of imidazolium-based ILs, some authors indicate that the π–cation interaction [19,20] is likely responsible, while others believe the Van der Waals forces-based interactions [21] cause the rupture of the π–π attractive forces between the CNTs. With respect to the phosphonium-based ILs, in the absence of aromatic groups in their structures, π–cation interactions are preferred [57]. Moreover, as these ILs contain cations with long aliphatic chains, a surfactant effect has also been suggested as being responsible for the dispersive effect [15,36,37].

3.3. Electrical Properties

Figure 4 shows the electrical conductivity vs the CNT content of the different epoxy/IL systems. It can be clearly seen that the addition of CNTs led to a dramatic increase in conductivity in all the epoxy/IL systems. For example, the conductivity values of the 0.2 wt.% CNT composition were 1.13×10^{-7}, 5.67×10^{-8}, 1.96×10^{-9}, and 9.99×10^{-9} S/cm for the IL-P-TMPP, IL-P-DCA, IL-I-DCA, and Aradur-based systems, respectively. The percolation thresholds are shown in Table 4, and the fitting parameters B and t (Equation (3)) along with the conductivity values corresponding to the p_c are presented in Table S1. As can be seen, the percolation threshold was reached at very low CNT fractions—0.025 wt.% or lower—in all cases. The IL-P-DCA-cured epoxy NC had the lowest p_c at 0.001 wt.%. This is consistent with the best dispersion levels on the TEM micrographs. Table 4 also shows how the similar CNT dispersion levels of the other two IL-based systems (and for the reference amine-based one) also resulted in similar percolation threshold values. Thus, at low CNT shares, tiny changes in the degree of dispersion can affect the percolation threshold, i.e., the minimum concentration required to create the conductive path necessary for the transition from an insulator to semiconductor to take place.

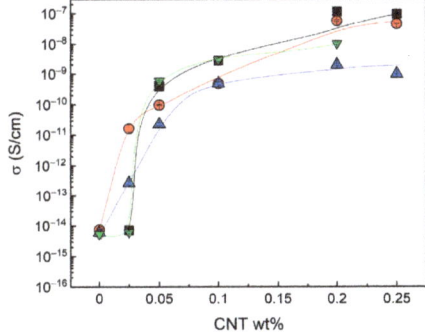

Figure 4. Electrical conductivity of IL-P-TMPP (■), IL-P-DCA (●), IL-I-DCA (▲), and the reference Aradur (▼) epoxy/IL NCs as a function of the CNT concentration.

Table 4. Percolation thresholds of the epoxy/IL and epoxy/Aradur NCs.

Curing Agent	p_c (CNT wt.%)
IL-P-TMPP	0.025
IL-P-DCA	0.001
IL-I-DCA	0.02
Aradur	0.025

At high CNT shares, the conductivity values of the systems cured with the phosphonium-based ILs (i.e., IL-P-TMPP and IL-P-DCA) were two orders of magnitude higher than those cured with the imidazolium-based IL (i.e., IL-I-DCA), suggesting that the long chains in the structures of the former two are more effective at preventing the formation of CNT aggregates.

In the literature, a wide range of percolation threshold concentrations have been reported for epoxy/CNT NCs. This variation is probably due to the myriad factors that affect the formation of percolated networks, namely, the aspect ratio of the CNTs. Usually, the reported p_c values do not exceed 1–2 wt.% [58], with most ranging from 0.1 to 1 wt.% [59–61]. However, figures as low as 0.04 wt.% [62,63] or 0.001 wt.% [64] have also been reported. Lopes Pereira et al. [48] studied the effect of adding ILs as nanofiller-dispersing agents and reported an increase of three orders of magnitude in electrical conductivity compared to the non-IL system. Regarding the conductivity of the Nanocyl NC7000 CNT-based percolated systems, a wide range of values (from 10^{-7} to 10^{-2} S/cm) can be found in our bibliography depending on the experimental technique or conditions used [15,36,37].

3.4. Mechanical Properties

Figure 5 shows the flexural modulus, flexural strength, and deformation at break of the epoxy/IL NCs as a function of the CNT content. The data for the reference unfilled epoxy/Aradur system have also been included (the green, solid line shows the average value, while the shaded area shows the standard deviation).

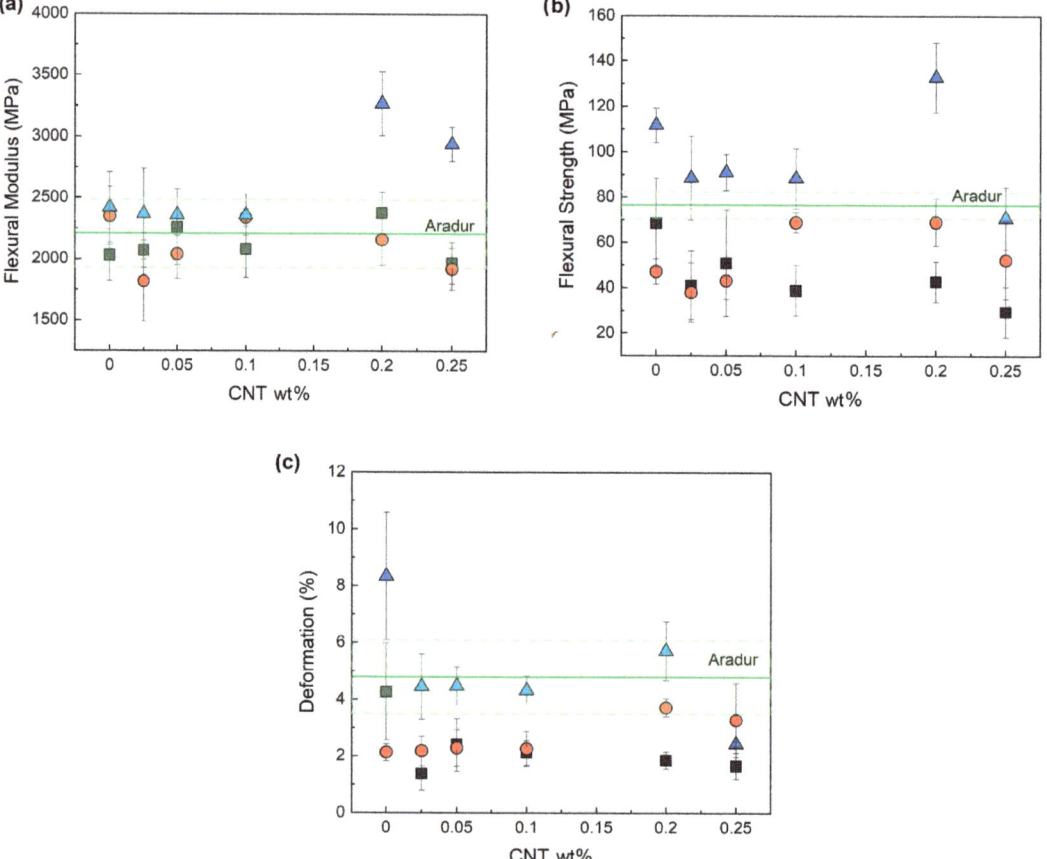

Figure 5. Flexural modulus (**a**), flexural strength (**b**), and deformation at break (**c**) of the epoxy/CNT systems cured with ILs (IL-P-TMPP (■), IL-P-DCA (●), and IL-I-DCA (▲)). The values of the unfilled epoxy resin cured with an amine-based curing agent (Aradur) are also shown as a reference (the solid green line marks the average value, and the shaded area shows the standard deviation).

As Figure 5a shows, overall, the flexural modulus of the epoxy resin was barely affected by the addition of the CNTs. Slight increases were only observed for the IL-I-DCA-based system at high CNT shares. These results suggest that neither the crosslinking density (Table 3) nor the degree of dispersion of the CNTs (Figure 3a–c) in these systems are directly related to their mechanical properties. Similar results were obtained in the reference epoxy/Aradur system, given that the addition of 0.2 wt.% CNTs did not lead to significant changes in the flexural modulus (2270 MPa) or in flexural strength (56.0 MPa). This is further discussed below.

No significant trends in flexural strength were observed in any of the systems (Figure 5b). As with the flexural modulus, the IL-I-DCA-cured samples with a high concentration of CNTs scored highest. It is worth noting that the flexural modulus and strength data of some of the compositions are similar or better than those of the amine-cured reference system.

With respect to high-strain mechanical properties, the deformation at break (Figure 5c) and the impact strength (Figure 6) both decreased when CNTs were added. This was to be expected because nanofillers in general—and CNTs in particular—are known to act as crack initiators and/or stress-concentration points, thus leading to decreases in ductility and toughness [1].

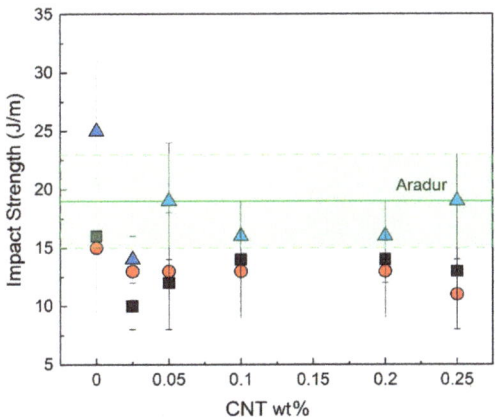

Figure 6. Impact strength of the epoxy/CNT systems cured with ILs (IL-P-TMPP (■), IL-P-DCA (●), and IL-I-DCA (▲)). The value of the unfilled epoxy resin cured with an amine-based curing agent (Aradur) is also shown as a reference (green solid line for the average value and shaded area for the standard deviation).

Epoxy resins that have been mechanically improved by adding CNTs have been extensively documented in the literature [1,2]. In amine-cured epoxy/CNT NCs with ILs as dispersing agents, the mechanical properties were further enhanced due to the enhanced levels of dispersion [6,7,9,10]. Nevertheless, the fact that ILs can also act as plasticizers also needs to be taken into account. Accordingly, depending on the amount of IL used, the plasticizing effect may be predominant, causing decreases in low-strain mechanical properties [7].

Few papers on the roles of ILs as both curing and dispersing agents in epoxy systems have examined their mechanical behavior. For those that have, organoclays [65], silica [47], core–shell particles [43], and CNTs [8] were used as nanofillers. In the epoxy/IL/CNT system [8], for example, the authors reported that Young's modulus increased with the CNT content with respect to the unfilled epoxy/IL-I-DCA system. However, this increase was not significant above 1 wt.% CNTs, which is probably due to the deficient dispersion associated with an excessive number of CNTs.

3.5. Adhesive Properties

Figure 7 shows the lap shear strength of the epoxy/IL NCs as a function of their CNT content. The corresponding result for the unfilled epoxy/Aradur system is also shown as a reference. As can be seen, regardless of the IL used, the addition of low concentrations of CNTs led to increases in the lap shear strength compared to the unfilled compositions. However, higher concentrations of CNTs did not lead to higher lap shear strength values. A maximum improvement in adhesive properties (30%) was attained with the IL-P-DCA-cured epoxy system containing 0.025 wt.% CNTs. Similar increases—close to 30% (from 6.8 to 8.8 MPa)—were also observed when 0.2% wt.% CNTs were added to the reference epoxy/Aradur system.

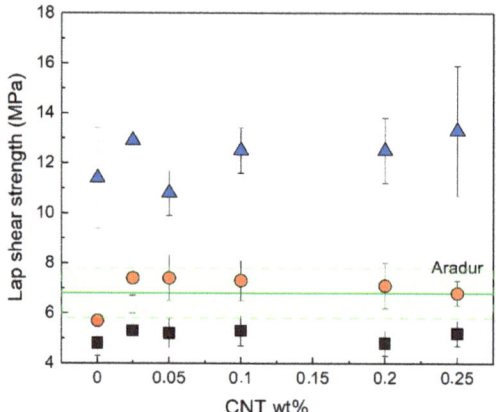

Figure 7. Lap shear strength of the epoxy/CNT systems cured with ILs (IL-P-TMPP (■), IL-P-DCA (●), and IL-I-DCA (▲)). The lap shear strength of the unfilled epoxy resin cured with an amine-based curing agent (Aradur) is also shown as a reference (the green, solid line marks the average value, and shaded area shows the standard deviation).

Our findings are consistent with the results reported in the literature for epoxy/CNT systems [4,5,11]. In an epoxy/CNT/amine system, Alves et al. [16] reported that the improvement in the dispersion of CNTs effected by a masterbatch led to enhanced adhesive properties. However, they also observed—in the same study—that even though the dispersion of the CNTs was enhanced when an IL was added, the lap shear strength decreased by over 50% due to the lubricating effect of the IL [13,66].

Considering the results in Figure 7, it is noteworthy that, regardless of the CNT content, all the epoxy/IL-I-DCA NCs showed significantly enhanced adhesive properties compared to both the reference unfilled epoxy/Aradur system and the CNT-filled epoxy/Aradur composition. So, in summary, electrically conductive epoxy adhesives with outstanding adhesive properties were obtained when ILs were used as both the curing and dispersing agents.

4. Conclusions

The dual role of ionic liquids (ILs) as effective curing and dispersing agents produced volatile-amine-free epoxy/CNT nanocomposites with a better balance of mechanical, electrical, and adhesive properties. Three different ILs were tested, and all three led to good dispersion of the nanofiller, featuring individually dispersed CNTs as well as some small aggregates. Overall, with a percolation threshold of 0.001 wt.%, the IL-P-DCA system was the most effective.

The addition of CNTs had no effect on the thermal or low-strain mechanical properties of the epoxy/IL systems. However, it did improve the systems' adhesive properties. The epoxy/IL-P-DCA system containing 0.025 wt.% CNTs improved by 30% and was the best of the three. This work proves that, using very small amounts of CNTs, it is possible to obtain electrically conductive, amine-free epoxy adhesives with similar mechanical properties but greater lap shear strength than the reference amine-cured epoxy system. ILs have a lower vapor pressure, and a significantly lower amount is needed to effectively cure epoxy resins. Therefore, replacing conventional epoxy resin curing agents (amines, anhydrides, etc.) with ILs is a major step forward in the development of more sustainable materials.

Supplementary Materials: The following are available online at https://www.mdpi.com/article/10.3390/nano13040725/s1, Table S1: Fitting parameters (B and t) and conductivity values corresponding to the p_c of the epoxy/CNT/IL NCs cured with IL-P-TMPP, IL-P-DCA and IL-I-DCA and of the reference epoxy/CNT/Aradur systems.

Author Contributions: Research, L.O., I.O., N.A. and G.G.-E.; writing—original draft preparation, L.O., I.O., N.A. and G.G.-E.; writing—review and editing, L.O., I.O., N.A. and G.G.-E. All authors have read and agreed to the published version of the manuscript.

Funding: This research was funded by Ministerio de Economía y Competitividad (MINECO, Spain) (MAT2017-84116-R), Basque Government (Eusko Jaurlaritza) (Project IT1309-19) and the University of the Basque Country (UPV/EHU) (pre-doctoral grant awarded to Lidia Orduna).

Acknowledgments: Lidia Orduna acknowledges the grant awarded by the University of the Basque Country (UPV/EHU).

Conflicts of Interest: The authors declare no conflict of interest.

References

1. Singh, N.P.; Gupta, V.; Singh, A.P. Graphene and carbon nanotube reinforced epoxy nanocomposites: A review. *Polymer* **2019**, *180*, 121724. [CrossRef]
2. Martin-Gallego, M.; Yuste-Sanchez, V.; Sanchez-Hidalgo, R.; Verdejo, R.; Lopez-Manchado, M. Epoxy Nanocomposites Filled with Carbon Nanoparticles. *Chem. Rec.* **2018**, *18*, 928–939. [CrossRef]
3. Shukla, M.K.; Sharma, K. Effect of Carbon Nanofillers on the Mechanical and Interfacial Properties of Epoxy Based Nanocomposites: A Review. *Polym. Sci. Ser. A* **2019**, *61*, 439–460. [CrossRef]
4. Jojibabu, P.; Jagannatham, M.; Haridoss, P.; Ram, G.J.; Deshpande, A.P.; Bakshi, S.R. Effect of different carbon nano-fillers on rheological properties and lap shear strength of epoxy adhesive joints. *Compos. Part A Appl. Sci. Manuf.* **2016**, *82*, 53–64. [CrossRef]
5. Kumar, A.; Kumar, K.; Ghosh, P.; Rathi, A.; Yadav, K. Raman MWCNTs toward superior strength of epoxy adhesive joint on mild steel adherent. *Compos. Part B Eng.* **2018**, *143*, 207–216. [CrossRef]
6. Kleinschmidt, A.C.; Almeida, J.H.S.; Donato, R.K.; Schrekker, H.S.; Marques, V.C.; Corat, E.J.; Amico, S.C. Functionalized-Carbon Nanotubes with Physisorbed Ionic Liquid as Filler for Epoxy Nanocomposites. *J. Nanosci. Nanotechnol.* **2016**, *16*, 9132–9140. [CrossRef]
7. Gholami, H.; Arab, H.; Mokhtarifar, M.; Maghrebi, M.; Baniadam, M. The effect of choline-based ionic liquid on CNTs' arrangement in epoxy resin matrix. *Mater. Des.* **2016**, *91*, 180–185. [CrossRef]
8. Watters, A.; Cuadra, J.; Kontsos, A.; Palmese, G. Processing-structure–property relationships of SWNT–epoxy composites prepared using ionic liquids. *Compos. Part A Appl. Sci. Manuf.* **2015**, *73*, 269–276. [CrossRef]
9. Hameed, N.; Salim, N.V.; Hanley, T.L.; Sona, M.; Fox, B.L.; Guo, Q. Individual dispersion of carbon nanotubes in epoxy via a novel dispersion–curing approach using ionic liquids. *Phys. Chem. Chem. Phys.* **2013**, *15*, 11696–11703. [CrossRef] [PubMed]
10. Chen, C.; Li, X.; Wen, Y.; Liu, J.; Li, X.; Zeng, H.; Xue, Z.; Zhou, X.; Xie, X. Noncovalent engineering of carbon nanotube surface by imidazolium ionic liquids: A promising strategy for enhancing thermal conductivity of epoxy composites. *Compos. Part A: Appl. Sci. Manuf.* **2019**, *125*, 105517. [CrossRef]
11. Srivastava, V. Effect of carbon nanotubes on the strength of adhesive lap joints of C/C and C/C–SiC ceramic fibre composites. *Int. J. Adhes. Adhes.* **2011**, *31*, 486–489. [CrossRef]
12. Zhao, L.; Li, Y.; Cao, X.; You, J.; Dong, W. Multifunctional role of an ionic liquid in melt-blended poly(methyl methacrylate)/multi-walled carbon nanotube nanocomposites. *Nanotechnology* **2012**, *23*, 255702. [CrossRef] [PubMed]
13. Carrión, F.; Espejo, C.; Sanes, J.; Bermúdez, M. Single-walled carbon nanotubes modified by ionic liquid as antiwear additives of thermoplastics. *Compos. Sci. Technol.* **2010**, *70*, 2160–2167. [CrossRef]
14. Ahmad, A.; Mahmood, H.; Mansor, N.; Iqbal, T.; Moniruzzaman, M. Ionic liquid assisted polyetheretherketone-multiwalled carbon nanotubes nanocomposites: An environmentally friendly approach. *J. Appl. Polym. Sci.* **2020**, *138*, 50159. [CrossRef]
15. Santos, D.F.; Carvalho, A.P.A.; Soares, B.G. Phosphonium-based ionic liquid as crosslinker/dispersing agent for epoxy/carbon nanotube nanocomposites: Electrical and dynamic mechanical properties. *J. Mater. Sci.* **2019**, *55*, 2077–2089. [CrossRef]
16. Alves, F.F.; Silva, A.A.; Soares, B.G. Epoxy-MWCNT composites prepared from master batch and powder dilution: Effect of ionic liquid on dispersion and multifunctional properties. *Polym. Eng. Sci.* **2017**, *58*, 1689–1697. [CrossRef]
17. Sanes, J.; Saurín, N.; Carrión, F.; Ojados, G.; Bermúdez, M. Synergy between single-walled carbon nanotubes and ionic liquid in epoxy resin nanocomposites. *Compos. Part B Eng.* **2016**, *105*, 149–159. [CrossRef]
18. Watters, A.L.; Palmese, G.R. Ultralow percolation threshold of single walled carbon nanotube-epoxy composites synthesized via an ionic liquid dispersant/initiator. *Mater. Res. Express* **2014**, *1*, 035013. [CrossRef]
19. Fukushima, T.; Kosaka, A.; Ishimura, Y.; Yamamoto, T.; Takigawa, T.; Ishii, N.; Aida, T. Molecular Ordering of Organic Molten Salts Triggered by Single-Walled Carbon Nanotubes. *Science* **2003**, *300*, 2072–2074. [CrossRef]

20. Kim, H.; Choi, J.; Lim, S.; Choi, H. Preparation and nanoscopic internal structure of single-walled carbon nanotube-ionic liquid gel. *Synth. Met.* **2005**, *154*, 189–192. [CrossRef]
21. Wang, J.; Chu, H.; Li, Y. Why Single-Walled Carbon Nanotubes Can Be Dispersed in Imidazolium-Based Ionic Liquids. *ACS Nano* **2008**, *2*, 2540–2546. [CrossRef]
22. Ke, K.; Pötschke, P.; Gao, S.; Voit, B. An Ionic Liquid as Interface Linker for Tuning Piezoresistive Sensitivity and Toughness in Poly(vinylidene fluoride)/Carbon Nanotube Composites. *ACS Appl. Mater. Interfaces* **2017**, *9*, 5437–5446. [CrossRef] [PubMed]
23. Pereira, E.C.L.; Soares, B.G.; Silva, A.A.; da Silva, J.M.F.; Barra, G.M.; Livi, S. Conductive heterogeneous blend composites of PP/PA12 filled with ionic liquids treated-CNT. *Polym. Test.* **2019**, *74*, 187–195. [CrossRef]
24. Wang, P.; Zhou, Y.; Hu, X.; Wang, F.; Chen, J.; Xu, P.; Ding, Y. Improved mechanical and dielectric properties of PLA/EMA-GMA nanocomposites based on ionic liquids and MWCNTs. *Compos. Sci. Technol.* **2020**, *200*, 108347. [CrossRef]
25. Soares da Silva, J.P.; Soares, B.G.; Livi, S.; Barra, G.M.O. Phosphonium-based ionic liquid as dispersing agent for MWCNT in melt-mixing polystyrene blends: Rheology, electrical properties and EMI shielding effectiveness. *Mater. Chem. Phys.* **2017**, *189*, 162–168. [CrossRef]
26. Xing, C.; Zhao, L.; You, J.; Dong, W.; Cao, X.; Li, Y. Impact of Ionic Liquid-Modified Multiwalled Carbon Nanotubes on the Crystallization Behavior of Poly(vinylidene fluoride). *J. Phys. Chem. B* **2012**, *116*, 8312–8320. [CrossRef] [PubMed]
27. Soares, B.G.; Calheiros, L.F.; Silva, A.A.; Indrusiak, T.; Barra, G.M.O.; Livi, S. Conducting melt blending of polystyrene and EVA copolymer with carbon nanotube assisted by phosphonium-based ionic liquid. *J. Appl. Polym. Sci.* **2017**, *135*, 45564. [CrossRef]
28. Hallett, J.P.; Welton, T. Room-Temperature Ionic Liquids: Solvents for Synthesis and Catalysis. 2. *Chem. Rev.* **2011**, *111*, 3508–3576. [CrossRef]
29. Wishart, J.F. Energy applications of ionic liquids. *Energy Environ. Sci.* **2009**, *2*, 956–961. [CrossRef]
30. Lee, J.K.; Kim, M.-J. Ionic Liquid-Coated Enzyme for Biocatalysis in Organic Solvent. *J. Org. Chem.* **2002**, *67*, 6845–6847. [CrossRef]
31. Lins, L.C.; Livi, S.; Duchet-Rumeau, J.; Gérard, J.-F. Phosphonium ionic liquids as new compatibilizing agents of biopolymer blends composed of poly(butylene-adipate-co-terephtalate)/poly(lactic acid) (PBAT/PLA). *RSC Adv.* **2015**, *5*, 59082–59092. [CrossRef]
32. Pereira, E.C.L.; da Silva, J.M.F.; Jesus, R.B.; Soares, B.G.; Livi, S. Bronsted acidic ionic liquids: New transesterification agents for the compatibilization of polylactide/ethylene-co-vinyl acetate blends. *Eur. Polym. J.* **2017**, *97*, 104–111. [CrossRef]
33. Leroy, E.; Jacquet, P.; Coativy, G.; Reguerre, A.L.; Lourdin, D. Compatibilization of starch–zein melt processed blends by an ionic liquid used as plasticizer. *Carbohydr. Polym.* **2012**, *89*, 955–963. [CrossRef] [PubMed]
34. Chen, B.-K.; Wu, T.-Y.; Chang, Y.-M.; Chen, A.F. Ductile polylactic acid prepared with ionic liquids. *Chem. Eng. J.* **2012**, *215-216*, 886–893. [CrossRef]
35. Scott, M.P.; Rahman, M.; Brazel, C.S. Application of ionic liquids as low-volatility plasticizers for PMMA. *Eur. Polym. J.* **2003**, *39*, 1947–1953. [CrossRef]
36. Soares, B.G.; Riany, N.; Silva, A.A.; Barra, G.M.O.; Livi, S. Dual-role of phosphonium—Based ionic liquid in epoxy/MWCNT systems: Electric, rheological behavior and electromagnetic interference shielding effectiveness. *Eur. Polym. J.* **2016**, *84*, 77–88. [CrossRef]
37. Maka, H.; Spychaj, T.; Pilawka, R. Epoxy resin/phosphonium ionic liquid/carbon nanofiller systems: Chemorheology and properties. *Express Polym. Lett.* **2014**, *8*, 723–732. [CrossRef]
38. Yin, Y.; Liu, M.; Wei, W.; Zheng, C.; Gao, J.; Zhang, W.; Zheng, C.; Deng, P.; Xing, Y. DGEBA/imidazolium ionic liquid systems: The influence of anions on the reactivity and properties of epoxy systems. *J. Adhes. Sci. Technol.* **2017**, *32*, 1114–1127. [CrossRef]
39. Da Silva, L.C.O.; Soares, B.G. New all solid-state polymer electrolyte based on epoxy resin and ionic liquid for high temperature applications. *J. Appl. Polym. Sci.* **2017**, *135*, 45838. [CrossRef]
40. Binks, F.C.; Cavalli, G.; Henningsen, M.; Howlin, B.J.; Hamerton, I. Examining the nature of network formation during epoxy polymerisation initiated with ionic liquids. *Polymer* **2018**, *150*, 318–325. [CrossRef]
41. Binks, F.C.; Cavalli, G.; Henningsen, M.; Howlin, B.J.; Hamerton, I. Investigating the mechanism through which ionic liquids initiate the polymerisation of epoxy resins. *Polymer* **2018**, *139*, 163–176. [CrossRef]
42. Silva, A.A.; Livi, S.; Netto, D.B.; Soares, B.G.; Duchet, J.; Gérard, J.-F. New epoxy systems based on ionic liquid. *Polymer* **2013**, *54*, 2123–2129. [CrossRef]
43. Nguyen, T.K.L.; Soares, B.G.; Duchet-Rumeau, J.; Livi, S. Dual functions of ILs in the core-shell particle reinforced epoxy networks: Curing agent vs dispersion aids. *Compos. Sci. Technol.* **2017**, *140*, 30–38. [CrossRef]
44. Leclère, M.; Livi, S.; Maréchal, M.; Picard, L.; Duchet-Rumeau, J. The properties of new epoxy networks swollen with ionic liquids. *RSC Adv.* **2016**, *6*, 56193–56204. [CrossRef]
45. Orduna, L.; Razquin, I.; Otaegi, I.; Aranburu, N.; Guerrica-Echevarría, G. Ionic Liquid-Cured Epoxy/PCL Blends with Improved Toughness and Adhesive Properties. *Polymers* **2022**, *14*, 2679. [CrossRef]
46. Orduna, L.; Razquin, I.; Aranburu, N.; Guerrica-Echevarría, G. Are ionic liquids effective curing agents for preparing epoxy adhesives? *Int. J. Adhes. Adhes.* **2022**, *under review for publication*.
47. Throckmorton, J.A.; Watters, A.L.; Geng, X.; Palmese, G.R. Room temperature ionic liquids for epoxy nanocomposite synthesis: Direct dispersion and cure. *Compos. Sci. Technol.* **2013**, *86*, 38–44. [CrossRef]
48. Pereira, E.C.L.; Soares, B.G. Conducting epoxy networks modified with non-covalently functionalized multi-walled carbon nanotube with imidazolium-based ionic liquid. *J. Appl. Polym. Sci.* **2016**, *133*, 43976. [CrossRef]

49. Hameed, A.; Islam, M.; Ahmad, I.; Mahmood, N.; Saeed, S.; Javed, H. Thermal and mechanical properties of carbon nanotube/epoxy nanocomposites reinforced with pristine and functionalized multiwalled carbon nanotubes. *Polym. Compos.* **2014**, *36*, 1891–1898. [CrossRef]
50. Gkikas, G.; Barkoula, N.-M.; Paipetis, A. Effect of dispersion conditions on the thermo-mechanical and toughness properties of multi walled carbon nanotubes-reinforced epoxy. *Compos. Part B Eng.* **2012**, *43*, 2697–2705. [CrossRef]
51. Gkikas, G.; Douka, D.-D.; Barkoula, N.-M.; Paipetis, A. Nano-enhanced composite materials under thermal shock and environmental degradation: A durability study. *Compos. Part B Eng.* **2015**, *70*, 206–214. [CrossRef]
52. Suave, J.; Coelho, L.A.; Amico, S.C.; Pezzin, S.H. Effect of sonication on thermo-mechanical properties of epoxy nanocomposites with carboxylated-SWNT. *Mater. Sci. Eng. A* **2009**, *509*, 57–62. [CrossRef]
53. Liao, Y.-H.; Marietta-Tondin, O.; Liang, Z.; Zhang, C.; Wang, B. Investigation of the dispersion process of SWNTs/SC-15 epoxy resin nanocomposites. *Mater. Sci. Eng. A* **2004**, *385*, 175–181. [CrossRef]
54. Rathore, D.K.; Prusty, R.K.; Ray, B.C. Mechanical, thermomechanical, and creep performance of CNT embedded epoxy at elevated temperatures: An emphasis on the role of carboxyl functionalization. *J. Appl. Polym. Sci.* **2017**, *134*, 44851. [CrossRef]
55. Abdalla, M.; Dean, D.; Robinson, P.; Nyairo, E. Cure behavior of epoxy/MWCNT nanocomposites: The effect of nanotube surface modification. *Polymer* **2008**, *49*, 3310–3317. [CrossRef]
56. Zheng, X.; Li, D.; Feng, C.; Chen, X. Thermal properties and non-isothermal curing kinetics of carbon nanotubes/ionic liquid/epoxy resin systems. *Thermochim. Acta* **2015**, *618*, 18–25. [CrossRef]
57. Soares, B.G. Ionic liquid: A smart approach for developing conducting polymer composites: A review. *J. Mol. Liq.* **2018**, *262*, 8–18. [CrossRef]
58. Pecastaings, G.; Delhaes, P.; Derre, A.; Saadaoui, H.; Carmona, F.; Cui, S. Role of interfacial effects in carbon nanotube/epoxy nanocomposite behavior. *J. Nanosci. Nanotechnol.* **2004**, *4*, 838–843. [CrossRef]
59. Li, J.; Ma, P.C.; Chow, W.S.; To, C.K.; Tang, B.Z.; Kim, J.-K. Correlations between percolation threshold, dispersion state, and aspect ratio of carbon nanotubes. *Adv. Funct. Mater.* **2007**, *17*, 3207–3215. [CrossRef]
60. Tang, L.-C.; Wan, Y.-J.; Peng, K.; Pei, Y.-B.; Wu, L.-B.; Chen, L.-M.; Shu, L.-J.; Jiang, J.-X.; Lai, G.-Q. Fracture toughness and electrical conductivity of epoxy composites filled with carbon nanotubes and spherical particles. *Compos. Part A Appl. Sci. Manuf.* **2013**, *45*, 95–101. [CrossRef]
61. Thostenson, E.T.; Chou, T.-W. Carbon Nanotube Networks: Sensing of Distributed Strain and Damage for Life Prediction and Self Healing. *Adv. Mater.* **2006**, *18*, 2837–2841. [CrossRef]
62. Chapartegui, M.; Markaide, N.; Florez, S.; Elizetxea, C.; Fernandez, M.; Santamaria, A. Curing of epoxy/carbon nanotubes physical networks. *Polym. Eng. Sci.* **2011**, *52*, 663–670. [CrossRef]
63. Sandler, J.; Shaffer, M.; Prasse, T.; Bauhofer, W.; Schulte, K.; Windle, A. Development of a dispersion process for carbon nanotubes in an epoxy matrix and the resulting electrical properties. *Polymer* **1999**, *40*, 5967–5971. [CrossRef]
64. Sandler, J.K.W.; Kirk, J.E.; Kinloch, I.A.; Shaffer, M.S.P.; Windle, A.H. Ultra-low electrical percolation threshold in carbon-nanotube-epoxy composites. *Polymer* **2003**, *44*, 5893–5899. [CrossRef]
65. Yu, Y.; Zhang, B.; Wang, Y.; Qi, G.; Tian, F.; Yang, J.; Wang, S. Co-continuous structural electrolytes based on ionic liquid, epoxy resin and organoclay: Effects of organoclay content. *Mater. Des.* **2016**, *104*, 126–133. [CrossRef]
66. Sanes, J.; Aviles, M.-D.; Saurin, N.; Espinosa, T.; Carrion, F.-J.; Bermudez, M.-D. Synergy between graphene and ionic liquid lubricant additives. *Tribol. Int.* **2017**, *116*, 371–382. [CrossRef]

Disclaimer/Publisher's Note: The statements, opinions and data contained in all publications are solely those of the individual author(s) and contributor(s) and not of MDPI and/or the editor(s). MDPI and/or the editor(s) disclaim responsibility for any injury to people or property resulting from any ideas, methods, instructions or products referred to in the content.

Review

Decadal Journey of CNT-Based Analytical Biosensing Platforms in the Detection of Human Viruses

Joydip Sengupta [1] and Chaudhery Mustansar Hussain [2,*]

[1] Department of Electronic Science, Jogesh Chandra Chaudhuri College, Kolkata 700033, India
[2] Department of Chemistry and Environmental Science, New Jersey Institute of Technology, Newark, NJ 07102, USA
* Correspondence: chaudhery.m.hussain@njit.edu

Abstract: It has been proven that viral infections pose a serious hazard to humans and also affect social health, including morbidity and mental suffering, as illustrated by the COVID-19 pandemic. The early detection and isolation of virally infected people are, thus, required to control the spread of viruses. Due to the outstanding and unparalleled properties of nanomaterials, numerous biosensors were developed for the early detection of viral diseases via sensitive, minimally invasive, and simple procedures. To that aim, viral detection technologies based on carbon nanotubes (CNTs) are being developed as viable alternatives to existing diagnostic approaches. This article summarizes the advancements in CNT-based biosensors since the last decade in the detection of different human viruses, namely, SARS-CoV-2, dengue, influenza, human immunodeficiency virus (HIV), and hepatitis. Finally, the shortcomings and benefits of CNT-based biosensors for the detection of viruses are outlined and discussed.

Keywords: carbon nanotube (CNT); biosensor; virus; SARS-CoV-2; toxicity; biocompatibility

1. Introduction

1.1. Carbon Nanotubes and Their Applicability in Biosensing

The carbon nanotube (CNT) is the 1D allotrope of carbon; the experimental evidence was first reported by L.V. Radushkevich and V.M. Lukyanovich [1] from the Institute of Physical Chemistry and Electrochemistry of the Russian Academy of Sciences, in 1952. However, after Ijima's paper [2] in 1991, research interest in CNT escalated rapidly. Structurally, CNTs can be divided into two major categories (Figure 1) based on the number of graphitic layers, namely, single-wall CNT (SWCNT) and multi-wall CNT (MWCNT). Depending on the direction of the roll-up, SWCNTs can have different structures, namely, the zigzag, armchair, or chiral formations. While depending on the nature of the wrapping i.e., whether a graphitic sheet is rolled around itself multiple times (Swiss roll) or if the graphitic sheets are arranged as concentric cylinders (Russian doll), MWCNT can also be categorized, as in Scheme 1.

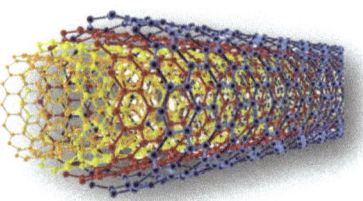 

Figure 1. The nanostructure of multi-wall carbon nanotube (MWCNT) (**left**) and single-wall carbon nanotube (SWCNT) (**right**) (adapted with permission from Ref. [3]).

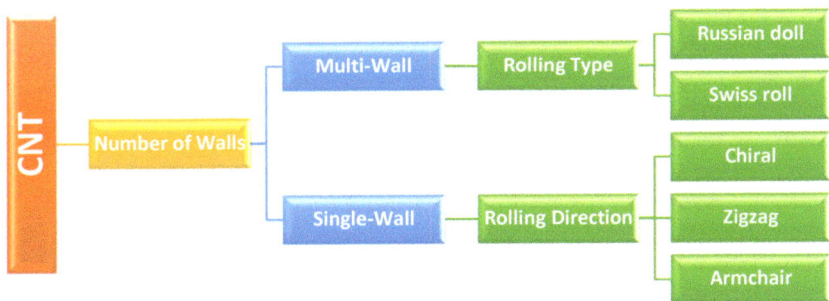

Scheme 1. Different categories of carbon nanotube (CNT).

CNT can be synthesized through various means, such as arc discharge, laser ablation, chemical vapor deposition (CVD), etc. However, because of the scalability, CVD-based approaches appear to be the most appropriate for large-scale CNT synthesis [4]. A wide range of prospective applications in important industrial fields, including nanoelectronics and biotechnology are promised by the distinctive mix of electrical, thermal, mechanical, and chemical characteristics that CNTs display. Additionally, among the many nanomaterials, CNTs are particularly intriguing because they provide an exceedingly tiny inner hollow core, virtually a one-dimensional space, for material storage. Thus, a novel structure can also be formed by filling the core of a CNT with the components necessary for specific applications.

Previous research has discovered a linkage between biomolecules in living beings and illnesses. The monitoring of aberrant physical parameters and the early diagnosis of illnesses help to minimize mortality and ensure organisms' physical health. Conventional laboratory approaches for assessing pathogenic variables are typically time-consuming, expensive, and complicated. Biosensors can facilitate the reliable and rapid analysis of metabolites in the body to help the current therapeutic procedure. CNTs have unique physicochemical and photoelectric qualities that can improve the performance of biosensors, such as a greater surface area for better catalyst adhesion; CNT-modified electrodes offer quicker electron transfer, resulting in enhanced sensitivity of detection for biosensors. CNT's unrivaled electronic features, such as quantum wire-like behavior, ballistic-type electronic conduction, remarkable thermal properties derived from phonon quantization, excellent flexibility, and high breaking stress despite its low density make CNT one of the best transducer materials for the transmission of signals related to the recognition of analytes, metabolites, or disease biomarkers. The curvature of the tube contributes to CNTs' high reactivity and sensitivity to chemical or environmental interactions. Moreover, as the carbon atom near the end of an open-ended tube has only two bonds, foreign molecules can easily enter the structure, thereby helping in the preferential addition of one or more species for functionalization. More importantly, from the viewpoint of biosensors, CNTs can act effectively as scaffolds for the immobilization of biomolecules at their surface. These fascinating characteristics have led to CNTs being widely used in biosensor applications.

1.2. Biosensors

According to the IUPAC, a biosensor (Figure 2) can be defined as "a device that uses specific biochemical reactions mediated by isolated enzymes, immunosystems, tissues, organelles, or whole cells to detect chemical compounds, usually by electrical, thermal, or optical signals" [4]. In 1962, the first biosensor for monitoring blood glucose was reported by Clark et al. [5]; later, a biosensor was also developed for the detection of the virus [6]. In 1998, Davis et al. [7] were able to immobilize the proteins on CNTs; afterward, in 1999, Balavoine et al. [8] were successful in developing the first biosensor using CNT. The timeline of biosensor development is represented in Figure 3.

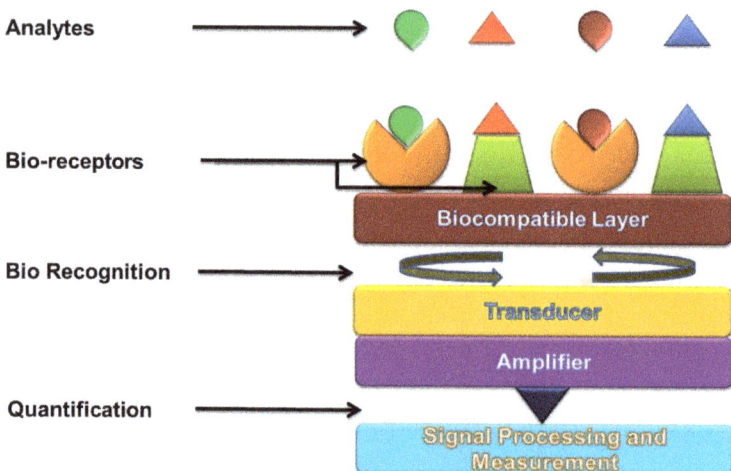

Figure 2. The schematic structure and operating principle of a biosensor.

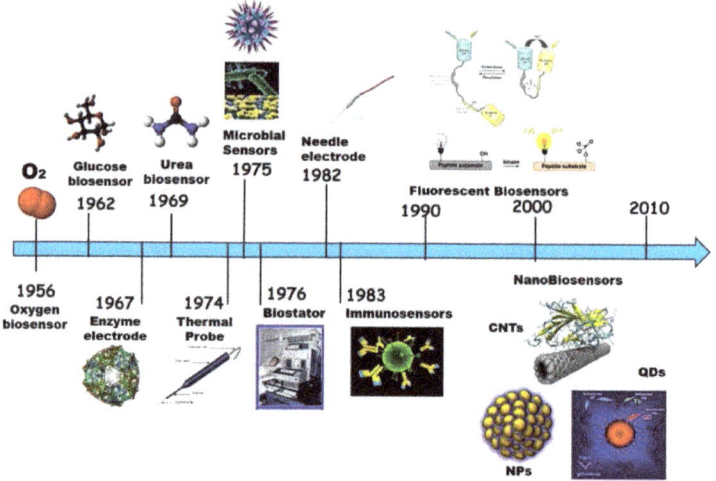

Figure 3. Biosensor development timeline (Reprinted with permission from Ref. [9]).

To date, different varieties of biosensors have been fabricated (Scheme 2) based on the analyte and transducer used. However, the type of biosensors that are used for human virus detection falls within the scope of discussion in this review.

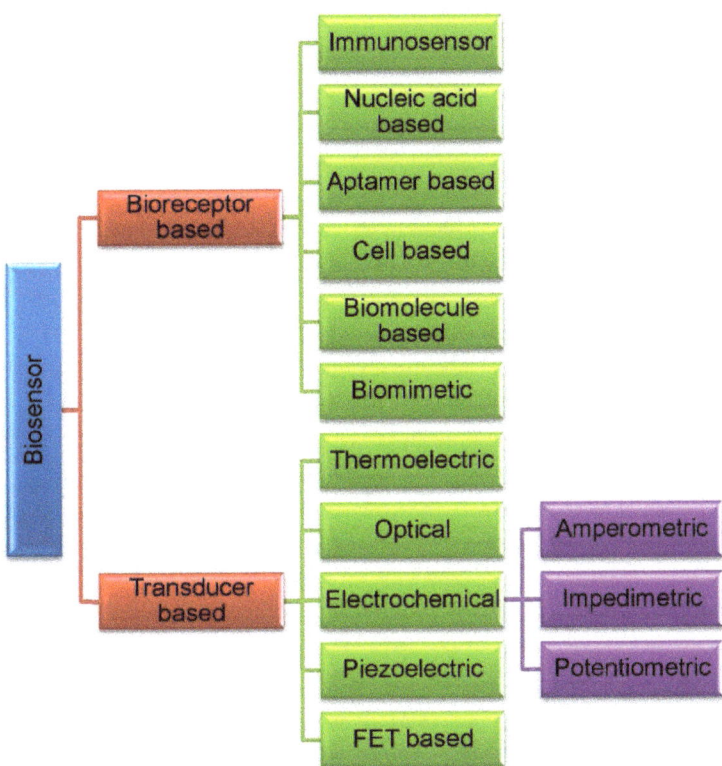

Scheme 2. Classification of biosensors, based on analyte and transducer use (adapted with permission from Ref. [10]).

2. Types of Biosensors Used for Virus Detection

2.1. Immunosensors

Because of its capacity to handle information, the immune system is an appealing subject in scientific studies. The major purpose of an immune system, as part of the system's defensive mechanism, is to accredit and ascertain all cells and molecules in the assembly and classify these biological substances as either toxic or non-toxic. When exposed to foreign substances (i.e., antigens), specialized immune system cells make immunoglobulins (i.e., antibodies) that attach to these antigens precisely. An immunosensor (Figure 4, top left), an affinity-based biosensing device, exploits the concept of immunology and employs an antibody for the specific molecular identification of antigens that are immobilized on a transducer surface, and then develops a stable immunocomplex. The immunocomplex is calculated and quantified by connecting the antibody and antigen interactions to the surface of a transducer. The transducer detects the response and transforms it into an electrical signal, which may then be processed, recorded, and examined. The detection of the target analyte in immunosensors might be direct, by witnessing the production of immunocomplexes, or indirect, by using a label. Immunosensors can be categorized into several categories, based on various methodologies, such as electrochemical, impedimetric, potentiometric, amperometric, voltammetric, conductometric, capacitive, and surface plasmon resonance (SPR)-based methodologies.

2.2. Optical Biosensor

An optical biosensor, a compact analytical instrument, combines an optical transducer system with a biorecognition-sensing element (Figure 4, bottom left). An optical biosensor's primary goal is to provide a signal that is proportional to the concentration of the material being analyzed (the analyte). Optical detection is made possible by using the interplay between the optical field and a biorecognition element. Label-free and label-based optical biosensing are the main two categories of optical biosensors. In a label-free mode, the interaction between the substances is analyzed, and the transducer directly generates the measured signal. In contrast, label-based sensing makes use of a label to assess the biorecognition event and generates an optical signal using a colorimetric, fluorescent, or luminescent approach. However, in some cases, such as antibody-antigen interactions, when a label is coupled with one of the bio-reactants, then this labeling might modify the binding characteristics, introducing systematic inaccuracy into biosensor analysis.

2.3. Electrochemical Biosensor

Due to the direct conversion of a biological event to an electrical signal, electrochemical biosensors provide an appealing technique for analyzing the content of a biological sample. The measurement of electrical characteristics in biosensing, for extracting information from biological systems, is generally electrochemical in nature, with a bio-electrochemical component serving as the major transduction aspect (Figure 4, bottom right). While biosensing devices use a variety of recognition components, electrochemical detection approaches mostly involve enzymes. This is mainly owing to their unique binding properties and bio-catalytic activity. In bio-electrochemistry, the reaction under examination would typically create a quantifiable current (amperometric), a measurable potential or charge buildup (potentiometric), or a measurable impedance (impedimetric). The electrodes are essential components for the operation of electrochemical biosensors since reactions are generally observed near the electrode's surface. Depending on the electrode's parameters, the material, the surface modification, or the electrode's size have a significant impact on the capability of detection. In general, three electrodes, namely, the reference electrode, counter or auxiliary electrode, and working electrode are needed for electrochemical sensing. To maintain a known and constant voltage, the reference electrode is kept away from the reaction site. The counter electrode creates a link to the electrolytic solution so that a current may be supplied to the working electrode, while the working electrode acts as the transduction element in the biological reaction. These electrodes ought to be chemically stable and conductible to achieve a faithful analysis.

2.4. Field-Effect Transistor (FET)-Based Biosensor

FET biosensors, which have the characteristics of being quick, inexpensive, and straightforward, stood out among a wide spectrum of electrical sensing devices as one of the most promising options for biosensing (Figure 4, top right). This cutting-edge technology, which has evolved since 1970 [11] in various forms, is the easiest method for the quick and accurate detection of numerous analytes. Specific probes on the conducting channel of FET-based biosensors can be embedded to provide real-time and label-free analysis. A FET is a type of solid-state device that controls the semiconductor's electron conductivity between its source and drain terminals by the application of a third gate electrode, via an insulator. To recognize specific analytes, biological receptors are immobilized on the sensing channels, which are linked to the source and drain electrodes. After exposing the biosensor to target analytes and forming specific biological complexes, the transducer system converts biochemical changes into a measurable signal. The addition of charged biomolecules to the surface of the gate dielectric is equivalent to the application of voltage by the use of a gate electrode and results in threshold voltage variations. Therefore, the FET biosensors' underlying method relies on the conductance of the species that have been adsorbed. The two main types of FETs are n-type and p-type devices, wherein electrons and holes, respectively, serve as the principal charge carriers. An n-type FET sensor will respond by increasing the conductance if the target molecule is positively charged as a result of electron aggregation. Conversely, the conductance will be reduced if the target is a molecule with a negative charge. When it comes to the p-type FET system, the opposite tendency is applicable.

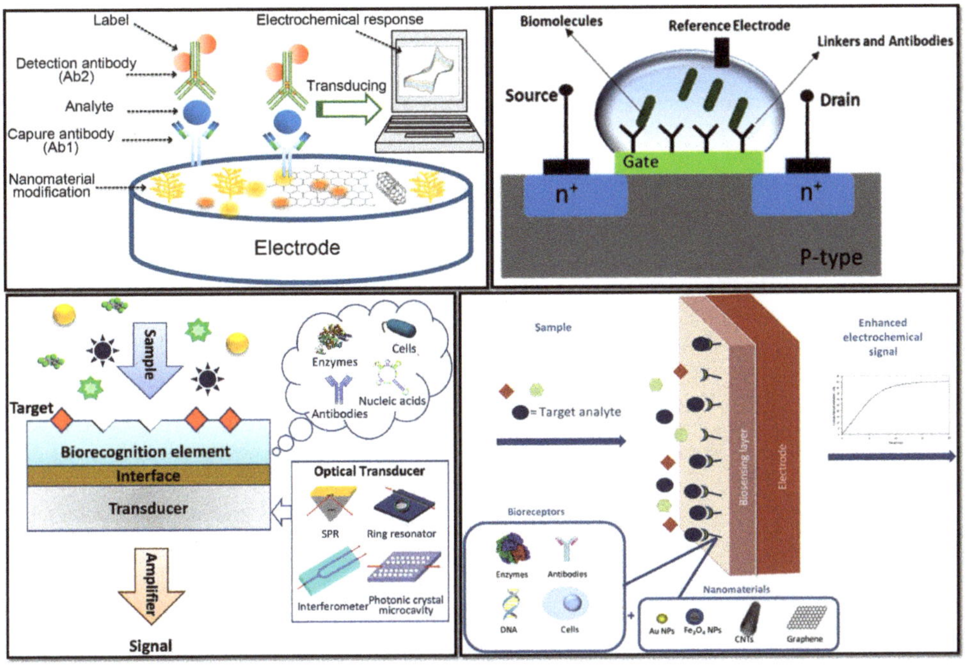

Figure 4. (**Top left**) A schematic representation of an electrochemical immunosensor (Reprinted with permission from Ref. [12]); (**top right**) schematic diagram of a field effect transistor (FET)-based biosensor with a source and drain (Reprinted with permission from Ref. [13]); (**bottom left**) schematic diagram of optical biosensor constitution (Reprinted with permission from Ref. [14]); (**bottom right**) the main constituents of a nanomaterial-based electrochemical biosensor (Reprinted with permission from Ref. [15]).

3. CNTs-Based Biosensors for the Detection of Human Viruses
3.1. SARS-CoV-2

The SWCNT-based optical sensing approach was employed by Pinals et al. [16] to detect the SARS-CoV-2 spike protein. The angiotensin-converting enzyme 2 (ACE2), which has a strong affinity with the SARS-CoV-2 spike protein, was used to functionalize SWCNTs. A 2-fold increase in fluorescence was observed within 90 min of SARS-CoV-2 spike protein exposure, which exhibited a limit of detection (LOD) of 12.6 nM for the device. Shao et al. [17] functionalized a high-purity semiconducting SWCNT surface with an anti-SARS-CoV-2 spike protein antibody and used it as a channel in an FET-type biosensor, to detect SARS-CoV-2 antigens in clinical nasopharyngeal samples. The fabricated device exhibited an LOD of 0.55 fg/mL (Figure 5). For efficient and precise identification of the SARS-CoV-2 S1 antigens in fortified saliva samples, Zamzami et al. [18] created a rapid, simple-to-use, inexpensive, and quantitative CNT-based antibody-functionalized p-type depletion FET biosensor. Through a non-covalent interaction with the linker 1-pyrenebutanoic acid succinimidyl ester (PBASE), the SARS-CoV-2 S1 antibody was bound to the CNT surface in the FET channel region. The CNT-FET biosensor successfully identified the SARS-CoV-2 S1 antigen in 10 mM AA buffer at a pH of 6.0, at concentrations ranging from 0.1 fg/mL to 5.0 pg/mL, with an LOD of 4.12 fg/mL. On a flexible Kapton substrate, Thanihaichelvan et al. [19] created CNT-FETs and immobilized the reverse sequence of the SARS-CoV-2

RNA-dependent RNA polymerase gene onto the CNT channel, to develop a biosensor for coronavirus detection with an LOD of 10 fM. The primary signal generation depended on RNA hybridization, while the main signal transducer was a liquid-gated CNT-FET. Cardoso et al. [20] used carboxylated CNTs and screen-printed (SP) them onto carbon electrodes (CNT-SPE). Later, the electrode was modified with EDC/NHS coupling chemistry to produce an amine layer, to adsorb the SARS-CoV-2 spike protein antibody. The sensor exhibited a linear response between 1.0 pg/mL and 10 ng/mL, with an LOD of ~0.7 pg/mL. A platform for electrochemical sensing was fabricated by Curti et al. [21] employing SWCNT-SPEs, which were functionalized with a DNA aptamer that was already redox-tagged. In the presence of the SARS-CoV-2 spike protein S1, the concentration-dependent folding of a DNA aptamer occurred, which resulted in a change in the amperometric signal with an LOD of 7 nM. Monoclonal antibodies were employed for the functionalization of SWCNT by Li et al. [22], to develop an extremely sensitive immuno-resistive sensor for the detection of SARS-CoV-2. To minimize the contact resistance, silver electrodes were screen-printed onto SWCNTs, and the complete arrangement was mounted on polyethylene terephthalate (PET) film. The LOD of the developed sensor was 350 genome equivalents/mL. Through systematic analysis, Kim et al. [23] examined the relationships between different thin-film characteristics and the sensitivity of CNT thin-film-based immunosensors for the rapid detection of the SARS-CoV-2 virus. They found that smaller surface roughness and better CNT alignment resulted in improved sensitivity at a given film thickness, with a LOD value of 5.62 fg/mL, because of the enhanced bioreceptor-binding surface area. Table 1 summarizes the earlier discussion of the creation of CNT-based biosensors to identify the SARS-CoV-2 virus.

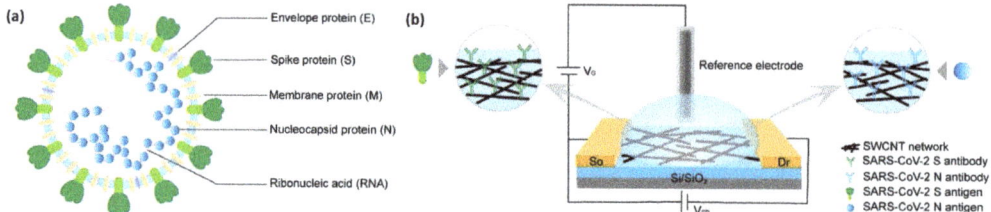

Figure 5. The detection of SARS-CoV-2 Ag, using SWCNT-based FET biosensors. (**a**) Schematic structure of SARS-CoV-2 to demonstrate the targeting proteins. (**b**) Schematic illustration of a liquid-gated SWCNT FET for the detection of SARS-CoV-2 SAg and NAg. Interdigitated gold electrodes (yellow blocks) are configured as the source (So) and drain (Dr) (Reprinted with permission from Ref. [17]).

3.2. Dengue Virus (DNV)

Based on CNT-SPE, an immunosensor for non-structural protein 1 (NS1) of the DNV was created by Dias et al. [24]. They employed a uniform mixture of carboxylated CNTs distributed in carbon ink to make the CNT-SPE and used an ethylenediamine film to covalently attach anti-NS1 antibodies to CNT-SPE. The developed biosensor was able to detect the DNV NS1 protein, with an LOD of 12 ng/mL. A robust poly (allylamine) (PAH) sandwich-based immunosensor was fabricated by Silva et al. [25] for detecting DNV NS1. A thin coating of PAH on carboxylated CNTs helped to immobilize anti-NS1 antibodies on the electrode surface. To strongly bond CNTs to the electrode surface, as well as anti-NS1 antibodies, through their Fc terminus to prevent random immobilization, PAH, a cationic polymer, was used. The fabricated immunosensor had a linear range of operation between 0.1 g/mL and 2.5 g/mL, with an LOD of 0.035 g/mL. An SWCNT-based, inexpensive, label-free chemiresistive biosensor was designed by Wasik et al. [26], where, for the first time, heparin was utilized as a biorecognition component as opposed to a conventional antibody. The biosensor revealed clinically significant sensitivity for people infected by *Aedes aegypti*, concerning the detection of whole DNV, with an LOD of 8 DNV/chip. Later, they modified the biosensor by employing a network of anti-dengue NS1 monoclonal antibodies for the functionalization of SWCNTs, instead of using herpin [27]. The modified biosensor exhibited a linear response in the range of 0.03–1200 ng/mL, with an LOD of 0.09 ng/mL. Almost all laboratory and commercial dengue NS1 diagnostic measures include a blood-drawing procedure, which limits the advantage of point-of-care (POC) diagnostics and reduces patient readiness. Instead of blood, NS1 can be extracted from human saliva for the early detection of dengue infection, which is a straightforward, non-invasive, pain-free, and economical process that can be performed by even untrained/less-trained workers. This pathway was also explored by Wasik et al. [28] through the fabrication of a label-free chemiresistive immunosensor employing a network of anti-dengue NS1 monoclonal antibody-functionalized SWCNTs. The biosensor exhibited a detection range of ~1 ng/mL to 1000 ng/mL for the DNV NS1 protein (Figure 6). To detect DNV antibodies, a high-performance impedimetric immunosensor was developed by Palomar et al. [29] via the deposition of CNT on electrodes and was later functionalized with polypyrrole-NHS to immobilize the DNV 2 NS1 glycoprotein via covalent amide coupling. This biosensor exhibited high linearity after optimization, in a broad range of concentrations (10^{-13} to 10^{-5} g/mL). For the detection of dengue toxin, the CNT/Au nanoparticle (AuNP) composite was deposited on a homemade Au electrode by Palomar et al. [30], to immobilize dengue antibodies on AuNPs through covalent bonding. The electrochemical signal enhancement and improvement in overall performance were achieved due to the porosity of the tri-dimensional network of Au-CNT, with an LOD of 3×10^{-13} g/mL. Mendonça et al. [31] fabricated a label-free immunosensor employing a thin film of CNT-ethylenediamine. The covalent immobilization of the anti-NS1 monoclonal antibodies on CNTs allowed for great measurement stability. Finally, differential pulse voltammetry (DPV) was used to analyze the responses to dengue NS1. The measurement showed that the linear range for the operation of the biosensor was 20 to 800 ng/mL, with an LOD value of 6.8 ng/mL. The preceding discussion on the development of CNT-based biosensors for DNV detection is summarized in Table 1.

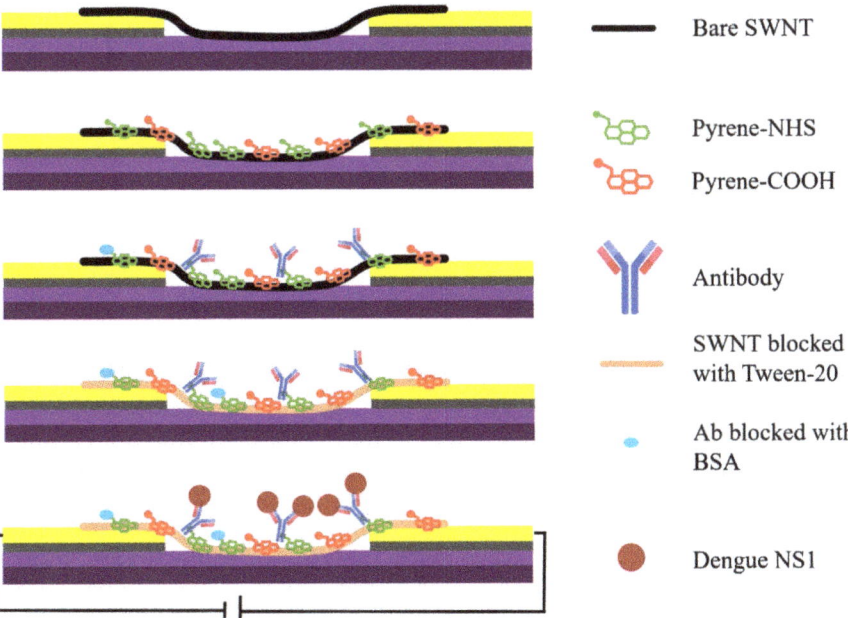

Figure 6. Schematic of the functionalization of SWNT networks with the Dengue Virus (DNV) NS1 antibody (Reprinted with permission from Ref. [28]).

3.3. Influenza Virus

An MWCNT-cobalt phthalocyanine nanocomposite and poly (amidoamine) (PAMAM) dendrimer were deposited on a glassy carbon electrode (GCE) by Zhu et al. [32] to electrochemically detect the avian influenza virus (AIV) genotype in a label-free format. The DNA probes were, then, effectively immobilized on the modified electrode using the coupling agent, G4 PAMAM dendrimer, and were monitored with DPV to achieve an LOD of 1.0 pg/mL. The surface of MWCNTs was utilized by Tam et al. [33] to immobilize the DNA probe by covalent interactions between the DNA sequence's amine and phosphate groups. Changes in conductance of the sensor surface were used to detect the hybridization of the DNA probe and the influenza viral DNA for the label-free detection of the influenza virus with an LOD of 0.5 nM. CNT-SPEs were used by Bonanni et al. [34] as an electrochemical-sensing platform for the detection of influenza viral DNA. The CNT-SPEs were functionalized with carboxylic molecules to covalently immobilize the oligonucleotide probe, employing facile carbodiimide chemistry. A direct coupling and sandwich scheme, two distinct techniques for the impedimetric detection of DNA hybridization, were employed and compared. The sandwich scheme revealed better results, with an LOD value of 7.5 fM. A self-assembled SWCNT thin film was prepared by Lee et al. [35] for the fabrication of an affordable, label-free biosensor to detect the swine influenza virus (SIV). The anti-SIV antibodies were bounded with the SWCNT thin film for the detection of SIV, with an LOD of 180 $TCID_{50}$/mL, by monitoring the change in resistance (up to 12%). Dielectrophoretic and electrostatic forces were used to deposit COOH-functionalized SWCNTs on a self-assembled monolayer of the polyelectrolyte, polydiallyl dimethyl-ammonium chloride (PDDA), by Singh et al. [36]. Viral antibodies were immobilized, utilizing biotin-avidin coupling, after avidin was coated on the PDDA-SWCNT channels (Figure 7). Changes in the channels' resistance were monitored to detect influenza viruses, with a detection limit of 1 PFU/mL. AuNP-decorated CNTs (Au-CNTs) were produced, utilizing phytochemical composites at room temperature, and were employed by Lee et al. to build a plasmon-assisted fluoro-immunoassay (PAFI) for the detection of the influenza virus [37]. The surfaces of Au-CNTs and CdTe quantum dots (CdTe-QDs), the photoluminescence intensity of which varied according to viral concentration, were conjugated with specific antibodies against the influenza virus to achieve an LOD of 1 ng/mL (for Beijing/262/95 (H1N1)), 0.1 pg/mL (for New Caledonia/20/99IvR116 (H1N1)), and 50 PFU/mL (for Yokohama/110/2009 (H3N2)). Later, a two-step method was adopted by Lee et al. [38] to decorate CNTs with Au/magnetic nanoparticles, which offered superior magnetic properties and high electrical conductivity. Later, the Au nanoparticle's surface was conjugated with thiol-group-functionalized probe DNA to detect influenza viral DNA with a detection limit of approximately 8.4 pM. AuNPs were bonded to the CNT surface via in situ accumulation under mild conditions by Ahmed et al. [39]. The improved peroxidase-like activity of the Au-CNT nanohybrid was employed to develop a supersensitive colorimetric optical sensor to detect the influenza virus. The sensor comprised a test system containing specific influenza antibodies, Au-CNT nanohybrids, 3, 3', 5, 5'-tetramethyl-benzidine (TMB), and H_2O_2. The color of the test system turned blue upon the addition of the influenza virus and the LOD was 3.4 PFU/mL. Fu et al. [40] fabricated chemiresistor-type biosensors to detect AIV, employing semiconducting SWCNTs or nitrogen-doped MWCNTs and non-covalently functionalizing them with DNA probe sequences. Complementary DNA target sequences of AIV, with concentrations ranging from 2 pM to 2 nM, could be detected by the constructed biosensor after 15 min at room temperature. To detect influenza type A viral DNA, a CNT-FETs-based DNA sensor was developed by Tran et al. [41]. The initial probe DNA, the hybridization period, and the reaction temperature were some of the aspects that were examined since they affected the sensing data. The DNA sensor demonstrated a quick response time of under a minute, with a very low (1 pM) detection limit and a broad linear detection range of 1 pM to 10 nM. Wang et al. [42] compared the effectiveness of the aptamer and antibody, concerning the detection of influenza A virus (California/07/2009 (pdmH1N1)), employing an MWCNT-Au conjugated sensing surface with a di-electrode.

They found that the electric response and affinity of aptamers were much stronger than those of antibodies. Wang et al. also reported that the aptamer could provide an LOD in the range of 10 fM, whereas the antibody showed an LOD of 1 pM in detecting the influenza A virus. Using flexible and stable SP-CNT-polydimethylsiloxane electrodes, a paper-based immunosensor for the label-free hemagglutinin antigen (HA) detection of several AIVs (H5N1, H7N9, and H9N2) was demonstrated by Lee et al. [43], whereby immune responses were measured via DPV. The LOD values for the different viruses were 55.7 pg/mL for H5N1 HA, 99.6 pg/mL for H7N9 HA, and 54.0 pg/mL for H9N2 HA. In Table 1, an overview of the CNT-based biosensors for influenza detection is provided.

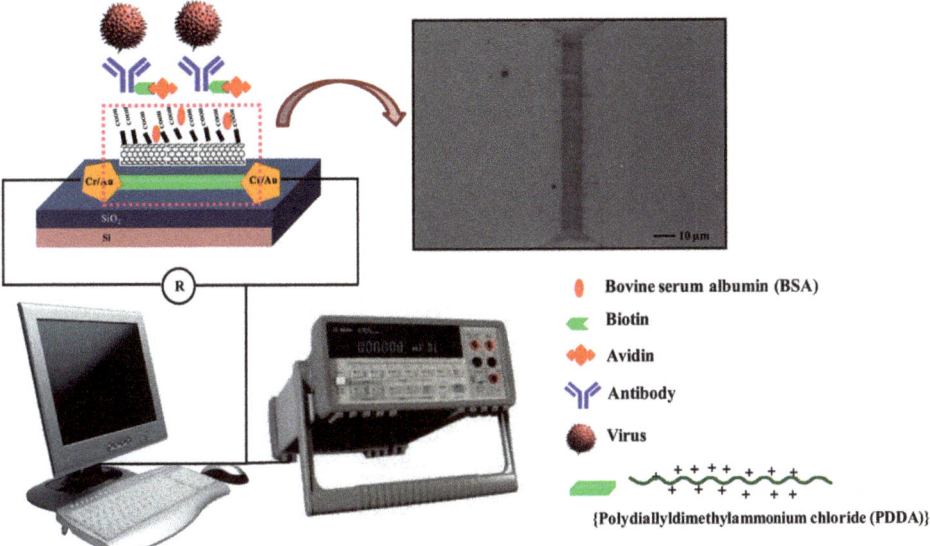

Figure 7. Schematic illustration of the SWCNT-based immunosensor system for H1N1 virus detection. The inset shows an optical image of di-electrophoretically deposited SWCNTs on a PDDA self-assembled monolayer (Reprinted with permission from Ref. [36]).

3.4. Human Immunodeficiency Virus (HIV)

Mahmoud et al. [44] immobilized thiol-terminated ferrocene-pepstatin (ThFcP) conjugate on an SWCNT/AuNP-modified Au electrode. Electrochemical impedance spectroscopy (EIS) was used to track the nature of the interaction between HIV-1 protease and the ThFcP conjugate, which resulted in the change of the interfacial characteristics of Au electrodes. The electrochemical biosensor was able to detect HIV-1 protease, even at a 10 pM level. Later, thiolated SWCNT/AuNPs were used to modify the disposable SP-Au electrode surface, while ThFcP was subsequently self-assembled on those surfaces by Mahmoud et al. [45] to fabricate a sensitive electrochemical biosensor to detect HIV-1 protease. A nanocomposite of AuNP, amino-functionalized MWCNT, and acetone-extracted propolis (AEP) was prepared by Kheiri et al. [46] and deposited in the same way on an Au electrode for immobilization of the p24 antibody (anti-p24 Ab) to create an immunosensor for the detection of the HIV antigen. The immunosensor demonstrated high electrochemical sensitivity in detecting p24 in a range of concentration from 0.01 to 60.00 ng/mL, with an LOD of 0.0064 ng/mL. An advanced molecularly imprinted polymers (MIPs) electrochemical sensor was fabricated on MWCNT-modified GCE by Ma et al. [47] through the polymerization of the surface, using acrylamide (AAM), N,N′-methylene bisacrylamide (MBA), and ammonium persulphate (APS) as the functional monomer, cross-linking agent, and initiator, respectively. The developed sensor was capable of detecting HIV-p24 in human serum samples, with an LOD of 0.083 pg/cm^3. A chitosan/glutaraldehyde crosslinking system was employed by Giannetto et al. [48] to immobilize the target protein on disposable CNT-SPE for the maximum exposure of p24, to interact with a mouse anti-p24 IgG1. The linear operating range of the immunosensor was 10 pM to 1 nM, with an LOD of 2 pM, for HIV-related p24 capsid protein in human serum. Harvey et al. [49] observed that denatured proteins can improve the optical responsiveness of CNTs to nucleic acids. Their study revealed that following hybridization, hydrophobic regions of the denatured protein interact with the surface of the CNTs, which results in a larger shift in the nanotube emission. They later employed this strategy for the detection of intact HIV in serum. A nickel-organic composite/AuNP/CNT/polyvinyl alcohol (PVA) substance was used to fabricate a flexible paper-based electrode by Lu et al. [50] for the detection of HIV DNA (Figure 8). The methylene blue was employed as a redox indicator for DNA hybridization on the electrode. The large surface area and the presence of π-electron, donated by the Ni-Au composite, facilitate the higher loading of target DNA. With a linear range of 10 nM–1 μM and an LOD of 0.13 nM, this flexible paper electrode demonstrated good sensing capability. The previous discussion on the creation of CNT-based biosensors to identify HIV is summarized in Table 1.

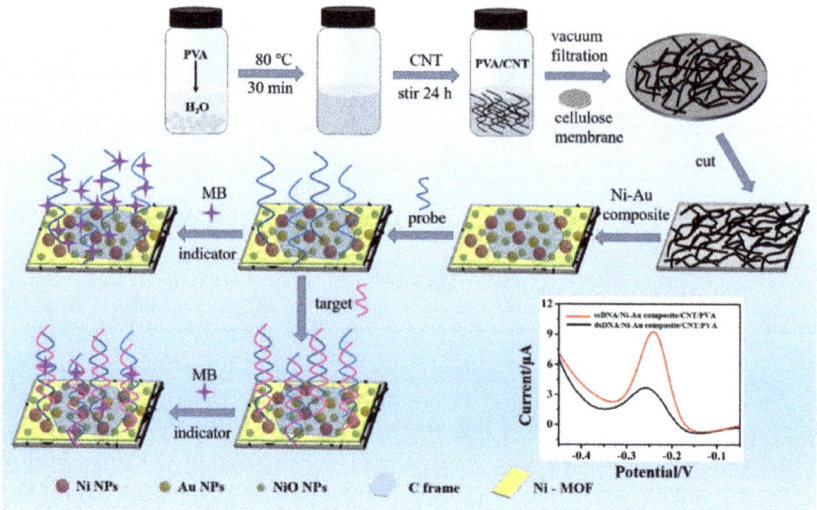

Figure 8. Schematic of the fabrication process for a flexible Ni-Au composite/CNT/PVA film electrode and the detection of the target Human Immunodeficiency Virus (HIV) DNA (Reprinted with permission from Ref. [50].

3.5. Hepatitis Virus

To detect the short DNA sequences associated with the Hepatitis B virus (HBV), a label-free electrochemical DNA biosensor was developed by Li et al. [51], employing 4,4′-diaminoazobenzene (4,4′-DAAB) and MWCNT-modified GCE. The CNT carboxyl groups were covalently linked to the oligonucleotides and the DPV was used to monitor the hybridization reaction, with an LOD of 1.1×10^{-8} M. To detect HBV, Oh et al. [52] created an FET-based biosensor, using CNTs consisting of a microfluidic channel with immobilized hepatitis B antibody on it. The electrical conductance changed over time, owing to the presence of the hepatitis B antigen in the channel. The change in the channel conductance was proportional to the hepatitis B antigen concentration. Ly et al. [53] immobilized bovine IgG, employing cyclic voltammetry on a DNA-linked CNT electrode to fabricate an electrochemical biosensor for the detection of human HBV in non-treated blood. The relative standard deviation of 0.2 mL HBV was 0.04 (n = 4) within the working limits of 0.035–0.242 mg/mL anti-bovine IgG. A CNT-conducting polymer (CP) network was created by Hu et al. [54] via drop-casting a CNT solution on a GCE, followed by the electrochemical polymerization of a poly (pyrrole propionic acid) (pPPA) film for crosslinking and stabilizing the CNTs. The CNTs served as the network's structural foundation and provided excellent specific surface areas for immobilizing antibodies. Moreover, owing to its self-limiting growth characteristic, the conducting film facilitated CNT in forming a stable network and offered ample carboxyl groups to immobilize the probe proteins for the detection of hepatitis B surface antigen in serum, with an LOD of 0.01 ng/mL. Amino-CNT and hyaluronic acid (HA) were bonded with amide groups and assembled onto the surface of GCE by Cabral et al. [55]; the response of the electrode in the presence of hepatitis B core protein antibodies was measured by square-wave voltammetry (SWV). The immunosensor response was linear up to 6.0 ng/mL, with an LOD of 0.03 ng/mL. For the detection of the core hepatitis B antigen, an electrochemical immunosensor based on polytyramine (PTy)-CNT composite was developed by Trindade et al. [56]. Because of the substantial creation of NH_3^+ ionic species, the composite possesses high catalytic activity. The HBV was electrochemically identified by SWV in a label-free and reagent-free manner. The immunosensor exhibited a LOD of 0.89 ng/mL while operating in a linear range of 1.0 to 5.0 ng/mL. An AuNPs/chitosan-ferrocene-ammoniated MWCNT (CS-Fc-AMWNT) nanocomposite was prepared by Chen et al. [57] via the Schiff base reaction to achieve a large specific surface area, adequate conductivity, and exceptional biocompatibility. The electrochemical deposition was used to modify the AuNPs for the screen-printed electrode (SPE), while physical adsorption was used to adhere the CS-Fc-AMWNTs composite to the electrode surface. Later, through glutaraldehyde cross-linking, hepatitis B surface antibodies were immobilized on the surface of the electrode. The biosensor can be operated in the range of 1–250 ng/mL for the detection of hepatitis B antigen, with an LOD of 0.26 ng/mL. The SP carbon electrode was modified by Upan et al. [58], through the addition of CNTs that were embellished with AuNP and AgNP (Figure 9). The AuNPs offered biocompatibility and a wide surface area for the immobilization of the hepatitis B surface antibody, which aided the signal improvement. Subsequently, in DPV detection, AgNPs served as a sensing probe to detect the target antigen in the linear range of 1–40 ng/mL with an LOD of 0.86 ng/mL. Using a peptide nucleic acid-functionalized SWCNT-FET biosensor, Dastagir et al. [59] also exhibited the explicit and label-free detection of a hepatitis C virus RNA sequence, with a detection limit of 0.5 pM. The deposition precipitation approach was used by Pusomjit et al. [60] to create Pt nanoparticles that were later coated onto SWCNTs. The generated nanocomposite was finally used as a substrate to immobilize antibodies on a paper-based, SP graphene electrode surface for the purposes of analyzing the hepatitis C virus. DPV was used to measure the target antigen in the range of 0.05 to 1000 pg/mL, with an LOD value of 0.015 pg/mL. Table 1 summarises the preceding discussion on the development of CNT-based biosensors and their use to detect several types of human viruses HBV.

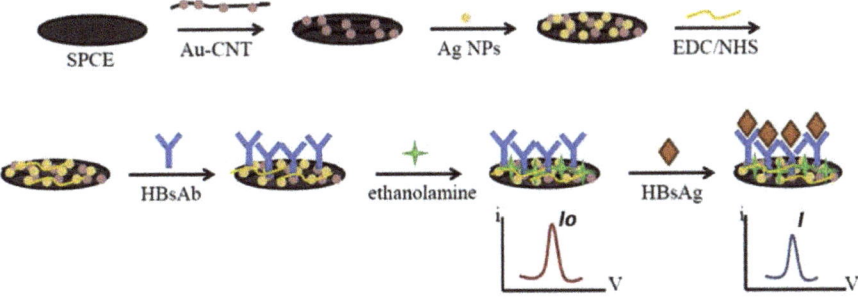

Figure 9. The stepwise fabrication process of the immunosensor for the detection of Hepatitis B virus (HBV) (Reprinted with permission from Ref. [58]).

Table 1. CNT-based biosensors for human virus detection.

Virus.	Target	Sensor Type	Limit of Detection (LOD)	Detection Range	Detection Platform	Ref. No
SARS-CoV-2	Antigen	FET-based	4.12 fg/mL	0.1 fg/mL to 5.0 pg/mL	CNT surface	[18]
	Antigen	FET-based	0.55 fg/mL	5.5 fg/mL to 5.5 pg/mL	CNT surface	[17]
	Gene	FET-based	10 fM	NA	CNT surface	[19]
	Protein	Optical	12.6 nM	NA	SWCNT substrate	[16]
	Antibody	Electrochemical	0.7 pg/mL	0.7 pg/mL and 10.0 ng/mL	Electrode	[20]
	Protein	Electrochemical	7 nM	NA	Electrode	[21]
	Protein	Immunosensor	350 genome equivalents/mL	NA	Electrode	[22]
	Protein	Immunosensor	5.62 fg/mL	NA	CNT thin-film	[23]
Dengue	Antibody	Immunosensor	10^{-13} g/mL	10^{-13} to 10^{-5} g/mL	Electrode	[29]
	Protein	Immunosensor	12 ng/mL	40 ng/mL to 2 µg/mL	Electrode	[24]
	Protein	Immunosensor	6.8 ng/mL	20 to 800 ng/mL	CNT surface	[31]
	Protein	Immunosensor	0.035 µg/mL	0.1 to 2.5 µg/mL	Electrode	[25]
	Virus	Immunosensor	8 DNV/chip	NA	Electrode	[26]
	Antibody	Immunosensor	0.09 ng/mL	0.03 to 1200 ng/mL	Electrode	[27]
	Protein	Immunosensor	1 ng/mL	1 to 1000 ng/mL	Electrode	[28]
	Toxin	Electrochemical	3×10^{-13} g/mL	0.001 to 2 µg/mL	Electrode	[30]
Influenza	Virus	Immunosensor	1 ng/mL (for Beijing/262/95 (H1N1)), 0.1 pg/mL (for New Caledonia/20/99IvR116 (H1N1)) and 50 PFU/mL (for Yokohama/110/2009 (H3N2))	50 to 10,000 PFU/mLm (for Yokohama/110/2009 (H3N2))	Au-CNT surface	[37]
	Aptamer and antibody	Immunosensor	10 fM	NA	Au-CNT surface	[42]
	Antibody	Immunosensor	180 TCID$_{50}$/mL	NA	CNT surface	[35]
	DNA	Immunosensor	2 pM	2 to 200 pM	Semiconducting SWCNTs or nitrogen-doped MWCNTs	[40]
	DNA	FET-based	1 pM	1 pM to 10 nM	CNT surface	[41]
	DNA	Immunosensor	8.4 pM	1 pM to 10 nM	Electrode	[38]
	DNA	Electrochemical	0.5 nM	NA	Electrode	[33]
	Virus	Immunosensor	1 PFU/mL	1 to 10,000 PFU/mL	CNT surface	[36]
	Virus	Optical	3.4 PFU/mL	3.4 to 10 PFU/mL	Au-CNT surface	[39]
	DNA	Electrochemical	7.5 fM	NA	Electrode	[34]
	Antigen	Immunosensor	55.7 pg/mL for H5N1 HA, 99.6 pg/mL for H7N9 HA, and 54.0 pg/mL for H9N2 HA	100 pg/mL to 100 ng/m	Electrode	[43]
	DNA	Electrochemical	1.0 pg/mL	0.01 to 500 ng/mL	Electrode	[32]

Table 1. Cont.

Virus.		Target	Sensor Type	Limit of Detection (LOD)	Detection Range	Detection Platform	Ref. No
HIV		Virus	Electrochemical	0.083 pg/cm^3	1.0×10^{-4} to 2 ng/cm^3	Electrode	[47]
		Antigen	Immunosensor	0.0064 ng/mL	0.01 to 60.00 ng/mL	Electrode	[46]
		Protease Enzyme	Electrochemical	NA	NA	Electrode	[45]
		DNA	Electrochemical	0.13 nM	10 nM to 1 μM	Electrode	[50]
		Protease Enzyme	Electrochemical	NA	10 pM	Electrode	[44]
		Viral nucleic acids	Optical	NA	NA	Fluorescence Detector	[49]
		Protein	Immunosensor	2 pM	10 pM to 1 nM	Electrode	[48]
Hepatitis	B	Antibody	Immunosensor	0.89 ng/mL	1.0 to 5.0 ng/mL	Electrode	[56]
		Antigen	FET-based	NA	NA	CNT surface	[52]
		Virus	Electrochemical	NA	NA	Electrode	[53]
		Antigen	Immunosensor	0.26 ng/mL	1–250 ng/mL	Electrode	[57]
		Antibody	Immunosensor	0.03 ng/mL	0.03 to 6.0 ng/mL	Electrode	[55]
		Antigen	Immunosensor	0.01 ng/mL	NA	Electrode	[54]
		DNA	Electrochemical	1.1×10^{-8} M	7.94×10^{-8} to 1.58×10^{-6} M	Electrode	[51]
		Antigen	Immunosensor	0.86 ng/mL	1–40 ng/mL	Electrode	[58]
	C	Antigen	Immunosensor	0.015 pg/mL	0.05 to 1000 pg/mL	Electrode	[60]
		RNA	FET-based	0.5 pM	NA	CNT surface	[59]

4. Current Challenges and Future Perspectives

Carbon nanomaterials are frequently employed in biomedical applications, due to their multifunctionality and minimal complexity in surface modification, which, in turn, improves their biophysical characteristics. CNT is the sp^2 hybridized allotrope of carbon, with a hollow cylindrical tubular structure, and has a high aspect ratio. The efficiency and precision of detection in sensing viral genomes, proteins, and other viral cellular biological components using CNT are governed by the exotic characteristics of the nanotubes. CNTs differ from conventional nanomaterials in that they possess many novel physiochemical properties that hold great promise for a variety of applications, including biosensing. It is envisaged that the utilization of CNT's special capabilities in a biological setting would lead to significant improvements in disease diagnosis, monitoring, and treatment. Consequently, CNT-based biosensors offer many benefits over other types of sensors, such as those based on metal oxides or silicon, including high sensitivity, quick response times, reduced redox reaction potential, and longer lifetimes with greater stability. Although CNTs have many desirable qualities and benefits, dispersion, which is triggered by the high surface energy of the CNTs, stands in the way of moving forward. Because of their extreme hydrophobicity, CNTs cannot be dissolved in water or other common solvents. In order to increase their solubility and other functional qualities, CNTs must be functionalized, depending on the application. The benefits and drawbacks of employing CNTs in biological applications are listed in Table 2.

Table 2. Pros and cons of using CNTs for biomedical applications. (Reprinted with permission from Ref. [61].

Pros	Cons
• Unique mechanical properties offer in vivo stability • Extremely large aspect ratio offers a template for the development of multimodal devices • Capacity to readily cross biological barriers; novel delivery systems • Unique electrical and semiconducting properties; constitute advanced components for in vivo devices • Hollow, fibrous, light structure with different flow dynamics properties; advantageous in vivo transport kinetics • Mass production-low cost; attractive for drug development	• Nonbiodegradable • Large available surface area for protein opsonization • As-produced material insoluble in most solvents; needs to be surface-treated preferably by covalent functionalization chemistries to confer aqueous solubility (i.e., biocompatibility) • Bundling; large structures with less than optimum biological behavior • Healthy tissue tolerance and accumulation; unknown parameters that require toxicological profiling of the material • A great variety of CNT types makes standardization and toxicological evaluation cumbersome

To establish a faithful nano-bio interface, CNTs may be engineered by means of covalent or non-covalent modification, EDC/NHS chemistry, click chemistry, and length-location tuning. Numerous viruses can go into a dormant stage called latency, wherein they remain inactive inside the host cell before activation [62–66]. Often, it is necessary to identify the virus, even in the latent stage, to eliminate the probability of infection/reinfection. Thus, the pathways may be searched through the functionalization or surface modification of CNTs, to enable CNTs to detect the virus in both the dominant and latent stages. Although the great sensitivity and extended durability of CNTs-based viral biosensors make them a promising candidate for viral detection, their accessibility and the economic perspective are essential for rapid diagnosis. The supervision of economic feasibility is an important component of commercialization. The manufacturing complexity levels should be kept to a minimum from the beginning of the design process, by means of the selection of facile fabrication methods that can be easily scaled up. The selection of materials is also very important, particularly in terms of balancing the cost and the desired material features for the required application. The CNT exhibits tremendous economic potential, especially in light of recent developments that demonstrate a substantial decrease in the cost of making high-quality CNT, which fosters a silver lining in CNT-based nanodevice fabrication. Wearable biosensors with wireless communication facilities should be introduced to address this problem [67], enabling the patient to benefit from swift analysis and report the findings, to ensure a timely diagnosis. The integration of biosensors with electronic gadgets possessing smart read-out capabilities is turning into a top requirement in contemporary state-of-the-art living, in line with the current technological evolution [68]. Conversely, the toxicity of CNT and the greenness of biosensors are currently the major challenges regarding biocompatibility and sustainability. The toxicity of CNT can be reduced with the use of several techniques, such as tuning the surface defects [69], utilizing native small-molecule drugs [70], and attenuating the CNT length [71]. The greenness of the biosensor may be achieved through the green synthesis of CNT; the entire sensing operation must be performed in a green manner, i.e., using green solvents, green waste management, green power management, etc. [72]. In many cases, it was observed that the viruses mutated very rapidly (e.g., SARS-CoV-2 [73]); thus, it is vital to create dependable and effective methods based on integrated multiple biosensor technology for the quick detection of several mutations of a virus at once [74]. Moreover, CNT can also be used as an antiviral agent to inhibit viruses. By adding protoporphyrin IX to acid-functionalized MWCNTs, Banerjee et al. [75] created porphyrin-conjugated MWCNTs, which significantly reduced the capacity of the Influenza A virus to infect mammalian cells when exposed to visible light. Iannazzo et al. [76] investigated the anti-HIV efficacy of several functionalized MWCNTs. The findings demonstrated that the antiviral activity of functionalized MWCNTs was regulated by their hydrophilic functionality and water dispersibility. Thus, CNT can also be employed in a bimodal "detection-inhibition" role to fight against viruses. Finally, further research into the nanotechnology used in virus detection is needed to achieve novel platforms that might completely revolutionize the current viral identification systems used in clinics.

Author Contributions: Conceptualization, C.M.H.; methodology, J.S.; writing-original draft preparation, J.S.; writing-review and editing, C.M.H. All authors have read and agreed to the published version of the manuscript.

Funding: This research received no external funding.

Institutional Review Board Statement: Not applicable.

Informed Consent Statement: Not applicable.

Data Availability Statement: Not applicable.

Conflicts of Interest: The authors declare no conflict of interest.

References

1. Radushkevich, L.V.; Lukyanovich, V.M. The Structure of Carbon Produced by Thermal Decomposition of Carbon Monoxide on an Iron Catalyst. *Russ. J. Phys. Chem.* **1952**, *26*, 88–95.
2. Iijima, S. Helical Microtubules of Graphitic Carbon. *Nature* **1991**, *354*, 56–58. [CrossRef]
3. Nurazzi Norizan, M.; Harussani Moklis, M.; Demon, S.Z.N.; Abdul Halim, N.; Samsuri, A.; Syakir Mohamad, I.; Feizal Knight, V.; Abdullah, N. Carbon Nanotubes: Functionalisation and Their Application in Chemical Sensors. *RSC Adv.* **2020**, *10*, 43704–43732. [CrossRef]
4. Chemistry (IUPAC), T.I.U. of P. and A. IUPAC-Biosensor (B00663). Available online: https://goldbook.iupac.org/terms/view/B00663 (accessed on 18 October 2022).
5. Clark, L.C., Jr.; Lyons, C. Electrode Systems for Continuous Monitoring in Cardiovascular Surgery. *Ann. N.Y. Acad. Sci.* **1962**, *102*, 29–45. [CrossRef] [PubMed]
6. Parry, R.P.; Love, C.; Robinson, G.A. Detection of Rubella Antibody Using an Optical Immunosensor. *J. Virol. Methods* **1990**, *27*, 39–48. [CrossRef]
7. Davis, J.J.; Green, M.L.H.; Allen, O.; Hill, H.; Leung, Y.C.; Sadler, P.J.; Sloan, J.; Xavier, A.V.; Chi Tsang, S. The Immobilisation of Proteins in Carbon Nanotubes. *Inorg. Chim. Acta* **1998**, *272*, 261–266. [CrossRef]
8. Balavoine, F.; Schultz, P.; Richard, C.; Mallouh, V.; Ebbesen, T.W.; Mioskowski, C. Helical Crystallization of Proteins on Carbon Nanotubes: A First Step towards the Development of New Biosensors. *Angew. Chem. Int. Ed.* **1999**, *38*, 1912–1915. [CrossRef]
9. Tîlmaciu, C.-M.; Morris, M.C. Carbon Nanotube Biosensors. *Front. Chem.* **2015**, *3*, 59. [CrossRef]
10. Sharma, A.; Agrawal, A.; Kumar, S.; Awasthi, K.K.; Awasthi, K.; Awasthi, A. 23-Zinc Oxide Nanostructures–Based Biosensors. In *Nanostructured Zinc Oxide*; Awasthi, K., Ed.; Metal Oxides; Elsevier: Amsterdam, The Netherlands, 2021; pp. 655–695, ISBN 978-0-12-818900-9.
11. Bergveld, P. Development of an Ion-Sensitive Solid-State Device for Neurophysiological Measurements. *IEEE Trans. Biomed. Eng.* **1970**, *BME-17*, 70–71. [CrossRef] [PubMed]
12. Zhang, Z.; Cong, Y.; Huang, Y.; Du, X. Nanomaterials-Based Electrochemical Immunosensors. *Micromachines* **2019**, *10*, 397. [CrossRef] [PubMed]
13. Masurkar, N.; Varma, S.; Mohana Reddy Arava, L. Supported and Suspended 2D Material-Based FET Biosensors. *Electrochem* **2020**, *1*, 260–277. [CrossRef]
14. Chen, Y.; Liu, J.; Yang, Z.; Wilkinson, J.S.; Zhou, X. Optical Biosensors Based on Refractometric Sensing Schemes: A Review. *Biosens. Bioelectron.* **2019**, *144*, 111693. [CrossRef] [PubMed]
15. Dridi, F.; Marrakchi, M.; Gargouri, M.; Saulnier, J.; Jaffrezic-Renault, N.; Lagarde, F. 5-Nanomaterial-Based Electrochemical Biosensors for Food Safety and Quality Assessment. In *Nanobiosensors*; Grumezescu, A.M., Ed.; Academic Press: Cambridge, MA, USA, 2017; pp. 167–204, ISBN 978-0-12-804301-1.
16. Pinals, R.L.; Ledesma, F.; Yang, D.; Navarro, N.; Jeong, S.; Pak, J.E.; Kuo, L.; Chuang, Y.-C.; Cheng, Y.-W.; Sun, H.-Y.; et al. Rapid SARS-CoV-2 Spike Protein Detection by Carbon Nanotube-Based Near-Infrared Nanosensors. *Nano Lett.* **2021**, *21*, 2272–2280. [CrossRef]
17. Shao, W.; Shurin, M.R.; Wheeler, S.E.; He, X.; Star, A. Rapid Detection of SARS-CoV-2 Antigens Using High-Purity Semiconducting Single-Walled Carbon Nanotube-Based Field-Effect Transistors. *ACS Appl. Mater. Interfaces* **2021**, *13*, 10321–10327. [CrossRef]

18. Zamzami, M.A.; Rabbani, G.; Ahmad, A.; Basalah, A.A.; Al-Sabban, W.H.; Nate Ahn, S.; Choudhry, H. Carbon Nanotube Field-Effect Transistor (CNT-FET)-Based Biosensor for Rapid Detection of SARS-CoV-2 (COVID-19) Surface Spike Protein S1. *Bioelectrochemistry* **2022**, *143*, 107982. [CrossRef]
19. Thanihaichelvan, M.; Surendran, S.N.; Kumanan, T.; Sutharsini, U.; Ravirajan, P.; Valluvan, R.; Tharsika, T. Selective and Electronic Detection of COVID-19 (Coronavirus) Using Carbon Nanotube Field Effect Transistor-Based Biosensor: A Proof-of-Concept Study. *Mater. Today Proc.* **2022**, *49*, 2546–2549. [CrossRef]
20. Cardoso, A.R.; Alves, J.F.; Frasco, M.F.; Piloto, A.M.; Serrano, V.; Mateus, D.; Sebastião, A.I.; Matos, A.M.; Carmo, A.; Cruz, T.; et al. An Ultra-Sensitive Electrochemical Biosensor Using the Spike Protein for Capturing Antibodies against SARS-CoV-2 in Point-of-Care. *Mater. Today Bio* **2022**, *16*, 100354. [CrossRef] [PubMed]
21. Curti, F.; Fortunati, S.; Knoll, W.; Giannetto, M.; Corradini, R.; Bertucci, A.; Careri, M. A Folding-Based Electrochemical Aptasensor for the Single-Step Detection of the SARS-CoV-2 Spike Protein. *ACS Appl. Mater. Interfaces* **2022**, *14*, 19204–19211. [CrossRef]
22. Li, T.; Soelberg, S.D.; Taylor, Z.; Sakthivelpathi, V.; Furlong, C.E.; Kim, J.-H.; Ahn, S.; Han, P.D.; Starita, L.M.; Zhu, J.; et al. Highly Sensitive Immunoresistive Sensor for Point-Of-Care Screening for COVID-19. *Biosensors* **2022**, *12*, 149. [CrossRef]
23. Kim, S.Y.; Lee, J.-C.; Seo, G.; Woo, J.H.; Lee, M.; Nam, J.; Sim, J.Y.; Kim, H.-R.; Park, E.C.; Park, S. Computational Method-Based Optimization of Carbon Nanotube Thin-Film Immunosensor for Rapid Detection of SARS-CoV-2 Virus. *Small Sci.* **2022**, *2*, 2100111. [CrossRef] [PubMed]
24. Dias, A.C.M.S.; Gomes-Filho, S.L.R.; Silva, M.M.S.; Dutra, R.F. A Sensor Tip Based on Carbon Nanotube-Ink Printed Electrode for the Dengue Virus NS1 Protein. *Biosens. Bioelectron.* **2013**, *44*, 216–221. [CrossRef]
25. Silva, M.M.S.; Dias, A.C.M.S.; Silva, B.V.M.; Gomes-Filho, S.L.R.; Kubota, L.T.; Goulart, M.O.F.; Dutra, R.F. Electrochemical Detection of Dengue Virus NS1 Protein with a Poly(Allylamine)/Carbon Nanotube Layered Immunoelectrode. *J. Chem. Technol. Biotechnol.* **2015**, *90*, 194–200. [CrossRef]
26. Wasik, D.; Mulchandani, A.; Yates, M.V. A Heparin-Functionalized Carbon Nanotube-Based Affinity Biosensor for Dengue Virus. *Biosens. Bioelectron.* **2017**, *91*, 811–816. [CrossRef] [PubMed]
27. Wasik, D.; Mulchandani, A.; Yates, M.V. Point-of-Use Nanobiosensor for Detection of Dengue Virus NS1 Antigen in Adult Aedes Aegypti: A Potential Tool for Improved Dengue Surveillance. *Anal. Chem.* **2018**, *90*, 679–684. [CrossRef] [PubMed]
28. Wasik, D.; Mulchandani, A.; Yates, M.V. Salivary Detection of Dengue Virus NS1 Protein with a Label-Free Immunosensor for Early Dengue Diagnosis. *Sensors* **2018**, *18*, 2641. [CrossRef]
29. Palomar, Q.; Gondran, C.; Marks, R.; Cosnier, S.; Holzinger, M. Impedimetric Quantification of Anti-Dengue Antibodies Using Functional Carbon Nanotube Deposits Validated with Blood Plasma Assays. *Electrochim. Acta* **2018**, *274*, 84–90. [CrossRef]
30. Palomar, Q.; Xu, X.; Gondran, C.; Holzinger, M.; Cosnier, S.; Zhang, Z. Voltammetric Sensing of Recombinant Viral Dengue Virus 2 NS1 Based on Au Nanoparticle–Decorated Multiwalled Carbon Nanotube Composites. *Microchim. Acta* **2020**, *187*, 363. [CrossRef]
31. Mendonça, P.D.; Santos, L.K.B.; Foguel, M.V.; Rodrigues, M.A.B.; Cordeiro, M.T.; Gonçalves, L.M.; Marques, E.T.A.; Dutra, R.F. NS1 Glycoprotein Detection in Serum and Urine as an Electrochemical Screening Immunosensor for Dengue and Zika Virus. *Anal. Bioanal. Chem.* **2021**, *413*, 4873–4885. [CrossRef]
32. Zhu, X.; Ai, S.; Chen, Q.; Yin, H.; Xu, J. Label-Free Electrochemical Detection of Avian Influenza Virus Genotype Utilizing Multi-Walled Carbon Nanotubes–Cobalt Phthalocyanine–PAMAM Nanocomposite Modified Glassy Carbon Electrode. *Electrochem. Commun.* **2009**, *11*, 1543–1546. [CrossRef]
33. Tam, P.D.; Van Hieu, N.; Chien, N.D.; Le, A.-T.; Anh Tuan, M. DNA Sensor Development Based on Multi-Wall Carbon Nanotubes for Label-Free Influenza Virus (Type A) Detection. *J. Immunol. Methods* **2009**, *350*, 118–124. [CrossRef]
34. Bonanni, A.; Pividori, M.I.; Valle, M. del Impedimetric Detection of Influenza A (H1N1) DNA Sequence Using Carbon Nanotubes Platform and Gold Nanoparticles Amplification. *Analyst* **2010**, *135*, 1765–1772. [CrossRef]
35. Lee, D.; Chander, Y.; Goyal, S.M.; Cui, T. Carbon Nanotube Electric Immunoassay for the Detection of Swine Influenza Virus H1N1. *Biosens. Bioelectron.* **2011**, *26*, 3482–3487. [CrossRef]
36. Singh, R.; Sharma, A.; Hong, S.; Jang, J. Electrical Immunosensor Based on Dielectrophoretically-Deposited Carbon Nanotubes for Detection of Influenza Virus H1N1. *Analyst* **2014**, *139*, 5415–5421. [CrossRef]
37. Lee, J.; Ahmed, S.R.; Oh, S.; Kim, J.; Suzuki, T.; Parmar, K.; Park, S.S.; Lee, J.; Park, E.Y. A Plasmon-Assisted Fluoro-Immunoassay Using Gold Nanoparticle-Decorated Carbon Nanotubes for Monitoring the Influenza Virus. *Biosens. Bioelectron.* **2015**, *64*, 311–317. [CrossRef]
38. Lee, J.; Morita, M.; Takemura, K.; Park, E.Y. A Multi-Functional Gold/Iron-Oxide Nanoparticle-CNT Hybrid Nanomaterial as Virus DNA Sensing Platform. *Biosens. Bioelectron.* **2018**, *102*, 425–431. [CrossRef] [PubMed]
39. Ahmed, S.R.; Kim, J.; Suzuki, T.; Lee, J.; Park, E.Y. Enhanced Catalytic Activity of Gold Nanoparticle-Carbon Nanotube Hybrids for Influenza Virus Detection. *Biosens. Bioelectron.* **2016**, *85*, 503–508. [CrossRef] [PubMed]
40. Fu, Y.; Romay, V.; Liu, Y.; Ibarlucea, B.; Baraban, L.; Khavrus, V.; Oswald, S.; Bachmatiuk, A.; Ibrahim, I.; Rümmeli, M.; et al. Chemiresistive Biosensors Based on Carbon Nanotubes for Label-Free Detection of DNA Sequences Derived from Avian Influenza Virus H5N1. *Sens. Actuators B Chem.* **2017**, *249*, 691–699. [CrossRef]
41. Tran, T.L.; Nguyen, T.T.; Huyen Tran, T.T.; Chu, V.T.; Thinh Tran, Q.; Tuan Mai, A. Detection of Influenza A Virus Using Carbon Nanotubes Field Effect Transistor Based DNA Sensor. *Phys. E Low-Dimens. Syst. Nanostruct.* **2017**, *93*, 83–86. [CrossRef]

42. Wang, F.; Gopinath, S.C.; Lakshmipriya, T. Aptamer-Antibody Complementation on Multiwalled Carbon Nanotube-Gold Transduced Dielectrode Surfaces to Detect Pandemic Swine Influenza Virus. *IJN* **2019**, *14*, 8469–8481. [CrossRef] [PubMed]
43. Lee, D.; Bhardwaj, J.; Jang, J. Paper-Based Electrochemical Immunosensor for Label-Free Detection of Multiple Avian Influenza Virus Antigens Using Flexible Screen-Printed Carbon Nanotube-Polydimethylsiloxane Electrodes. *Sci. Rep.* **2022**, *12*, 2311. [CrossRef] [PubMed]
44. Mahmoud, K.A.; Luong, J.H.T. Impedance Method for Detecting HIV-1 Protease and Screening for Its Inhibitors Using Ferrocene-Peptide Conjugate/Au Nanoparticle/Single-Walled Carbon Nanotube Modified Electrode. *Anal. Chem.* **2008**, *80*, 7056–7062. [CrossRef] [PubMed]
45. Mahmoud, K.A.; Luong, J.H.T. A Sensitive Electrochemical Assay for Early Detection of HIV-1 Protease Using Ferrocene-Peptide Conjugate/Au Nanoparticle/Single Walled Carbon Nanotube Modified Electrode. *Anal. Lett.* **2010**, *43*, 1680–1687. [CrossRef]
46. Kheiri, F.; Sabzi, R.E.; Jannatdoust, E.; Shojaeefar, E.; Sedghi, H. A Novel Amperometric Immunosensor Based on Acetone-Extracted Propolis for the Detection of the HIV-1 P24 Antigen. *Biosens. Bioelectron.* **2011**, *26*, 4457–4463. [CrossRef] [PubMed]
47. Ma, Y.; Shen, X.-L.; Zeng, Q.; Wang, H.-S.; Wang, L.-S. A Multi-Walled Carbon Nanotubes Based Molecularly Imprinted Polymers Electrochemical Sensor for the Sensitive Determination of HIV-P24. *Talanta* **2017**, *164*, 121–127. [CrossRef] [PubMed]
48. Giannetto, M.; Costantini, M.; Mattarozzi, M.; Careri, M. Innovative Gold-Free Carbon Nanotube/Chitosan-Based Competitive Immunosensor for Determination of HIV-Related P24 Capsid Protein in Serum. *RSC Adv.* **2017**, *7*, 39970–39976. [CrossRef]
49. Harvey, J.D.; Baker, H.A.; Ortiz, M.V.; Kentsis, A.; Heller, D.A. HIV Detection via a Carbon Nanotube RNA Sensor. *ACS Sens.* **2019**, *4*, 1236–1244. [CrossRef]
50. Lu, Q.; Su, T.; Shang, Z.; Jin, D.; Shu, Y.; Xu, Q.; Hu, X. Flexible Paper-Based Ni-MOF Composite/AuNPs/CNTs Film Electrode for HIV DNA Detection. *Biosens. Bioelectron.* **2021**, *184*, 113229. [CrossRef] [PubMed]
51. Li, X.-M.; Zhan, Z.-M.; Ju, H.-Q.; Zhang, S.-S. Label-Free Electrochemical Detection of Short Sequences Related to the Hepatitis B Virus Using 4,4′-Diaminoazobenzene Based on Multiwalled Carbon Nanotube-Modified GCE. *Oligonucleotides* **2008**, *18*, 321–328. [CrossRef]
52. Oh, J.; Yoo, S.; Chang, Y.W.; Lim, K.; Yoo, K.-H. Carbon Nanotube-Based Biosensor for Detection Hepatitis B. *Curr. Appl. Phys.* **2009**, *9*, e229–e231. [CrossRef]
53. Ly, S.Y.; Cho, N.S. Diagnosis of Human Hepatitis B Virus in Non-Treated Blood by the Bovine IgG DNA-Linked Carbon Nanotube Biosensor. *J. Clin. Virol.* **2009**, *44*, 43–47. [CrossRef]
54. Hu, Y.; Zhao, Z.; Wan, Q. Facile Preparation of Carbon Nanotube-Conducting Polymer Network for Sensitive Electrochemical Immunoassay of Hepatitis B Surface Antigen in Serum. *Bioelectrochemistry* **2011**, *81*, 59–64. [CrossRef] [PubMed]
55. Cabral, D.G.A.; Lima, E.C.S.; Moura, P.; Dutra, R.F. A Label-Free Electrochemical Immunosensor for Hepatitis B Based on Hyaluronic Acid–Carbon Nanotube Hybrid Film. *Talanta* **2016**, *148*, 209–215. [CrossRef]
56. Trindade, E.K.G.; Dutra, R.F. A Label-Free and Reagentless Immunoelectrode for Antibodies against Hepatitis B Core Antigen (Anti-HBc) Detection. *Colloids Surf. B Biointerfaces* **2018**, *172*, 272–279. [CrossRef]
57. Chen, Z.; Bai, Y.; Zhao, F.; Cao, L.; Han, G.; Yin, S. Disposable Amperometric Immunosensor for Hepatitis B Antigen Detection Based on Multiwalled Carbon Nanotubes and Ferrocene Decorated Screen Printed Electrode. *J. Biomed. Nanotechnol.* **2019**, *15*, 930–938. [CrossRef]
58. Upan, J.; Banet, P.; Aubert, P.-H.; Ounnunkad, K.; Jakmunee, J. Sequential Injection-Differential Pulse Voltammetric Immunosensor for Hepatitis B Surface Antigen Using the Modified Screen-Printed Carbon Electrode. *Electrochim. Acta* **2020**, *349*, 136335. [CrossRef]
59. Dastagir, T.; Forzani, E.S.; Zhang, R.; Amlani, I.; Nagahara, L.A.; Tsui, R.; Tao, N. Electrical Detection of Hepatitis C Virus RNA on Single Wall Carbon Nanotube-Field Effect Transistors. *Analyst* **2007**, *132*, 738–740. [CrossRef] [PubMed]
60. Pusomjit, P.; Teengam, P.; Chuaypen, N.; Tangkijvanich, P.; Thepsuparungsikul, N.; Chailapakul, O. Electrochemical Immunoassay for Detection of Hepatitis C Virus Core Antigen Using Electrode Modified with Pt-Decorated Single-Walled Carbon Nanotubes. *Microchim. Acta* **2022**, *189*, 339. [CrossRef]
61. Lacerda, L.; Raffa, S.; Prato, M.; Bianco, A.; Kostarelos, K. Cell-Penetrating CNTs for Delivery of Therapeutics. *Nano Today* **2007**, *2*, 38–43. [CrossRef]
62. Traylen, C.M.; Patel, H.R.; Fondaw, W.; Mahatme, S.; Williams, J.F.; Walker, L.R.; Dyson, O.F.; Arce, S.; Akula, S.M. Virus Reactivation: A Panoramic View in Human Infections. *Future Virol.* **2011**, *6*, 451–463. [CrossRef]
63. Suzich, J.B.; Cliffe, A.R. Strength in Diversity: Understanding the Pathways to Herpes Simplex Virus Reactivation. *Virology* **2018**, *522*, 81–91. [CrossRef] [PubMed]
64. Kerr, J.R. Epstein-Barr Virus (EBV) Reactivation and Therapeutic Inhibitors. *J. Clin. Pathol.* **2019**, *72*, 651–658. [CrossRef]
65. Cook, C.H.; Trgovcich, J. Cytomegalovirus Reactivation in Critically Ill Immunocompetent Hosts: A Decade of Progress and Remaining Challenges. *Antivir. Res.* **2011**, *90*, 151–159. [CrossRef]
66. Cary, D.C.; Fujinaga, K.; Peterlin, B.M. Molecular Mechanisms of HIV Latency. *J. Clin. Investig.* **2016**, *126*, 448–454. [CrossRef] [PubMed]
67. Ozer, T.; Henry, C.S. Paper-Based Analytical Devices for Virus Detection: Recent Strategies for Current and Future Pandemics. *TrAC Trends Anal. Chem.* **2021**, *144*, 116424. [CrossRef] [PubMed]
68. Beduk, T.; Beduk, D.; Hasan, M.R.; Guler Celik, E.; Kosel, J.; Narang, J.; Salama, K.N.; Timur, S. Smartphone-Based Multiplexed Biosensing Tools for Health Monitoring. *Biosensors* **2022**, *12*, 583. [CrossRef]

69. Requardt, H.; Braun, A.; Steinberg, P.; Hampel, S.; Hansen, T. Surface Defects Reduce Carbon Nanotube Toxicity in Vitro. *Toxicol. Vitr.* **2019**, *60*, 12–18. [CrossRef]
70. Qi, W.; Tian, L.; An, W.; Wu, Q.; Liu, J.; Jiang, C.; Yang, J.; Tang, B.; Zhang, Y.; Xie, K.; et al. Curing the Toxicity of Multi-Walled Carbon Nanotubes through Native Small-Molecule Drugs. *Sci. Rep.* **2017**, *7*, 2815. [CrossRef]
71. Kobayashi, N.; Izumi, H.; Morimoto, Y. Review of Toxicity Studies of Carbon Nanotubes. *J. Occup. Health* **2017**, *59*, 394–407. [CrossRef]
72. Sengupta, J.; Hussain, C.M. Prospective Pathways of Green Graphene-Based Lab-on-Chip Devices: The Pursuit toward Sustainability. *Microchim. Acta* **2022**, *189*, 177. [CrossRef]
73. SARS-CoV-2 Variants of Concern as of 27 October 2022. Available online: https://www.ecdc.europa.eu/en/covid-19/variants-concern (accessed on 3 November 2022).
74. Xi, H.; Jiang, H.; Juhas, M.; Zhang, Y. Multiplex Biosensing for Simultaneous Detection of Mutations in SARS-CoV-2. *ACS Omega* **2021**, *6*, 25846–25859. [CrossRef]
75. Banerjee, I.; Douaisi, M.P.; Mondal, D.; Kane, R.S. Light-Activated Nanotube–Porphyrin Conjugates as Effective Antiviral Agents. *Nanotechnology* **2012**, *23*, 105101. [CrossRef]
76. Iannazzo, D.; Pistone, A.; Galvagno, S.; Ferro, S.; De Luca, L.; Monforte, A.M.; Da Ros, T.; Hadad, C.; Prato, M.; Pannecouque, C. Synthesis and Anti-HIV Activity of Carboxylated and Drug-Conjugated Multi-Walled Carbon Nanotubes. *Carbon* **2015**, *82*, 548–561. [CrossRef]

Article

Bioinspired Spinosum Capacitive Pressure Sensor Based on CNT/PDMS Nanocomposites for Broad Range and High Sensitivity

Yanhao Duan [1,2], Jian Wu [1,2,*], Shixue He [1,2], Benlong Su [1,2], Zhe Li [2] and Youshan Wang [1,2]

1 National Key Laboratory of Science and Technology on Advanced Composites in Special Environments, Harbin Institute of Technology, Harbin 150090, China
2 Center for Rubber Composite Materials and Structures, Harbin Institute of Technology, Weihai 264209, China
* Correspondence: wujian@hitwh.edu.cn

Abstract: Flexible pressure sensors have garnered much attention recently owing to their prospective applications in fields such as structural health monitoring. Capacitive pressure sensors have been extensively researched due to their exceptional features, such as a simple structure, strong repeatability, minimal loss and temperature independence. Inspired by the skin epidermis, we report a high-sensitivity flexible capacitive pressure sensor with a broad detection range comprising a bioinspired spinosum dielectric layer. Using an abrasive paper template, the bioinspired spinosum was fabricated using carbon nanotube/polydimethylsiloxane (CNT/PDMS) composites. It was observed that nanocomposites comprising 1 wt% CNTs had excellent sensing properties. These capacitive pressure sensors allowed them to function at a wider pressure range (~500 kPa) while maintaining sensitivity (0.25 kPa^{-1}) in the range of 0–50 kPa, a quick response time of approximately 20 ms and a high stability even after 10,000 loading–unloading cycles. Finally, a capacitive pressure sensor array was created to detect the deformation of tires, which provides a fresh approach to achieving intelligent tires.

Keywords: flexible capacitive sensors; bioinspired spinosum; CNT/PDMS nanocomposite; tires

Citation: Duan, Y.; Wu, J.; He, S.; Su, B.; Li, Z.; Wang, Y. Bioinspired Spinosum Capacitive Pressure Sensor Based on CNT/PDMS Nanocomposites for Broad Range and High Sensitivity. *Nanomaterials* **2022**, *12*, 3265. https://doi.org/10.3390/nano12193265

Academic Editor: Muralidharan Paramsothy

Received: 17 August 2022
Accepted: 15 September 2022
Published: 20 September 2022

Publisher's Note: MDPI stays neutral with regard to jurisdictional claims in published maps and institutional affiliations.

Copyright: © 2022 by the authors. Licensee MDPI, Basel, Switzerland. This article is an open access article distributed under the terms and conditions of the Creative Commons Attribution (CC BY) license (https://creativecommons.org/licenses/by/4.0/).

1. Introduction

Flexible pressure sensors have become a rapidly growing research area with significant progress in 5G communication and artificial intelligence [1]. Currently, flexible pressure sensors are widely used in the monitoring field, such as in examining human health [2,3] and structural health detection [4,5]. Compared with rigid sensors, flexible pressure sensors can withstand various deformations, such as bend, compression and torsion [6,7]. With the introduction of modern technologies, such as network technology and cloud computing, intelligence has become a significant trend in automotive technology development and tires intelligence. The tires are the only part that contacts the vehicle and the road and have attracted considerable attention in improving vehicle safety and reducing fuel consumption. Measuring or estimating tire force is vital for analyzing and controlling vehicle behavior [8]. The contact patch features and the contact pressure of the tires are significant for the force estimation [9]. However, measuring exact forces in the real world has become a concern for researchers. The sensors built into the tires to form intelligent tires can provide reliable quantities of tire variables compared with traditional test methods based on onboard sensors [10]. The use of flexible pressure sensors for tires is a promising method for detecting the contact pressure of tires. Matsuzaki et al. fabricated a novel rubber-based sensor and attached it to the inner surface of a tire. The experiment showed that the sensor could accurately monitor the tire's behavior [11]. Son et al. examined a reliable triboelectric bicycle tire by inserting a dielectric electrode layer between the tire tread and the inner tube [12]. The pressure and deformation of the tire can be monitored in real time by the sensor. According to working mechanisms, flexible pressure sensors are mainly based on piezoresistive [13,14], capacitive [15,16], piezoelectric [17,18] and triboelectric effects [19,20]. Capacitive pressure sensors have attracted considerable research attention owing to their simple structure, good repeatability, low loss and temperature independence, which are considered the ideal choice for structural health testing [21]. These pressure sensors adopt a sandwich structure consisting of two electrode layers at the top and bottom and one sensing layer in the middle. Polydimethylsiloxane (PDMS) has been employed widely as a flexible substrate because of its excellent characteristics, including flexibility, biocompatibility and mechanical properties [22]. With the progress of materials science, materials with different microstructures have been used extensively to fabricate pressure sensors. In particular, carbon-based functional materials, such as nanotubes [23], graphene [24], carbon black [25] and graphene oxide [26], have been used broadly as conductive fillers in sensors because of their low cost and outstanding electrical and mechanical properties. Adding conductive fillers to the dielectric layer can produce a higher dielectric constant and improve the signal magnitude. Furthermore, to enhance the sensitivity of capacitive sensors, the compressibility and effective dielectric constant change value can be improved by patterning the dielectric layer, such as by adding microstructures [27,28]. In 2010, Mannsfeld et al. first added a micropyramid structure to capacitive sensors, significantly improving their sensitivity [29]. In addition to the micropyramid structure, some other microstructures, such as micropillar [30,31], microdome [32,33], wrinkle [34] and microporous [35,36], have been explored. Despite the considerable progress on capacitive sensors, the large-scale preparation of low-cost, high-sensitivity and broad-range capacitive pressure sensors presents enormous challenges. Many structures in nature provide humans with examples of good design, which have inspired researchers in bioinspired pressure sensors [37]. Skin plays an important role in the perception of force, and the dermis with a spinosum microstructure makes skin sensitive to external forces [38]. Bioinspired pressure sensors with a spinosum surface have been used for the development of low-cost pressure sensors with a high sensitivity and a large linearity [39]. Little research has been conducted on the spinosum microstructure for capacitive pressure sensor design. Flexible sensor arrays can achieve spatial pressure distribution monitoring and real-time trajectory mapping, which are used widely in various fields [40]. Using flexible sensor arrays on tires may achieve real-time detection of their contact pressure and provide more accurate quantities of the tire variables.

This article reports a bioinspired spinosum capacitive pressure sensor based on carbon nanotube/polydimethylsiloxane (CNT/PDMS) nanocomposites for detecting the contact pressure of tires. CNT/PDMS-based spinosum pressure sensors inspired by the epidermis tissue structure in human skin were fabricated. The spinosum microstructure of the dermis has a high similarity in topography with abrasive paper. A capacitive pressure sensor with a high sensitivity and a large sensing range can be achieved using abrasive papers with different surface roughness as a template. The flexible pressure sensor exhibited a high sensitivity of 0.25 kPa^{-1} in the range of 0–50 kPa, a wide effective working range of 0–500 kPa, ultrafast response and relaxation times of 20 ms and excellent cycling stability (>10,000 cycles). As shown in Table 1, our sensor exhibits a higher sensitivity and a fast response time compared with reported capacitive pressure sensors in the literature. CNT/PDMS-based spinosum pressure sensors are excellent candidates for health monitoring. Based on the novel design of our sensor, a sensor array was used successfully in the quantitative monitoring of contact pressure on a tire.

Table 1. Comparison of the performance between our work and recently reported flexible capacitive pressure sensors.

Dielectric Layer	Materials	Sensing Range (kPa)	Sensitivity (kPa^{-1})	Response Time (ms)	Ref
pillars	SU8	0–30	0.0065	70	[41]
porous	PS/graphene/MWCNTs	0–4.5	0.062	45	[42]
microdome	PDMS	0.5–10	0.055	200	[43]
pyramid	PDMS	0.5–6	0.185	-	[44]
microridge	PDMS	0–10	0.148	20	[45]
spinosum	CNT/PDMS	0–500	0.25	20	This study

2. Materials and Methods

2.1. Preparation of Spinosum Microstructure CNT/PDMS Films as the Dielectric Layer

Commercial abrasive papers (Hubei Yuli Abrasive Belt Group, Hubei, China) with three roughness values, no. 80, 320 and 600, were utilized as templates. Nanocomposites with different compositions (mass ratio of PDMS/CNTs = 10:X, X = 0, 0.05, 0.1 and 0.15) were prepared to fabricate the composite dielectric layer. CNTs (purity: 95%, diameter: 3–15 nm, length: 15–30 μm, Shenzhen Suiheng Technology Co., Ltd., Guangdong, China) were dispersed in alcohol (10 g, 99.7%, Ziyansheng Fine Chemical Co., Ltd., Shanghai, China) for 2 h with ultrasonic assistance. Then, a PDMS base (9.1 g) (Sylgard 184, Dow Corning Co., Ltd., Midland, MI, United States) was dispersed in the alcohol/CNT solution for 2 h with magnetic stirring. Then, the alcohol was allowed to evaporate at 150 °C. After the solution cooled to room temperature (22 °C), a curing agent (0.9 g) (Sylgard 184, Dow Corning Co., Ltd., Midland, MI, United States) was added and the solution was stirred for 1 h. Then, the degassed CNT/PDMS solution was poured onto the abrasive paper and cured at room temperature for 12 h. Finally, the CNT/PDMS films were carefully peeled off to obtain the spinosum microstructure. The thickness of the dielectric layer was 1 mm.

2.2. Preparation of the Sensor Array

The bottom and top electrodes of the sensor array, both containing strip electrodes, were orthogonally mounted face to face with the CNT/PDMS spinosum dielectric layer, forming a sandwich structure. A capacitive pressure sensor array was then created.

2.3. Characterization of the Morphology and Performance of the CNT/PDMS-Based Spinosum Pressure Sensors

The morphology and structure of the fabricated sensors were characterized via a field emission scanning electron microscope (SEM, MERLIN, Zeiss, Jena, Germany). The fabricated sensors' 3D morphology and structure were characterized via an optical microscope (DSX 510, Olympus, Tokyo, Japan). The contact area of the dielectric layer was characterized with a high-speed CCD camera (Yvision Technology Company, Guangdong, China). The loading of applied force was carried out with a testing machine (WDW-02, STAR Testing Technology Co., Ltd., Shandong, China), while the electrical signals of the pressure sensors and the permittivity were recorded at the same time via an LCR meter (TH2829A, Changzhou Tonghui Electronic Co. Ltd., Jiangsu, China) at a 100 kHz frequency with a 0–1 V alternating voltage bias. A fatigue-testing machine (FLPL203E, FULETEST Instrument Technology Co., Ltd., Shanghai, China) was used to test the response/release time, and the depressing speed and rising speed were both 1000 mm/min. The stability test also used the fatigue testing machine (FLPL203E, FULETEST, Instrument Technology Co., Ltd., Shanghai, China), and the depressing speed and rising speed were both 1000 mm/min.

2.4. Finite Element Analysis

Finite element analysis (FEA) was performed using the commercial package ABAQUS. The CNT/PDMS-based spinosum microstructure dielectric layer was modeled as an incompressible material with Young's modulus E~1.4 MPa. The Cu electrode was simply treated as a rigid plate and compressed downward. All contact interactions were assumed to have friction with a friction coefficient of 0.15 without penetration.

3. Results and Discussion

3.1. Designs for Spinosum Capacitive Pressure Sensors

So far, nature has evolved to optimize structures to adapt to complicated environmental conditions, inspiring many engineers and scientists. Skin, one of the largest organs of the human body, can interact with the surrounding environment. Because of the spinosum of the dermis, the skin can sensitively sense stimulation from the environment (Figure 1a). The sensitivity and detection accuracy of pressure sensors can be improved through introducing the spinosum microstructures, which are attributed to the concentrated local stress at the contact region. As shown in Figure 1b, the surface of abrasive paper displays smoothly interconnected ridges and dispersive holes. Because the spinosum microstructure has a similar morphology to that of abrasive paper, we used abrasive papers as a temple to fabricate the spinosum microstructure. Figure 1c depicts the schematic illustration of the fabrication process of the spinosum pressure sensor. The spinosum pressure sensor was fabricated as follows: first, the prepared CNT/PDMS nanocomposite was coated on the abrasive paper template through a blading method and cured at room temperature. Second, the CNT/PDMS film was peeled off from the abrasive paper to obtain the spinosum microstructure. Third, the as-prepared spinosum microstructure CNT/PDMS films were cut into regular sizes (10 mm × 10 mm).. Subsequently, a piece of copper foil tape (Dongguan Xinshi Packaging Materials Co., Ltd. Guangdong, China) was attached to the side of the polyethylene (PE) substrate as bottom electrode. Then another piece of copper foil tape was attached to the flat side of the CNT/PDMS film as top electrode. The CNT/PDMS film was sandwiched between two copper electrodes. Finally, the copper wires were attached onto the copper foil to complete the sensor fabrication. The multilayer was packaged between two transparent and soft 3M tape. As we know, the traditional lithography template method to fabricate pressure sensors is expensive. Compared with the lithography template method, using abrasive paper as a template to fabricate pressure sensors is very cost-effective. The above strategy has the advantages of simplicity, economy and ease of operation, and the filling materials are not expensive.

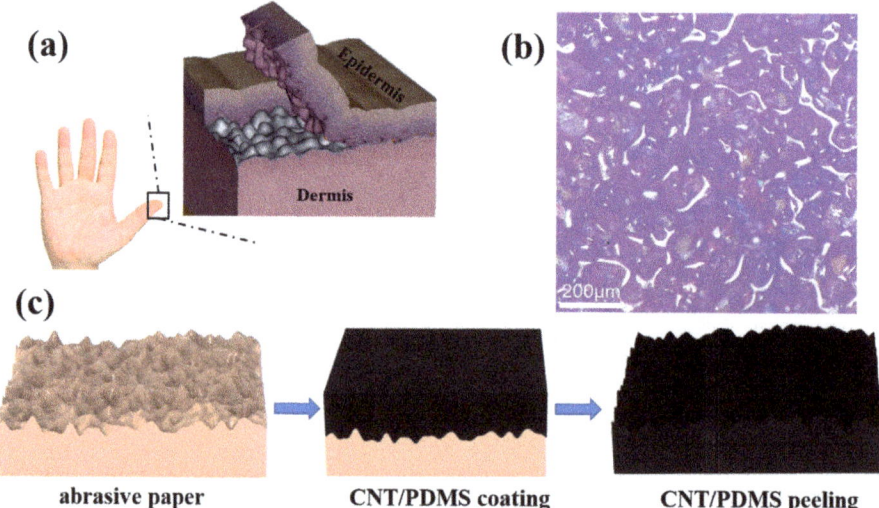

Figure 1. (a) Schematics of the microstructure of human epidermis. (b) Optical microscopy image of the no. 400 abrasive paper. (c) The fabrication process of the spinosum microstructure pressure sensor.

3.2. The Performance of Spinosum Capacitive Pressure Sensors

The pressure sensitivity (S) of capacitive pressure sensors can be defined as $S = (\Delta C/C_0)/\Delta P$, where ΔC is the relative change in capacitance, C_0 represents the initial capacitance when no pressure is applied and ΔP is the change in applied pressure. To investigate the effect of the spinosum microstructure on capacitive sensors, the distributed spinosum microstructure PDMS with increasing roughness was fabricated via abrasive papers no. 80–600. The as-prepared spinosum PDMS films were cut into 10 mm × 10 mm sizes and sandwiched between two copper electrodes. In order to characterize the capacitive responses of the sensors to external pressures, a high-precision test system containing a motorized test stand connected with a force gauge, an LCR meter and computers was used. The pressure load on the pressure sensors was precisely controlled via the force gauge, and the capacitive signals were collected through the LCR meter. In Figure 2a, the $\Delta C/C_0$ versus pressure change of the PDMS-based spinosum pressure sensors with different degrees of roughness using abrasive paper templates no. 80–600 is shown. The result indicates that the spinosum pressure sensor using abrasive paper template no. 600 exhibits the highest sensitivity compared to sensors of other degrees of roughness. However, the sensitivity is relatively low. To further improve the performance of the spinous pressure sensor, the dielectric layer was doped with CNTs to increase the dielectric constant.

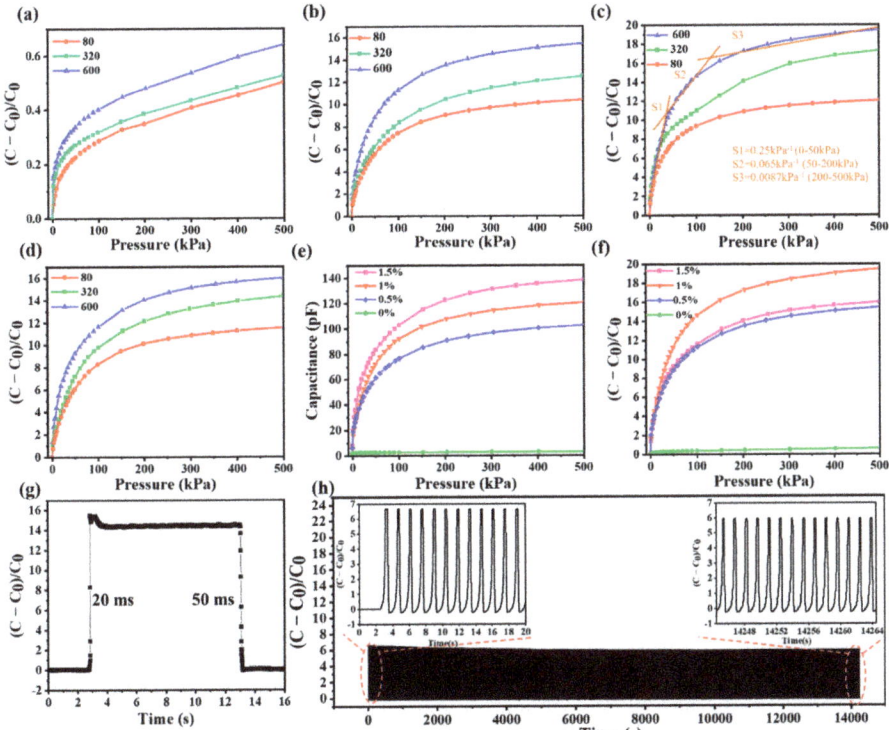

Figure 2. Sensing performances of the pressure sensors. (**a**) Pressure response of relative capacitance change in the sensor with (**a**) 0, (**b**) 0.5, (**c**) 1 and (**d**) 1.5 wt% CNT doping content using different abrasive papers. (**e**) Pressure response of the absolute capacitance of the sensor using abrasive paper no. 600 under different doping contents of CNTs. (**f**) Pressure response of the relative capacitance of the sensor using abrasive paper no. 600 under different doping contents of CNTs. (**g**) Response time and recovery time of the spinosum pressure sensor. (**h**) Durability test of the pressure sensor for 10,000 loading/unloading cycles at the pressure of 30 kPa.

The influence of the CNT doping content and the roughness of the abrasive surface on the performance of spinosum microstructure sensors have been investigated in this study. By using abrasive papers no. 80–600 as templates, the spinosum pressure sensors with different CNT doping content (0.5, 1 and 1.5 wt%) were fabricated and tested the performance of the sensors via the test system.

The CNT/PDMS-based spinosum pressure sensor exhibits a high sensitivity and a wide detection range. As shown in Figure 2b–d, similarly to the previous result, the $\Delta C/C_0$ versus pressure change for different CNT doping content (0.5, 1 and 1.5 wt%), the CNT/PDMS-based spinosum pressure sensor shows a similar regular, which is that the sensor using abrasive paper template no. 600 exhibits the highest sensitivity. The capacitance changes of spinosum pressure sensors fabricated with abrasive paper no. 600 with different amounts of CNTs (0, 0.5, 1 and 1.5 wt%) are plotted in Figure 2e. The regular is obvious: the absolute capacitive of the CNT/PDMS-based spinosum pressure sensors increase with the doping content of CNTs. However, the regular is different when the capacitance change is normalized by initial capacitance. The CNT/PDMS-based spinosum pressure sensors with 1 wt% CNT doping content show the highest sensitivity among the samples tested (Figure 2f). The sensitivity of 1 wt% doping CNT/PDMS-based spinosum pressure sensors prepared via abrasive paper no. 600 is 0.25 kPa^{-1} in the low-pressure range of 0–50 kPa, 0.065 kPa^{-1} for 50–200 kPa, and 0.0087 kPa^{-1} in the high-pressure range of 200–500 kPa.

Furthermore, the optimized design parameters (1 wt% CNT doping and fabricated from the abrasive paper no. 600) were selected from the above experiments and used to study the response time and the stability of the sensor. To investigate the response speed, the response and relaxation time of the pressure sensor were estimated in the process of loading/unloading. As described in Figure 2g, a response time of 20 ms and a relaxation time of 50 ms were observed, which indicate its fast pressure response and potential application in the monitoring field. Moreover, as shown in Figure 2h, the pressure sensor exhibits robust durability in the process of loading/unloading a massive pressure of 30 kPa up to 10,000 cycles. As we can see in the insets of the magnified view, the capacitance signal amplitude did not decrease significantly throughout the 10,000 load/unload test cycles.

3.3. The Effect of the Mesh Number of Abrasive Papers and CNT Doping Content on the Sensing Property

Figure 3a–i show the typical optical image of a spinosum microstructure CNT/PDMS dielectric layer using abrasive papers no. 80, 320 and 600, indicating that the spinosum microstructure of the abrasive paper surface is well-replicated. Generally, the height of the spinosum microstructure decreases as the mesh number of the abrasive paper increases, and the density of the spinosum increases as the mesh number of the abrasive paper increases. The successful fabrication of CNT/PDMS-based spinosum dielectric layers was confirmed through the SEM image, as shown in Figure 3j–l. The microstructures are distributed all over the surface, and as the roughness of the abrasive paper decreases, the microstructures' density increases, and the dimension of these microstructures decreases.

Figure 4 shows that the relative permittivity of CNT/PDMS nanocomposites with different CNT doping content increases as the pressure increases. When the doping content of CNTs reaches 1 wt%, its relative permittivity changes most obviously. In fact, the relative permittivity increase can be explained according to the percolation theory: $\varepsilon \propto (\rho$-$\rho_0)s$, where ε is the relative permittivity, ρ is the percolation threshold and ρ_0 is the CNT concentration [46]. The distance of CNTs decreases during the compression process, which leads to a decline in ρ and an increase in ε.

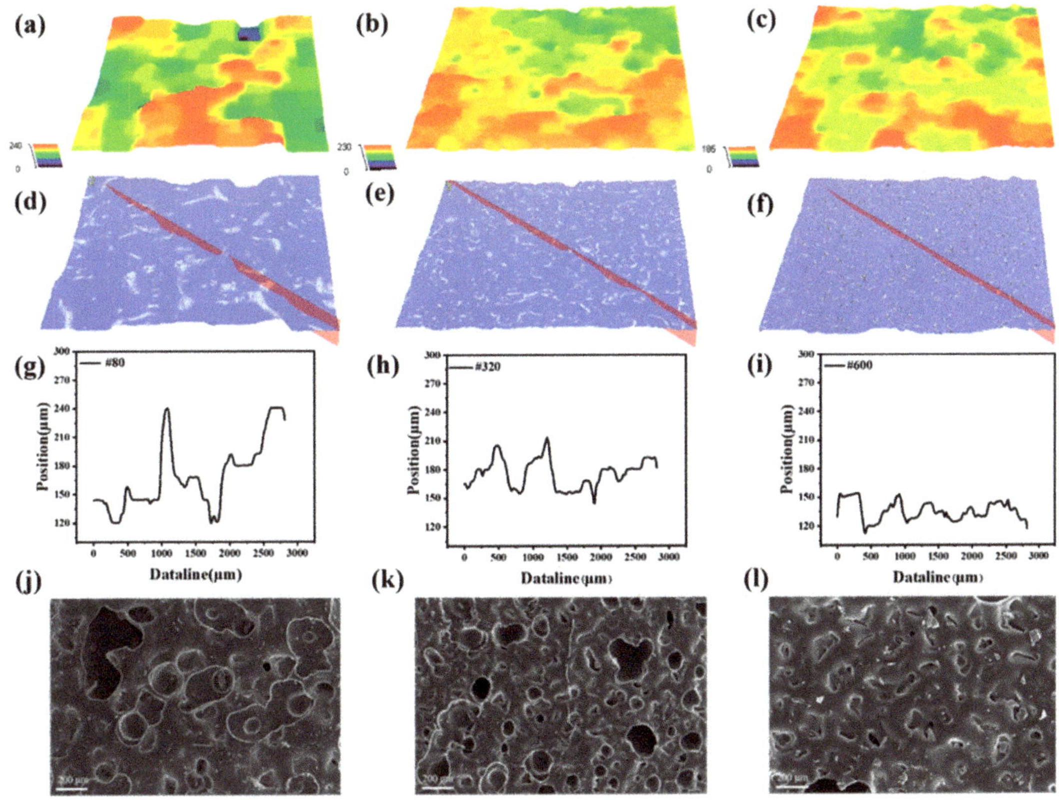

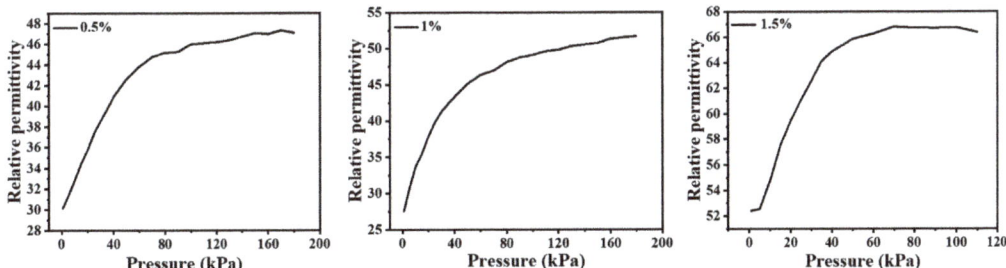

Figure 3. Characterization of the pressure sensors. The 3D morphology of spinosum CNT/PDMS dielectric layer using abrasive papers no. (**a**) 80, (**b**) 320 and (**c**) 600. The position of the marker line of the CNT/PDMS dielectric layer using abrasive papers no. (**d**) 80, (**e**) 320 and (**f**) 600. Height profile corresponding to the marked line on the diagonals using abrasive papers no. (**g**) 80, (**h**) 320 and (**i**) 600. The SEM images of the spinosum CNT/PDMS dielectric layer using abrasive papers no. (**j**) 80, (**k**) 320 and (**l**) 600.

Figure 4. Relative permittivity of the PDMS/CNT nanocomposites with 0.5, 1, and 1.5 wt% CNT under different pressures.

To further reveal the response mechanism of the sensor, the force distribution and deformation process of the spinosum microstructure were studied through finite element analysis. Two models with different roughness values have been conducted to simulate the compression process of the sensor. Model 1 (Figure 5a) has a low density of spinosum, but the height of the spinosum is large, which can represent abrasive papers with a low mesh number (the surface height is represented by color). Model 2 (Figure 5b) has a high density of spinosum but the height of the spinosum is low, which is similar to the abrasive papers with a high mesh number. Figure 5c,d show the stress distribution after the loading of 100 kPa for model 1 and model 2, respectively. As we can see, when under pressure, the stress of the spinosum structure is concentrated at the contact peak. The spinosum microstructure with a lower roughness has a more homogeneous pressure distribution than that with a greater roughness, and the stress concentration area of the spinosum microstructure with a lower roughness is much larger than that with a greater roughness. Thus, the relative permittivity of the underlying dielectric layer will be dramatically increased, which contributes to the stress concentration of the spinosum microstructure. Because model 2's stress concentration area is more extensive than model 1's, the relative permittivity changed obviously. Thus, the CNT/PDMS-based spinosum pressure sensor using abrasive paper no. 600 has a better performance.

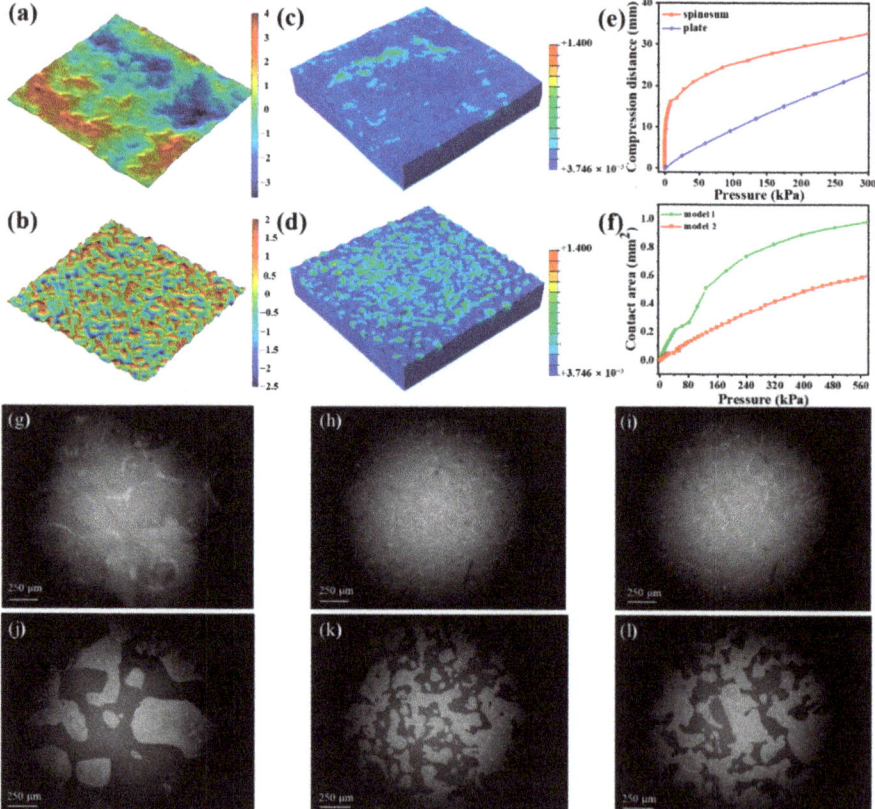

Figure 5. Sensing mechanisms of the pressure sensors. Schematic of the simulation (**a**) model 1 and (**b**) model 2 with different spinosum microstructures. FEA simulation showing the stress distribution of (**c**) model 1 and (**d**) model 2 at a pressure of 100 kPa. (**e**) The FEA simulation result of the compression distance variation for the spinosum and plate dielectric layer. (**f**) The FEA simulation result of the contact area variation between the dielectric layers with different architectures. The image of the spinosum dielectric layer using different abrasive papers no. (**g**) 80, (**h**) 320 and (**i**) 600 without pressure. The image of the contact area of the spinosum dielectric layer using different abrasive papers no. (**j**) 80, (**k**) 320 and (**l**) 600 at a pressure of 180 kPa.

By introducing the spinosum microstructure, the dielectric layer can more readily deform with applied pressure and be more easily compressed than without the microstructure (Figure 5e). As shown in Figure 5f, when under the same pressure, the contact area with electrodes in model 2 is smaller than in model 1. The roughness will decrease as the mesh number of the abrasive paper template increases; the spinosum microstructure with an extensive range of height differences is easily deformed due to stress concentration. Thus, the contact area with electrodes in model 1 is larger than in model 2. The simulation results are consistent with the experimental results. A transparent glass sheet was utilized to replace the electrodes for easy observation, and a 1.8 kg weight was placed above the sensor. A high-speed CCD camera was used to characterize the contact area change in the spinosum dielectric layer. As depicted in Figure 5g–l, when under the same pressure, the contact area (shaded area) gradually becomes smaller as the number of the abrasive papers' mesh increases. When under compression progress, the contact area of the spinosum dielectric layer gradually increases. The contact area of the spinosum microstructures with low roughness saturates quickly, contributing to a low sensitivity level.

In summary, the capacitance change in the pressure sensor is a synergistic effect of the spinosum microstructure and CNT doping. The spinosum microstructure leads to local stress amplification and the contact area change in the dielectric layer with electrodes. The spinosum microstructure can provide additional compressibility by accommodating compressed protrusions. Moreover, the viscoelasticity of the elastomer is reduced because of the microstructure, resulting in a faster sensor response. In addition, introducing the CNTs to the dielectric layer can significantly improve the capacitive magnitude because the dielectric constant is improved and the relative permittivity can change with the external pressure. The relative permittivity of the dielectric layer will be dramatically amplified by the spinosum microstructure. It is through the above-mentioned synergistic effect that the excellent performance of the sensor can be caused.

3.4. Application of Spinosum Capacitive Pressure Sensor

As shown in Figure 6a, a 3 × 3 sensor array has been fabricated to map pressure distribution, which consists of three strip copper foils as top and bottom electrodes and a 1 wt% CNT doping spinosum CNT/PDMS film using the abrasive paper template no. 600 as the dielectric layer. As shown in Figure 6b,c, this array can clearly distinguish the position and weight of different things, such as a pen and a rectangular mass. A new methodology is proposed in order to solve the tire–road contact pressure. The sensor-array-embedded tires can provide more reliable and exact tire information to the vehicle than traditional indirect estimation methods. A rubber tire with a diameter of 80 mm was selected as the research object to use flexible pressure sensor arrays to detect the contact pressure of tires. Because of its excellent performance, the 1 wt% CNT doping spinosum pressure sensor using the abrasive paper template no. 600 was used to detect the contact pressure of the tire. In order to fabricate an intelligent tire, as shown in Figure 6d, a 2 × 3 pixel sensing array sandwiched between cross-arrays of copper electrodes was designed to detect the deformation of the tire. This 2 × 3 pixel sensing array was adhered to the tread of the tire using 3M tape. Then, the tire was installed on a test platform (Figure 6e) and compressed for ten cycles with a downward distance of 0.7 mm. Cross-locating technology was used to obtain the changes in the 6-pixel signal of the sensor array under the compression progress. Figure 6f shows the signal of 6 pixels. It can be seen that the signal changes of 5 and 6 pixels are most apparent, and the signal changes of 1–4 pixels are relatively small. From the experimental results, the pressure distribution at different positions of the tire can be detected in real time by coupling the sensor array into the tire, which is of great significance for the realization of intelligent tires.

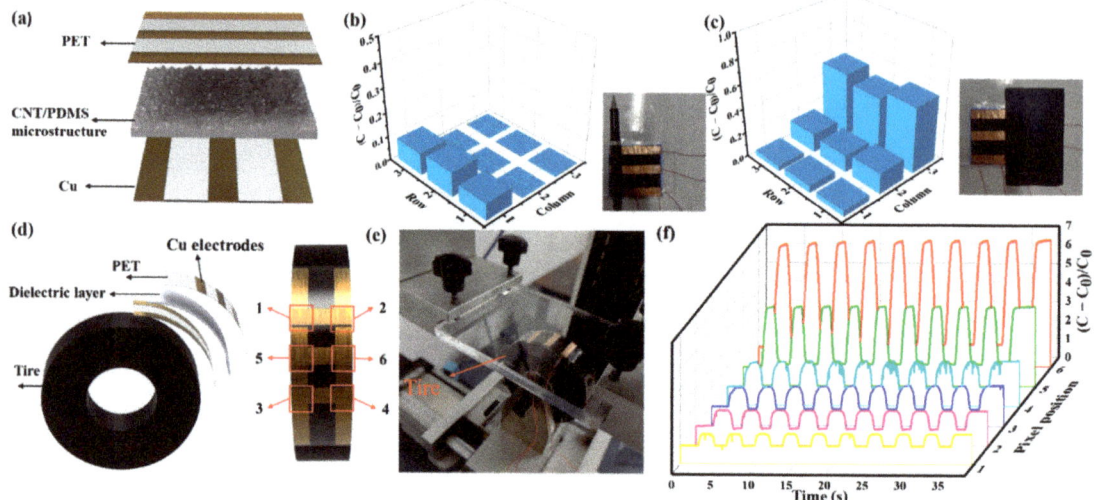

Figure 6. Application of the spinosum pressure sensors. (**a**) Schematic of the 3 × 3 pixelated pressure sensor array. Mapping of pressure distribution on the 3 × 3 sensor array upon (**b**) placing a pen and (**c**) placing a rectangular mass. (**d**) Schematic of the structure of the intelligent tire and the pixel position. (**e**) Photograph of the intelligent tire and test platform. (**f**) The capacitance signal at each pixel position of the intelligent tire under ten decompression cycles.

4. Conclusions

In summary, we have proposed a sensitive capacitive pressure sensor with a broad detection range inspired by the skin epidermis. A simple and low-cost fabrication process was proposed for CNT/PDMS-based spinosum pressure sensors through using the abrasive paper templates. Notably, the spinosum microstructure and doping content of CNTs can effectively improve the performance of pressure sensors: high sensitivity (0.25 kPa^{-1}), wider pressure range (~500 kPa), fast response time (20 ms) and excellent stability over 10,000 cycles. In addition, the effects of the mesh number of abrasive papers and CNT doping content on the sensing property are theoretically analyzed through simulations and experiments. Furthermore, a sensor array was manufactured for mapping the spatial distribution of pressure, which shows great potential for intelligent monitoring. Finally, a new methodology is proposed in order to solve the tire–road contact pressure and estimate related parameters by introducing a sensor array with a tire. Practically, this paper fabricated a bioinspired, cost-effective, broad-range, high-sensitivity, flexible sensor and opened a new patch to intelligently monitor the contact pressure of tires.

Author Contributions: Conceptualization, Y.D. and J.W.; methodology, Y.D. and B.S.; formal analysis, Y.D. and S.H.; investigation, Y.D., S.H. and J.W.; resources, Y.D.; data curation, Y.D. and B.S.; writing—original draft preparation, Y.D. and S.H.; writing—review and editing, J.W. and Y.W.; supervision, J.W., Z.L. and Y.W.; project administration, J.W.; funding acquisition, J.W. All authors have read and agreed to the published version of the manuscript.

Funding: This work was supported by the National Natural Science Foundation of China (Grant No. 52075119).

Data Availability Statement: The data presented in this study are available on request from the corresponding author.

Conflicts of Interest: The authors declare no conflict of interest.

References

1. Liu, Y.; Bao, R.; Tao, J.; Li, J.; Dong, M.; Pan, C. Recent progress in tactile sensors and their applications in intelligent systems. *Sci. Bull.* **2020**, *65*, 70–88. [CrossRef]
2. Tao, J.; Dong, M.; Li, L.; Wang, C.; Li, J.; Liu, Y.; Bao, R.; Pan, C. Real-time pressure mapping smart insole system based on a controllable vertical pore dielectric layer. *Microsyst. Nanoeng.* **2020**, *6*, 62. [CrossRef] [PubMed]
3. Huang, Y.; Wang, Z.; Zhou, H.; Guo, X.; Zhang, Y.; Wang, Y.; Liu, P.; Liu, C.; Ma, Y.; Zhang, Y. Highly sensitive pressure sensor based on structurally modified tissue paper for human physiological activity monitoring. *J. Appl. Polym. Sci.* **2020**, *137*, 48973. [CrossRef]
4. Xiong, W.; Guo, D.; Yang, Z.; Zhu, C.; Huang, Y. Conformable, programmable and step-linear sensor array for large-range wind pressure measurement on curved surface. *Sci. China Technol. Sci.* **2020**, *63*, 2073–2081. [CrossRef]
5. Liu, W.; Gong, L.; Yang, H. Integrated conductive rubber composites for contact deformation detection of tubular seals. *Polym. Test.* **2021**, *96*, 107089. [CrossRef]
6. Xiang, X.; Li, H.; Zhu, Y.; Xia, S.; He, Q. The composite hydrogel with " 2D flexible crosslinking point" of reduced graphene oxide for strain sensor. *J. Appl. Polym. Sci.* **2021**, *138*, 50801. [CrossRef]
7. Zeng, Q.; Lai, D.; Ma, P.; Lai, X.; Zeng, X.; Li, H. Fabrication of conductive and superhydrophobic poly(lactic acid) nonwoven fabric for human motion detection. *J. Appl. Polym. Sci.* **2022**, *139*, e52453. [CrossRef]
8. Lee, H.; Taheri, S. Intelligent tires? A review of tire characterization literature. *IEEE Intell. Transp. Syst. Mag.* **2017**, *9*, 114–135. [CrossRef]
9. Zhu, B.; Han, J.; Zhao, J. Tire-pressure identification using intelligent tire with three-axis accelerometer. *Sensors* **2019**, *19*, 2560. [CrossRef] [PubMed]
10. Lee, H.; Taheri, S. A novel approach to tire parameter identification. *Proc. Inst. Mech. Eng. Part D J. Automob. Eng.* **2018**, *233*, 55–72. [CrossRef]
11. Matsuzaki, R.; Keating, T.; Todoroki, A.; Hiraoka, N. Rubber-based strain sensor fabricated using photolithography for intelligent tires. *Sens. Actuators A* **2008**, *148*, 1–9. [CrossRef]
12. Son, J.-h.; Heo, D.; Song, Y.; Chung, J.; Kim, B.; Nam, W.; Hwang, P.T.J.; Kim, D.; Koo, B.; Hong, J.; et al. Highly reliable triboelectric bicycle tire as self-powered bicycle safety light and pressure sensor. *Nano Energy* **2022**, *93*, 106797. [CrossRef]
13. Luo, C.L.; Jiao, J.Y.; Su, X.J.; Zheng, L.X.; Yan, W.G.; Zhong, D.Z. Interlinked microcone resistive sensors based on self-assembly carbon nanotubes film for monitoring of signals. *Nanomaterials* **2022**, *12*, 2325. [CrossRef]
14. Xue, B.; Xie, H.; Zhao, J.; Zheng, J.; Xu, C. Flexible piezoresistive pressure sensor based on electrospun rough polyurethane nanofibers film for human motion monitoring. *Nanomaterials* **2022**, *12*, 723. [CrossRef]
15. Su, M.; Li, P.; Liu, X.; Wei, D.; Yang, J. Textile-based flexible capacitive pressure sensors: A review. *Nanomaterials* **2022**, *12*, 1495. [CrossRef]
16. Choi, J.; Kwon, D.; Kim, K.; Park, J.; Orbe, D.D.; Gu, J.; Ahn, J.; Cho, I.; Jeong, Y.; Oh, Y.; et al. Synergetic effect of porous elastomer and percolation of carbon nanotube filler toward high performance capacitive pressure sensors. *ACS Appl. Mater. Interfaces* **2020**, *12*, 1698–1706. [CrossRef]
17. Mu, J.; Xian, S.; Yu, J.; Zhao, J.; Song, J.; Li, Z.; Hou, X.; Chou, X.; He, J. Synergistic enhancement properties of a flexible integrated pan/pvdf piezoelectric sensor for human posture recognition. *Nanomaterials* **2022**, *12*, 1155. [CrossRef] [PubMed]
18. Wang, Z.L.; Song, J. Piezoelectric nanogenerators based on zinc oxide nanowire arrays. *Science* **2006**, *312*, 242–246. [CrossRef] [PubMed]
19. Goh, Q.L.; Chee, P.S.; Lim, E.H.; Ng, D.W.-K. An AI-assisted and self-powered smart robotic gripper based on eco-egain nanocomposite for pick-and-place operation. *Nanomaterials* **2022**, *12*, 1317. [CrossRef] [PubMed]
20. Fan, F.-R.; Tian, Z.-Q.; Lin Wang, Z. Flexible triboelectric generator. *Nano Energy* **2012**, *1*, 328–334. [CrossRef]
21. Mishra, R.B.; El-Atab, N.; Hussain, A.M.; Hussain, M.M. Recent progress on flexible capacitive pressure sensors: From design and materials to applications. *Adv. Mater. Technol.* **2021**, *6*, 2001023. [CrossRef]
22. Liu, S.-Y.; Lu, J.-G.; Shieh, H.-P.D. Influence of permittivity on the sensitivity of porous elastomer-based capacitive pressure sensors. *IEEE Sens. J.* **2018**, *18*, 1870–1876. [CrossRef]
23. Ha, K.H.; Zhang, W.; Jang, H.; Kang, S.; Wang, L.; Tan, P.; Hwang, H.; Lu, N. Highly sensitive capacitive pressure sensors over a wide pressure range enabled by the hybrid responses of a highly porous nanocomposite. *Adv. Mater.* **2021**, *33*, e2103320. [CrossRef] [PubMed]
24. He, Z.; Chen, W.; Liang, B.; Liu, C.; Yang, L.; Lu, D.; Mo, Z.; Zhu, H.; Tang, Z.; Gui, X. Capacitive pressure sensor with high sensitivity and fast response to dynamic interaction based on graphene and porous nylon networks. *ACS Appl. Mater. Interfaces* **2018**, *10*, 12816–12823. [CrossRef] [PubMed]
25. Zhang, P.; Zhang, J.; Li, Y.; Huang, L. Flexible and high sensitive capacitive pressure sensor with microstructured electrode inspired by ginkgo leaf. *J. Phys. D Appl. Phys.* **2021**, *54*, 465401. [CrossRef]
26. Wan, S.; Bi, H.; Zhou, Y.; Xie, X.; Su, S.; Yin, K.; Sun, L. Graphene oxide as high-performance dielectric materials for capacitive pressure sensors. *Carbon* **2017**, *114*, 209–216. [CrossRef]
27. Yang, J.; Luo, S.; Zhou, X.; Li, J.; Fu, J.; Yang, W.; Wei, D. Flexible, tunable, and ultrasensitive capacitive pressure sensor with microconformal graphene electrodes. *ACS Appl. Mater. Interfaces* **2019**, *11*, 14997–15006. [CrossRef] [PubMed]

28. Xiong, Y.; Shen, Y.; Tian, L.; Hu, Y.; Zhu, P.; Sun, R.; Wong, C.-P. A flexible, ultra-highly sensitive and stable capacitive pressure sensor with convex microarrays for motion and health monitoring. *Nano Energy* **2020**, *70*, 104436. [CrossRef]
29. Mannsfeld, S.C.; Tee, B.C.; Stoltenberg, R.M.; Chen, C.V.; Barman, S.; Muir, B.V.; Sokolov, A.N.; Reese, C.; Bao, Z. Highly sensitive flexible pressure sensors with microstructured rubber dielectric layers. *Nat. Mater.* **2010**, *9*, 859–864. [CrossRef]
30. Guo, Y.; Gao, S.; Yue, W.; Zhang, C.; Li, Y. Anodized aluminum oxide-assisted low-cost flexible capacitive pressure sensors based on double-sided nanopillars by a facile fabrication method. *ACS Appl. Mater. Interfaces* **2019**, *11*, 48594–48603. [CrossRef] [PubMed]
31. Pang, C.; Koo, J.H.; Nguyen, A.; Caves, J.M.; Kim, M.G.; Chortos, A.; Kim, K.; Wang, P.J.; Tok, J.B.-H.; Bao, Z. Highly skin-conformal microhairy sensor for pulse signal amplification. *Adv. Mater.* **2015**, *27*, 634–640. [CrossRef] [PubMed]
32. Miller, S.; Bao, Z. Fabrication of flexible pressure sensors with microstructured polydimethylsiloxane dielectrics using the breath figures method. *J. Mater. Res.* **2015**, *30*, 3584–3594. [CrossRef]
33. Lee, Y.; Park, J.; Cho, S.; Shin, Y.-E.; Lee, H.; Kim, J.; Myoung, J.; Cho, S.; Kang, S.; Baig, C.; et al. Flexible ferroelectric sensors with ultrahigh pressure sensitivity and linear response over exceptionally broad pressure range. *ACS Nano* **2018**, *12*, 4045–4054. [CrossRef] [PubMed]
34. Zeng, X.; Wang, Z.; Zhang, H.; Yang, W.; Xiang, L.; Zhao, Z.; Peng, L.-M.; Hu, Y. Tunable, ultrasensitive, and flexible pressure sensors based on wrinkled microstructures for electronic skins. *ACS Appl. Mater. Interfaces* **2019**, *11*, 21218–21226. [CrossRef]
35. Hwang, J.; Kim, Y.; Yang, H.; Oh, J.H. Fabrication of hierarchically porous structured PDMS composites and their application as a flexible capacitive pressure sensor. *Composites Part B Eng.* **2021**, *211*, 108607. [CrossRef]
36. Yang, L.; Liu, Y.; Filipe, C.D.M.; Ljubic, D.; Luo, Y.; Zhu, H.; Yan, J.; Zhu, S. Development of a highly sensitive, broad-range hierarchically structured reduced graphene oxide/polyhipe foam for pressure sensing. *ACS Appl. Mater. Interfaces* **2019**, *11*, 4318–4327. [CrossRef]
37. Li, T.; Li, Y.; Zhang, T. Materials, structures, and functions for flexible and stretchable biomimetic sensors. *Acc. Chem. Res.* **2019**, *52*, 288–296. [CrossRef]
38. Menon, G.K. New insights into skin structure: Scratching the surface. *Adv. Drug Deliv. Rev.* **2002**, *54*, S3–S17. [CrossRef]
39. Pang, Y.; Zhang, K.; Yang, Z.; Jiang, S.; Ju, Z.; Li, Y.; Wang, X.; Wang, D.; Jian, M.; Zhang, Y.; et al. Epidermis microstructure inspired graphene pressure sensor with random distributed spinosum for high sensitivity and large linearity. *ACS Nano* **2018**, *12*, 2346–2354. [CrossRef]
40. Duan, Y.; He, S.; Wu, J.; Su, B.; Wang, Y. Recent progress in flexible pressure sensor arrays. *Nanomaterials* **2022**, *12*, 2495. [CrossRef]
41. Pyo, S.; Choi, J.; Kim, J. Flexible, transparent, sensitive, and crosstalk-free capacitive tactile sensor array based on graphene electrodes and air dielectric. *Adv. Electron. Mater.* **2018**, *4*, 1700427. [CrossRef]
42. Qiu, J.; Guo, X.; Chu, R.; Wang, S.; Zeng, W.; Qu, L.; Zhao, Y.; Yan, F.; Xing, G. Rapid-response, low detection limit, and high-sensitivity capacitive flexible tactile sensor based on three-dimensional porous dielectric layer for wearable electronic skin. *ACS Appl. Mater. Interfaces* **2019**, *11*, 40716–40725. [CrossRef] [PubMed]
43. Mahata, C.; Algadi, H.; Lee, J.; Kim, S.; Lee, T. Biomimetic-inspired micro-nano hierarchical structures for capacitive pressure sensor applications. *Measurement* **2020**, *151*, 107095. [CrossRef]
44. Pignanelli, J.; Schlingman, K.; Carmichael, T.B.; Rondeau-Gagné, S.; Ahamed, M.J. A comparative analysis of capacitive-based flexible PDMS pressure sensors. *Sens. Actuators A* **2019**, *285*, 427–436. [CrossRef]
45. Kim, J.; Chou, E.F.; Le, J.; Wong, S.; Chu, M.; Khine, M. Soft wearable pressure sensors for beat-to-beat blood pressure monitoring. *Adv. Healthc Mater.* **2019**, *8*, e1900109. [CrossRef] [PubMed]
46. Lin, C.; Wang, H.; Yang, W. Variable percolation threshold of composites with fiber fillers under compression. *J. Appl. Phys.* **2010**, *108*, 013509. [CrossRef]

 *nanomaterials*

Article

High Permittivity Polymer Composites on the Basis of Long Single-Walled Carbon Nanotubes: The Role of the Nanotube Length

Shamil Galyaltdinov [1], Ivan Lounev [1,2], Timur Khamidullin [1], Seyyed Alireza Hashemi [3], Albert Nasibulin [4] and Ayrat M. Dimiev [1,*]

1. Laboratory for Advanced Carbon Nanomaterials, Chemical Institute, Kazan Federal University, 18 Kremlyovskaya Street, 420008 Kazan, Russia
2. Institute of Physics, Kazan Federal University, 18 Kremlyovskaya Street, 420008 Kazan, Russia
3. Nanomaterials and Polymer Nanocomposites Laboratory, School of Engineering, University of British Columbia, Kelowna, BC V1V 1V7, Canada
4. Skolkovo Institute of Science and Technology, Nobel Str. 3, 143026 Moscow, Russia
* Correspondence: amdimiev@kpfu.ru

Abstract: Controlling the permittivity of dielectric composites is critical for numerous applications dealing with matter/electromagnetic radiation interaction. In this study, we have prepared polymer composites, based on a silicone elastomer matrix and Tuball carbon nanotubes (CNT) via a simple preparation procedure. The as-prepared composites demonstrated record-high dielectric permittivity both in the low-frequency range (10^2–10^7 Hz) and in the X-band (8.2–12.4 GHz), significantly exceeding the literature data for such types of composite materials at similar CNT content. Thus, with the 2 wt% filler loading, the permittivity values reach 360 at 10^6 Hz and >26 in the entire X-band. In similar literature, even the use of conductive polymer hosts and various highly conductive additives had not resulted in such high permittivity values. We attribute this phenomenon to specific structural features of the used Tuball nanotubes, namely their length and ability to form in the polymer matrix percolating network in the form of neuron-shaped clusters. The low cost and large production volumes of Tuball nanotubes, as well as the ease of the composite preparation procedure open the doors for production of cost-efficient, low weight and flexible composites with superior high permittivity.

Keywords: carbon nanotubes; dielectric polymer composites; permittivity

1. Introduction

Controlling dielectric permittivity of polymer composites is critical for thin-film transistors [1], photovoltaic devices [2] and more broadly for materials aimed at absorption/reflection of electromagnetic radiation, especially in the X-band [3–6]. The most commonly used polymer matrices are silicon rubber [7–15], epoxy resin [16–20], polyvinylidene fluoride (PVDF) [21–24], and thermo-polyurethane (TPU) [25,26]. Correspondingly, the most broadly used conductive fillers are carbon nanotubes (CNTs), metal particles and graphene derivatives [4–28]. Among the fillers for this aim, CNTs are of special interest since they possess high aspect ratio, high electrical conductivity and superior mechanical strength [29]. In the literature, there are more papers on the use of multi-walled carbon nanotubes (MWCNTs) [6–10,13–17,21–24] rather than single-walled carbon nanotubes (SWCNT) [30–34]. This is because MWCNTs are more available, significantly cheaper and can be more easily and uniformly distributed in the polymer matrices.

Among all the SWCNTs commercially available on the market, nanotubes sold under the brand Tuball manufactured by OCSiAl are of particular interest [33–35]. This is because they are currently manufactured at a tonnage scale and are offered at the cheapest price. A structural feature of these nanotubes is their large diameter (1.4–2.2 nm), compared to other

SWCNTs such as HiPCo, CoMoCat, etc., and very large length [36]. Such unique features make them attractive as fillers [37,38] and additives to polymers and concrete [39].

The two-component composites, in which a dielectric polymer host contains only CNTs as a conductive filler, exhibit relatively low dielectric permittivity values (ε') even in the low-frequency range [7,9,11–15]. Higher dielectric permittivity values are normally achieved using either a conductive polymer host [22–24] or/and special additives [31]. The maximum permittivity value was achieved in [21] with PVDF matrix and modified CNTs; at the 3.5 wt% CNT content, the resulting composite exhibiting permittivity values of about 250 at 10^4 Hz.

In the X-band, for the two-component systems made of epoxy resin and MWCNTs, in most studies, the reported permittivity values do not exceed 13 [18,26,40]. The higher value of $\varepsilon' > 20$ was obtained when using a higher MWCNT content (5 wt%) [26]. To the best of our knowledge, the highest reported value of $\varepsilon' = 23$ at 3% CNT content was attained when using MWCNTs and a polar dielectric matrix, such as PVDF [41]. Alternative ways to increase the dielectric permittivity include addition of a third highly conductive component such as noble metals [6,42], which increases the cost of the final material. At the same time, for the two-component composites, consisting of a non-polar dielectric matrix and SWCNTs as conductive filler, high dielectric permittivity values have not been reported yet.

In this study, for the first time, we report the dielectric properties of the silicone elastomer-based composites reinforced with Tuball SWCNTs. The composites demonstrate tremendously high dielectric permittivity values both in the low-frequency range and in the X-band. Importantly, such values were obtained using affordable components by a simple preparation method without high-cost additives.

2. Experimental

2.1. Materials

The SWCNTs were of the Tuball brand manufactured by OCSiAl (Luxembourg) (01RW03.N1, batch no. 819); SWCNTs were purified by the manufacturer. According to the manufacturer, the content of nanotubes was ≥93% and the content of metallic impurities was <1%. The ToolDecor T 20–137, a two-part molding silicone elastomer, was supplied from Wacker (Munchen, Germany); it consists of two components, ToolDecor T 20–1 Base (Part A) and Catalyst T 37 Hardener (Part B).

Methylene chloride was purchased from Tatkhimprodukt LLC (Kazan, Russia) and used without additional purification.

2.2. Characterization

Raman spectra of nanotubes were acquired from the nanotube films using an ARS-3000 Raman microscope (NanoScanTechnologies, Russia) with the 532 nm excitation laser. The scanning electron microscopy (SEM) images were acquired with a Merlin field-emission high resolution scanning electron microscope (Carl Zeiss, Oberkochen, Germany) at accelerating voltage of incident electrons of 5 kV and current probe of 300 pA.

2.3. Preparation of Polymer Composites

To prepare reinforced composites, nanotubes in precalculated quantities were dispersed in CH_2Cl_2 by sonication with a tip sonicator Sonic-Vibra 750 (Sonics, Newtown, CT, USA) for 1 h at 30% amplitude. Methylene chloride was chosen as a solvent, because it can dissolve silicone and has a low boiling point, facilitating its subsequent evaporation. Then the resulting dispersion of nanotubes was mixed with part A of the silicone elastomer and agitated manually with a glass rod until homogeneous condition. Next, the mixture was mildly sonicated for 30 min at 20% amplitude. After that, the as-obtained paste was heated in a water bath at T = 60–70 °C with manual agitation until complete evaporation of CH_2Cl_2. The resulting paste was thoroughly mixed with part B in a weight ratio of 100A:4B and placed into silicone molds. The samples were cured for 12 h at room temperature. As a result of these operations, samples of composites in the form of disks were obtained: for the low-frequency measurements, the disks were fabricated with a diameter of 29.0 mm and a thickness of 3.5 mm; for the high-frequency measurements, the disks were fabricated with a diameter of 29.0 mm and a thickness of 7 mm.

2.4. Electrical Measurements

The permittivity and loss values for the low-frequency range were calculated from the capacitance, measured with the Novocontrol BDS Concept-80 impedance analyzer, (Novocontrol Technologies GmbH & Co. KG, Montabaur, Germany) with the automatic temperature control provided by the QUATRO cryo-system (the temperature uncertainty is ±0.5 °C). A sample was placed between two gold-plated electrodes of the capacitor. The capacitor was attached to the thermostated testing head. The measurements were conducted in the frequency range of 0.1 Hz–10 MHz. The data for the ultra-high frequency (UHF) range (0.1–70 GHz) were measured with the PNA-X Network Analyzer N5247A (Agilent Technologies, Santa Clara, CA, USA). Samples in the form of disks with a diameter of 29 mm and a thickness of 7 mm were placed at the end of a coaxial measuring probe (Performance Probe) with a diameter of 10 mm. When measuring on the PNA-X Agilent N5247A, the results were recorded using the Agilent 85070 built-in licensed software package (Santa Clara, CA, USA). The temperature was set at 25 °C. The processing of the experimental data was carried out with the WinFit software [43].

3. Results and Discussion

The used Tuball CNTs were characterized by SEM and Raman spectroscopy (Figure 1). The SEM images (Figure 1a) reveal that nanotubes exist mainly in the form of thick bundles consisting of tens and hundreds of individual nanotubes. According to the higher magnification SEM and TEM images [38], the primary bundles with diameters of 5–10 nm are assembled into secondary and even tertiary bundles with diameters of up to 50 nm. The existence of Tuball nanotubes in the form of bundles is a consequence of their longer length compared to other commercially available SWCNTs, such as HiPCo, CoMoCat, etc. [36].

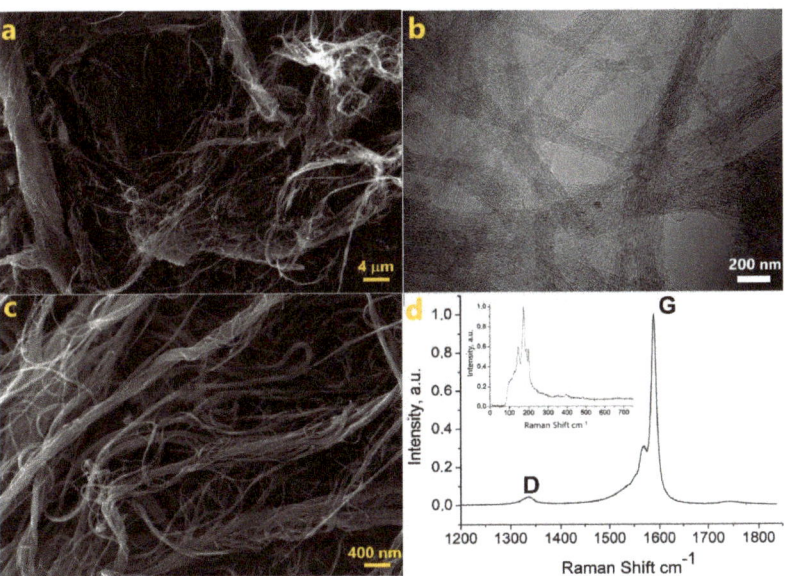

Figure 1. Characterization of as-received Tuball SWCNTs. (**a,c**) SEM images at different magnification. (**b**) TEM image. (**d**) Raman spectrum of CNTs. Tangential mode. The inset is the radial breathing mode. The spectra were acquired with the 532 nm excitation laser.

The Raman spectrum is typical for SWCNTs. In addition to the two-component G-band, there is a small D-band in the vicinity of ~1336 cm^{-1} (Figure 1b), which shows the presence of defects in the crystal lattice. The D-peak intensity is slightly higher than that in the as-received SWCNTs. Some additional defects were introduced during the sonication of the CNT dispersions. Radial breathing mode (RBM) is sensitive to the nanotube diameter [44,45] and coincides with the literature data for Tuball CNTs. The provided characteristics (Figure 1) are sufficient for the purposes of this study. For a more detailed description of the used CNTs, we refer to our previous publication [35].

Figure 2 shows the real (ε') and imaginary (ε'') parts of the complex permittivity of the composites with different filling fractions as a function of frequency. Real permittivity values in the range of 0.1–300 Hz are not shown in Figure 2, and are not considered in the approximation of the curves since the material in this frequency range is very sensitive to mechanical deformations that arise upon its contact with the electrodes of the measuring capacitor. The full range dielectric spectra are shown in the Supplementary Materials, Figure S1.

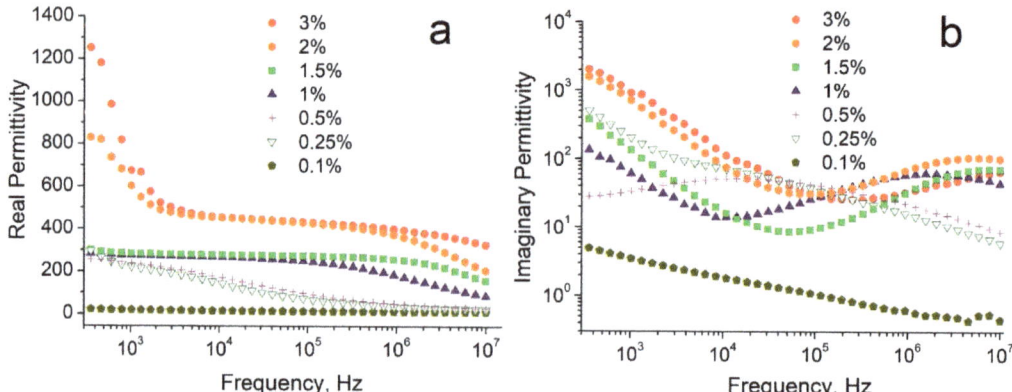

Figure 2. Frequency dependences of the (**a**) real and (**b**) imaginary parts of the complex permittivity for composites with different filler contents.

In general, an increase in the CNT content leads to an increase in the dielectric permittivity in the whole tested frequency range. The composites can be divided into three groups according to the proximity of the permittivity values:

(A) Group I—composites with a filler content of 0.1–0.5%;
(B) Group II—composites with a filler content of 1 and 1.5%;
(C) Group III—composites with a filler content of 2 and 3%.

For the composite with a filler content of 0.1%, ε' practically does not change over the entire frequency range. For the composites with a filler content of 0.25% and 0.5%, a smooth increase in ε' in the range of 1×10^7–3×10^2 Hz is observed without reaching a plateau at the low frequency end. For the Group II composites with a filler content of 1–1.5%, the growth of ε' values is observed in the range of 1×10^5–1×10^7 Hz, followed by a plateau below 5×10^4 Hz with ε' values around 280. Finally, for the Group III composites, a smooth increase in the ε' values is observed in the range of 3×10^5–1×10^7 Hz with a plateau starting at 2.5×10^5 Hz with ε' values reaching 470 at 1×10^4 Hz.

The imaginary parts of the complex permittivity (Figure 2b) specify the dielectric losses. In the tested frequency range, no loss peak is registered for the composite with a filler content of 0.1%. For the 0.25% sample, a broad peak in the range of 3×10^6–1×10^6 Hz is observed. For the composite with a 0.5% CNT content, a broad peak is centered at 1.4×10^5 Hz. The increase in the filler content leads to the shift of the loss peak position to the higher frequency region. Thus, for the 1% content, the loss peak is at 1.6×10^6 Hz, for 2% content, the peak is at 6×10^6 Hz, and for 3% content, the peak is located above

1×10^7 Hz. In the curves of the composites with 3% CNT content, only the left shoulder of the loss peak is visible in the entire frequency range. The peak of the dielectric losses is originated by the polarization delay with an increase in the frequency of external electric field, and is normally indicative of so-called dipolar relaxation [46].

Table 1 compares the published literature data with the values of the dielectric permittivity obtained in this study [14,15,21–24,31]. According to the presented data, the composites prepared in this work significantly outperform the known literature data for the polymer/carbon-nanotubes systems with the same CNT content. Importantly, the attained permittivity values are much higher than the values reported in studies [14,15] on similar systems made from silicon elastomer matrix and multi-wall carbon nanotubes. Even the use of conductive matrices such as PVDF [21–24], the modification of nanotubes [21], and addition of polyaniline (PANI) [31] do not afford such high values of ε', as we attain in this study.

Table 1. The literature data on the permittivity of different polymer composites, comprising CNTs as a conductive filler, in the frequency range of 10^4–10^6 Hz.

Polymer Host/Filler	CNT (wt%)	ε' at 10^4 Hz	ε' at 10^6 Hz	Ref.
PDMS/MWCNTs	3	~4.5	~4.3	[14]
Silicon rubber/MWCNTs	2.5	~4.5	~4.4	[15]
PVDF/functionalized MWCNTs	3.5	~250	~200	[21]
PVDF/MWCNTs	3.7	160	~100	[22]
PVDF/MWCNTs	2	~225	~50	[23]
PVDF/MWCNTs	4	~30	~25	[24]
TPU-PANI/SWCNTs	0.5	~100	~80	[31]
Silicon rubber/SWCNTs	2	~450	~360	This work

Normally, permittivity increases with increasing the conductive filler fraction. Subsequently, at very high CNT content (>7%), expectedly higher permittivity values have been reported in literature. However, for the CNT content range, tested in this wok (1–3%), we are reporting the record high values (Table 1).

Such a high permittivity values must be related to the special structural features of these CNTs. Namely, due to their length, Tuball CNTs form bundles, and do not easily unbundle even with prolonged high power sonication [35,36]. After unbundling they tend to quickly rebundle, if are not stabilized by surfactants or by other means. Respectively, in the composites, these CNTs would exist in the form of bundles. In addition to bundles, in polymers, CNTs form aggregates, consisting of many bundles. To investigate the structure of our composites, we used optical microscopy. Figure 3 represents optical microphotographs of the liquid CNT/silicon elastomer formulations with different CNT contents taken before the formulations have been cured. It is clear that, even at the 0.25% CNT content, the CNT bundles are joined into clusters, which are well visible in the fully transparent silicon elastomer. The shape of the clusters resemble neurons with the tails of the CNT bundles sticking out from the main cluster body. With increasing the CNT content, not only the number, but also the size of the clusters increase. The size of the clusters vary from 20 μm through 200 μm.

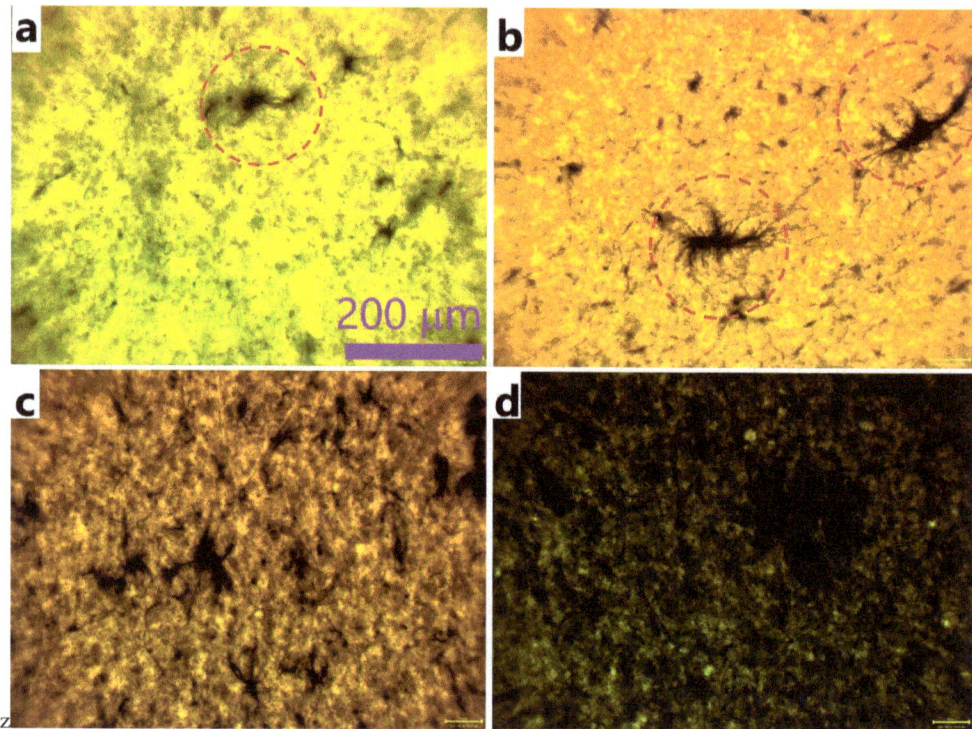

Figure 3. Optical microphotographs of liquid CNT/silicon elastomer formulations with different CNT content: 0.25% (**a**), 0.5% (**b**), 1.0% (**c**), and 2.0% (**d**). The scale bar is the same for all the four images. The red-line circles on (**a**,**b**) show the largest CNT clusters, present in the sample.

The sample regions containing isolated clusters of nanotubes form conductive inclusions, randomly distributed in the polymer matrix. According to the knowledge in the field, the pertinent physics behind the polarization mechanism in composite materials remains poorly understood [46]. Considering the structure of the composite, namely the size of the CNT clusters and their distribution, as well as the presence of the loss peaks on the imaginary part functions (Figure 2b), the registered phenomenon might be least contradictorily explained in the terms of the Maxwell-Wagner polarization [20,22,24,33,46,47]. The larger the size of the clusters and their number, the larger the interface with the polymer matrix. Respectively, the higher charge can be accumulated at the interface. Such polarization in the larger clusters will fully manifest at lower frequencies.

To extract additional relaxation parameters, the measured dielectric spectra were approximated by the superposition of the Cole-Cole function [48] and the Jonsher parameter [46] according to the following equation:

$$\varepsilon^*(\omega) = \varepsilon'(\omega) - i\varepsilon''(\omega) = \varepsilon_\infty + \frac{\Delta\varepsilon}{1 + (i\omega\tau)^\alpha} + B(i\omega)^{n-1} + i\frac{\sigma_0}{\varepsilon_0\omega} \qquad (1)$$

where $\varepsilon'(\omega)$ and $\varepsilon''(\omega)$ are the real and the imaginary parts of the complex permittivity; i is the imaginary unit; $\Delta\varepsilon = \varepsilon_s - \varepsilon_\infty$, ε_s is the static dielectric permittivity; ε_∞ is the permittivity at high frequency; ω is the cyclic frequency, $\Delta\varepsilon$, τ, α are the magnitudes of the dielectric strength, relaxation time, and Cole-Cole broadening parameter, respectively; B is the magnitude of the Jonscher correction; $0 < n \leq 1$ is the Jonscher parameter; σ_0 is the DC-conductivity; $\varepsilon_0 = 8.85 \times 10^{-12}$ F/m is the permittivity of vacuum. The approximation of the dielectric spectrum by Equation (1) for a sample with a filler content of 1% is shown in Figure S2.

Figure 4a shows that the DC-conductivity σ_0 smoothly increases with increasing the filler content in the composites that is in accordance with the literature values [7]. The dependence of the relaxation times on the filler concentration in the samples is shown in Figure 4b. In general, relaxation time decreases with an increase in the filler concentration.

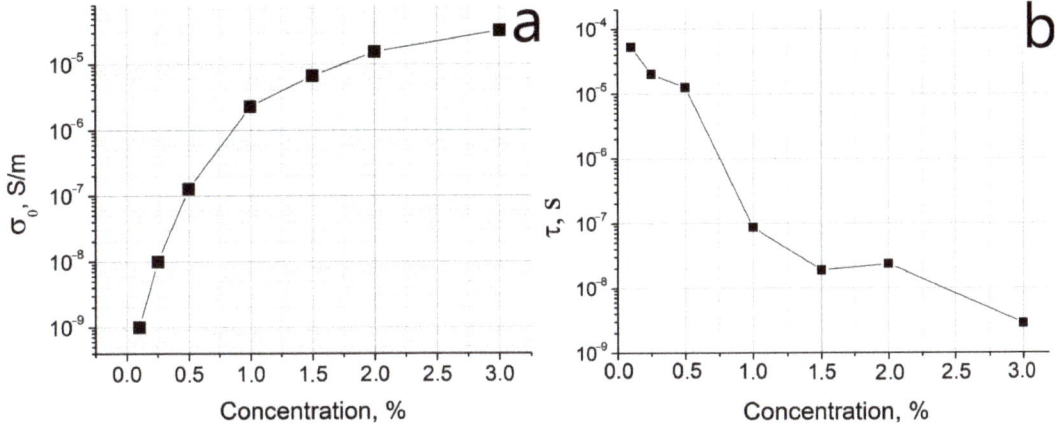

Figure 4. The approximation derived dependence of (**a**) DC-conductivity σ_0, and (**b**) relaxation times τ on the content of CNTs for a sample with a filler content of 1% in the frequency range of 300–10^7 Hz.

Figure 5 shows the frequency dependence of dielectric permittivity of the same materials in the X-band region. The original dielectric spectra of the composites in the whole UHF region are presented in the SI section (Figure S3). There is no significant change in the dielectric permittivity with frequency within the 8–12 GHz range (Figure 5a). Expectedly, composites with a higher filler content have higher ε' values. Based on the attained permittivity, the tested materials can be again divided into three groups with close ε' values:

I composites with filler content of 0.1–1%;
II composite with filler content of 1.5%;
III composites with filler content of 2–3%.

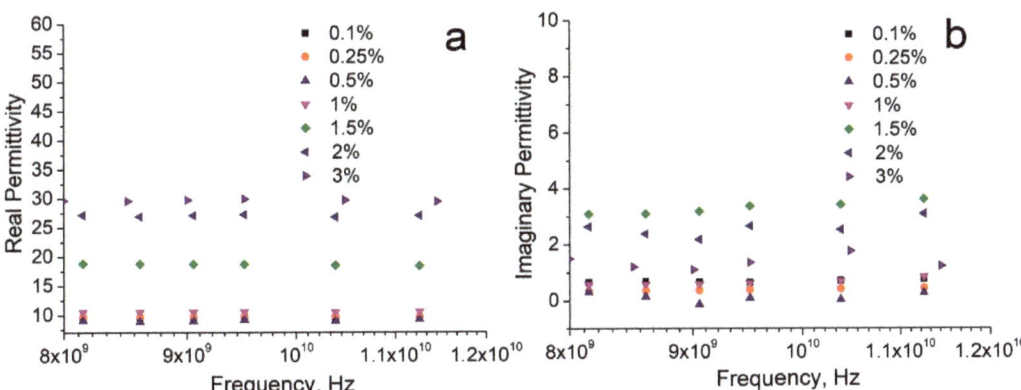

Figure 5. Frequency dependence of the (**a**) real and (**b**) imaginary parts of the complex permittivity for different filler contents in the X-band.

Comparison of our data with the published literature data in the X-band is presented in Table 2. Again, the permittivity values registered in this work notably exceed the literature data [18,26,40,41]. Even the use of the polar polymer host, such as PVDF [41], does not enable the values, attained in our study. At the same time, the values of the imaginary part of the dielectric permittivity, registered in this study (Figure 5b), are among the lowest in the literature [49].

Table 2. The literature data for the permittivity (ε') of different composites made from nanotubes and polymer matrix in the X-band.

Polymer Host/Filler	CNT Loading Fraction, %	ε'	Ref.
Epoxy resin/MWCNTs	1	~4.75	[18]
TPU/MWCNTs	3	~13	[26]
Epoxy resin/MWCNTs	2	3.0	[40]
PVDF/MWCNTs	3	~23	[41]
Silicone/SWCNTs	2	~27	This work
Silicone/SWCNTs	3	~30	This work

Figure 6a presents an approximation of the real and imaginary parts of the complex permittivity for a sample with a nanotube content of 2%. The spectra were approximated by the Cole-Cole function with the Jonsher parameter as was shown above.

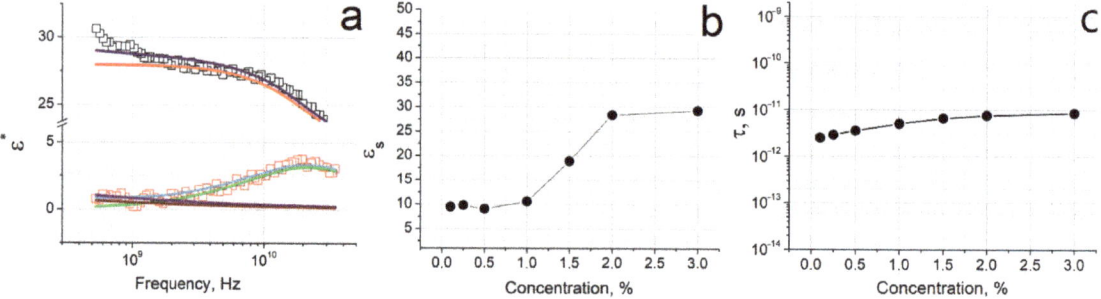

Figure 6. Approximation of the dielectric spectra for a sample with a nanotube content of 2% in the UHF region. (**a**) The empty black squares are the real part, and the empty red squares are the imaginary part of the complex permittivity. Blue and light blue lines are fitting functions for the real and imaginary parts of the spectrum, respectively; red and green lines are Cole-Cole functions for the real and imaginary parts of the spectrum, respectively; violet and brown lines are Jonsher corrections for the real and imaginary parts of the spectrum, respectively. (**b**) Dependence of the static dielectric permittivity on the filler content. (**c**) The function of the relaxation time on the filler content in the UHF region.

In contrast to the low-frequency spectra (Figures 2 and 4), no signs of DC-conductivity are observed in the UHF region, since the processes characteristic of DC-conductivity do not have time to occur during this short time intervals. Apparently, the periods of the field oscillations are rather short, and relaxation processes might reflect the polarization occurring inside the conducting CNT bundle. We hypothesize that the size of the CNT bundles of the Tuball nanotubes is the main factor, responsible for the record high permittivity values in the X-band.

Figure 6b shows the dependence of the static dielectric permittivity (ε_s) on the filler content in the UHF region. The ε_s values were obtained by approximating the dielectric spectra using the Cole-Cole equation [46,48]. In the range of 0.1–1%, the dielectric permittivity changes insignificantly with the filler content. After 1%, a sharp increase in ε_s is observed, followed by a plateau after >2%.

Figure 6c represents the function of the relaxation time τ on the filler content in the UHF region. The values of τ insignificantly change with increasing the filler content; that they have the same order of magnitude as opposed to the situation in the low-frequency region in which they differ ~10^5 times in the tested range of the filler contents (Figure 4b).

Apparently, there must be a difference in the polarization mechanisms in the two tested frequency regions. At low frequencies, high permittivity can be explained by the Maxwell-Wagner polarization with possible contribution of the charge transfer by hopping electrons through thin polymer layers. However, at high frequencies, we most likely face another polarization mechanism.

4. Conclusions

In this study, polymer composites have been prepared by incorporating the Tuball single-walled carbon nanotubes into a silicone matrix via a simple cost-efficient preparation procedure. The registered dielectric permittivity values in the low-frequency range ($3 \times 10^2 - 10^7$ Hz) and in the X-band significantly exceed the previously published literature data for the similar systems at similar CNT content. Even the use of polar polymer hosts and various highly conductive additives in the literature had not resulted in such high dielectric permittivity values. The high permittivity values registered in this study are the consequence of the specific features of the Tuball nanotubes, namely their high length and ability to join into long bundles, forming percolative clusters. The polarization mechanism in the two tested frequency ranges is suggested to be different.

Supplementary Materials: The following supporting information can be downloaded at: https://www.mdpi.com/article/10.3390/nano12193538/s1, Figure S1: Frequency dependences of the real (a) and imaginary (b) parts of the complex permittivity for composites with different filler contents in the low-frequency region; Figure S2: Approximation of the dielectric spectrum by the Equation (1) for a sample with a filler content of 1%, (a) real part, and (b) imaginary part. The violet line is the fitting function; the green line is the Johnsher parameter; the brown line is DC-conductivity; the red line is the Cole-Cole function; Figure S3: in the ultra-high frequency region. Frequency dependences of the real (a) and imaginary (b) parts of the complex permittivity for composites with different filler contents.

Author Contributions: Investigation, S.G., I.L., T.K. and S.A.H.; writing—original draft preparation, S.G.; conceptualization, A.M.D.; supervision, A.M.D.; writing—review and editing, A.N. and A.M.D. All authors have read and agreed to the published version of the manuscript.

Funding: This research was funded by the Russian Science Foundation; grant number 21-73-20024.

Data Availability Statement: The data presented in this study are available on request from the corresponding author.

Conflicts of Interest: The authors declare that they have no conflict of interest.

References

1. Boughias, O.; Belkaid, M.S.; Zirmi, R.; Trigaud, T.; Ratier, B.; Ayoub, N. Field Effect Transistors Based on Composite Films of Poly(4-Vinylphenol) with ZnO Nanoparticles. *J. Electron. Mater.* **2018**, *47*, 2447. [CrossRef]
2. James, D.K.; Tour, J.M. Graphene: Powder, Flakes, Ribbons, and Sheets. *Acc. Chem. Res.* **2013**, *46*, 2307. [CrossRef]
3. Younes, H.; Li, R.; Lee, S.E.; Kim, Y.K.; Choi, D. Gradient 3D-Printed Honeycomb Structure Polymer Coated with a Composite Consisting of Fe_3O_4 Multi-Granular Nanoclusters and Multi-Walled Carbon Nanotubes for Electromagnetic Wave Absorption. *Synth. Met.* **2021**, *275*, 116731. [CrossRef]
4. Ismail, M.M.; Rafeeq, S.N.; Sulaiman, J.M.A.; Mandal, A. Electromagnetic Interference Shielding and Microwave Absorption Properties of Cobalt Ferrite $CoFe_2O_4$/Polyaniline Composite. *Appl. Phys. A Mater. Sci. Process.* **2018**, *124*, 1–12. [CrossRef]
5. Pawar, S.P.; Bose, S. Extraordinary Synergy in Attenuating Microwave Radiation with Cobalt-Decorated Graphene Oxide and Carbon Nanotubes in Polycarbonate/Poly(Styrene-Co-Acrylonitrile) Blends. *ChemNanoMat* **2015**, *1*, 603. [CrossRef]
6. Yusof, Y.; Moosavi, S.; Johan, M.R.; Badruddin, I.A.; Wahab, Y.A.; Hamizi, N.A.; Rahman, M.A.; Kamangar, S.; Khan, T.M.Y. Electromagnetic Characterization of a Multiwalled Carbon Nanotubes-Silver Nanoparticles-Reinforced Polyvinyl Alcohol Hybrid Nanocomposite in X-Band Frequency. *ACS Omega* **2021**, *6*, 4184. [CrossRef]

7. Panda, S.; Goswami, S.; Acharya, B. Polydimethylsiloxane-Multiwalled Carbon Nanotube Nanocomposites as Dielectric Materials: Frequency, Concentration, and Temperature-Dependence Studies. *J. Electron. Mater.* **2019**, *48*, 2853. [CrossRef]
8. Cai, C.; Chen, T.; Chen, X.; Zhang, Y.T.; Gong, X.H.; Wu, C.G.; Hu, T. Enhanced Electromechanical Properties of Three-Phased Polydimethylsiloxane Nanocomposites via Surface Encapsulation of Barium Titanate and Multiwalled Carbon Nanotube with Polydopamine. *Macromol. Mater. Eng.* **2021**, *306*, 2100046. [CrossRef]
9. Jiang, M.J.; Dang, Z.M.; Xu, H.P. Giant Dielectric Constant and Resistance-Pressure Sensitivity in Carbon Nanotubes/Rubber Nanocomposites with Low Percolation Threshold. *Appl. Phys. Lett.* **2007**, *90*, 042914. [CrossRef]
10. Lopes, B.H.K.; Portes, R.C.; Amaral Junior, M.A.D.; Florez-Vergara, D.E.; Gama, A.M.; Silva, V.A.; Quirino, S.F.; Baldan, M.R. X Band Electromagnetic Property Influence of Multi-Walled Carbon Nanotube in Hybrid MnZn Ferrite and Carbonyl Iron Composites. *J. Mater. Res. Technol.* **2020**, *9*, 2369. [CrossRef]
11. Dimiev, A.; Lu, W.; Zeller, K.; Crowgey, B.; Kempel, L.C.; Tour, J.M. Low-Loss, High-Permittivity Composites Made from Graphene Nanoribbons. *ACS Appl. Mater. Interfaces* **2011**, *3*, 4657. [CrossRef] [PubMed]
12. Dimiev, A.; Zakhidov, D.; Genorio, B.; Oladimeji, K.; Crowgey, B.; Kempel, L.; Rothwell, E.J.; Tour, J.M. Permittivity of Dielectric Composite Materials Comprising Graphene Nanoribbons. the Effect of Nanostructure. *ACS Appl. Mater. Interfaces* **2013**, *5*, 7567. [CrossRef] [PubMed]
13. Lounev, I.V.; Musin, D.R.; Dimiev, A.M. New Details to Relaxation Dynamics of Dielectric Composite Materials Comprising Longitudinally Opened Carbon Nanotubes. *J. Phys. Chem. C* **2017**, *121*, 22995. [CrossRef]
14. Zhang, Z.; Sun, S.; Liu, L.; Yu, K.; Liu, Y.; Leng, J. Dielectric Properties of Carbon Nanotube/Silicone Elastomer Composites. In *Proceedings of SPIE*; SPIE: Bellingham, WA, USA, 2009; Volume 7493, p. 749315-1.
15. Pantazi, A.G.; Oprea, O.; Pantazi, A.; Palade, S.; Berbecaru, C.; Purica, M.; Matei, A.; Oprea, O.; Dragoman, D. Dielectric Properties of Multiwall Carbon Nanotube-Red Silicone Rubber Composites. *J. Optoelectron. Adv. Mater.* **2015**, *17*, 1319.
16. Wang, F.; Wang, J.W.; Li, S.Q.; Xiao, J. Dielectric Properties of Epoxy Composites with Modified Multiwalled Carbon Nanotubes. *Polym. Bull.* **2009**, *63*, 101. [CrossRef]
17. Zhao, K.; Gupta, S.; Chang, C.; Wei, J.; Tai, N.H. Layered Composites Composed of Multi-Walled Carbon Nanotubes/Manganese Dioxide/Carbon Fiber Cloth for Microwave Absorption in the X-Band. *RSC Adv.* **2019**, *9*, 19217. [CrossRef]
18. Sotiropoulos, A.; Koulouridis, S.; Masouras, A.; Kostopoulos, V.; Anastassiu, H.T. Carbon Nanotubes Films in Glass Fiber Polymer Matrix Forming Structures with High Absorption and Shielding Performance in X-Band. *Compos. Part B Eng.* **2021**, *217*, 108896. [CrossRef]
19. Dimiev, A.M.; Surnova, A.; Lounev, I.; Khannanov, A. Intrinsic Insertion Limits of Graphene Oxide into Epoxy Resin and the Dielectric Behavior of Composites Comprising Truly 2D Structures. *J. Phys. Chem. C* **2019**, *123*, 3461. [CrossRef]
20. Dimiev, A.M.; Lounev, I.; Khamidullin, T.; Surnova, A.; Valimukhametova, A.; Khannanov, A. Polymer Composites Comprising Single-Atomic-Layer Graphenic Conductive Inclusions and Their Unusual Dielectric Properties. *J. Phys. Chem. C* **2020**, *124*, 13715. [CrossRef]
21. Li, Q.; Xue, Q.; Hao, L.; Gao, X.; Zheng, Q. Large Dielectric Constant of the Chemically Functionalized Carbon Nanotube/Polymer Composites. *Compos. Sci. Technol.* **2008**, *68*, 2290. [CrossRef]
22. Shang, S.; Tang, C.; Jiang, B.; Song, J.; Jiang, B.; Zhao, K.; Liu, Y.; Wang, X. Enhancement of Dielectric Permittivity in Carbon Nanotube/Polyvinylidene Fluoride Composites by Constructing of Segregated Structure. *Compos. Commun.* **2021**, *25*, 100745. [CrossRef]
23. Wang, L.; Dang, Z.M. Carbon Nanotube Composites with High Dielectric Constant at Low Percolation Threshold. *Appl. Phys. Lett.* **2005**, *87*, 042903. [CrossRef]
24. Yuan, J.K.; Yao, S.H.; Dang, Z.M.; Sylvestre, A.; Genestoux, M.; Bai, J. Giant Dielectric Permittivity Nanocomposites: Realizing True Potential of Pristine Carbon Nanotubes in Polyvinylidene Fluoride Matrix through an Enhanced Interfacial Interaction. *J. Phys. Chem. C* **2011**, *115*, 5515. [CrossRef]
25. Jun, Y.S.; Habibpour, S.; Hamidinejad, M.; Park, M.G.; Ahn, W.; Yu, A.; Park, C.B. Enhanced Electrical and Mechanical Properties of Graphene Nano-Ribbon/Thermoplastic Polyurethane Composites. *Carbon* **2021**, *174*, 305. [CrossRef]
26. Kasgoz, A.; Korkmaz, M.; Durmus, A. Compositional and Structural Design of Thermoplastic Polyurethane/Carbon Based Single and Multi-Layer Composite Sheets for High-Performance X-Band Microwave Absorbing Applications. *Polymer* **2019**, *180*, 121672. [CrossRef]
27. Wang, P.; Yang, L.; Zhou, Y.; Gao, S.; Cao, T.; Feng, S.; Xu, P.; Ding, Y. Synergistic Effect of EVA-GMA and Nanofillers on Mechanical and Dielectric Properties of Polyamide. *Compos. Commun.* **2021**, *25*, 100738. [CrossRef]
28. Grimes, C.A.; Dickey, E.C.; Mungle, C.; Ong, K.G.; Qian, D. Effect of Purification of the Electrical Conductivity and Complex Permittivity of Multiwall Carbon Nanotubes. *J. Appl. Phys.* **2001**, *90*, 4134. [CrossRef]
29. Baughman, R.H.; Zakhidov, A.A.; De Heer, W.A. Carbon Nanotubes—The Route toward Applications. *Science* **2002**, *297*, 787. [CrossRef] [PubMed]
30. Grimes, C.A.; Mungle, C.; Kouzoudis, D.; Fang, S.; Eklund, P.C. The 500 MHz to 5.50 GHz Complex Permittivity Spectra of Single-Wall Carbon Nanotube-Loaded Polymer Composites. *Chem. Phys. Lett.* **2000**, *319*, 460. [CrossRef]
31. Dash, K.; Hota, N.K.; Sahoo, B.P. Fabrication of Thermoplastic Polyurethane and Polyaniline Conductive Blend with Improved Mechanical, Thermal and Excellent Dielectric Properties: Exploring the Effect of Ultralow-Level Loading of SWCNT and Temperature. *J. Mater. Sci.* **2020**, *55*, 12568. [CrossRef]

32. Badawi, A.; Alharthi, S.S.; Assaedi, H.; Alharbi, A.N.; Althobaiti, M.G. $Cd_{0.9}Co_{0.1}S$ Nanostructures Concentration Study on the Structural and Optical Properties of SWCNTs/PVA Blend. *Chem. Phys. Lett.* **2021**, *775*, 138701. [CrossRef]
33. Puértolas, J.A.; García-García, J.F.; Pascual, F.J.; González-Domínguez, J.M.; Martínez, M.T.; Ansón-Casaos, A. Dielectric Behavior and Electrical Conductivity of PVDF Filled with Functionalized Single-Walled Carbon Nanotubes. *Compos. Sci. Technol.* **2017**, *152*, 263. [CrossRef]
34. Abdelrazek, E.M.; Elashmawi, I.S.; Hezma, A.M.; Rajeh, A.; Kamal, M. Effect of an Encapsulate Carbon Nanotubes (CNTs) on Structural and Electrical Properties of PU/PVC Nanocomposites. *Phys. B Condens. Matter* **2016**, *502*, 48. [CrossRef]
35. Khamidullin, T.; Galyaltdinov, S.; Valimukhametova, A.; Brusko, V.; Khannanov, A.; Maat, S.; Kalinina, I.; Dimiev, A.M. Simple, Cost-Efficient and High Throughput Method for Separating Single-Wall Carbon Nanotubes with Modified Cotton. *Carbon* **2021**, *178*, 157. [CrossRef]
36. Krestinin, A.V.; Dremova, N.N.; Knerel'Man, E.I.; Blinova, L.N.; Zhigalina, V.G.; Kiselev, N.A. Characterization of SWCNT Products Manufactured in Russia and the Prospects for Their Industrial Application. *Nanotechnologies Russ.* **2015**, *10*, 537. [CrossRef]
37. Nano, A.; Chortos, A.; Pochorovski, I.; Lin, P.; Pitner, G.; Yan, X.; Gao, T.Z.; To, J.W.F.; Lei, T.; Will, J.W.; et al. Universal Selective Dispersion of Semiconducting Carbon Nanotubes from Commercial Sources Using a Supramolecular Polymer. *ACS Nano* **2017**, *11*, 5660.
38. Salamatov, I.N.; Yatsenko, D.A.; Khasin, A.A. Determination of the Diameter Distribution Function of Single-Wall Carbon Nanotubes by the X-Ray Diffraction Data. *J. Struct. Chem.* **2019**, *60*, 2001. [CrossRef]
39. Maus, C. TUBALL Single Wall Carbon Nanotubes: A New Additive for Thermoplastics. *Soc. Plast. Eng.* **2017**, 2193.
40. da Silva, V.A.; Rezende, M.C. S-Parameters, Electrical Permittivity, and Absorbing Energy Measurements of Carbon Nanotubes-Based Composites in X-Band. *J. Appl. Polym. Sci.* **2021**, *138*, e49843. [CrossRef]
41. Shayesteh Zeraati, A.; Mende Anjaneyalu, A.; Pawar, S.P.; Abouelmagd, A.; Sundararaj, U. Effect of Secondary Filler Properties and Geometry on the Electrical, Dielectric, and Electromagnetic Interference Shielding Properties of Carbon Nanotubes/Polyvinylidene Fluoride Nanocomposites. *Polym. Eng. Sci.* **2021**, *61*, 959. [CrossRef]
42. Bizhani, H.; Katbab, A.A.; Lopez-Hernandez, E.; Miranda, J.M.; Lopez-Manchado, M.A.; Verdejo, R. Preparation and Characterization of Highly Elastic Foams with Enhanced Electromagnetic Wave Absorption Based on Ethylene-Propylene-Diene-Monomer Rubber Filled with Barium Titanate/Multiwall Carbon Nanotube Hybrid. *Polymers* **2020**, *12*, 2278. [CrossRef]
43. Novocontrol Technologies Presents WinFIT. Available online: https://www.novocontrol.de/brochures/WinFIT.pdf (accessed on 12 December 2021).
44. Kataura, H. Bundle Effects of Single-Wall Carbon Nanotubes. *AIP Conf. Proc.* **2000**, *544*, 262.
45. Alvarez, L.; Righi, A.; Guillard, T.; Rols, S.; Anglaret, E.; Laplaze, D.; Sauvajol, J.-L.; Francé, F. Resonant Raman Study of the Structure and Electronic Properties of Single-Wall Carbon Nanotubes. *Chem. Phys. Lett.* **2000**, *316*, 186. [CrossRef]
46. Jonscher, A.K. Dielectric Relaxation in Solids. *J. Phys. D Appl. Phys.* **1999**, *32*, R57. [CrossRef]
47. Prodromakis, T.; Papavassiliou, C. Engineering the Maxwell–Wagner Polarization Effect. *Appl. Surf. Sci.* **2009**, *255*, 6989. [CrossRef]
48. Cole, R.H. On the Analysis of Dielectric Relaxation. *J. Chem. Phys.* **1955**, *23*, 493. [CrossRef]
49. Lee, S.E.; Kang, J.H.; Kim, C.G. Fabrication and Design of Multi-Layered Radar Absorbing Structures of MWNT-Filled Glass/Epoxy Plain-Weave Composites. *Compos. Struct.* **2006**, *76*, 397. [CrossRef]

Article

PMMA/SWCNT Composites with Very Low Electrical Percolation Threshold by Direct Incorporation and Masterbatch Dilution and Characterization of Electrical and Thermoelectrical Properties

Ezgi Uçar [1,2], Mustafa Dogu [3], Elcin Demirhan [2] and Beate Krause [1,*]

1. Leibniz-Institut für Polymerforschung Dresden e.V. (IPF), Hohe Str. 6, 01069 Dresden, Germany; ezgiucar@gmail.com
2. Chemical Engineering Department, Yildiz Technical University, Davutpasa Campus, Esenler, 34220 Istanbul, Türkiye; demirhan@yildiz.edu.tr
3. Mir Ar-Ge Inc., Research Department, Esenyurt, 34522 Istanbul, Türkiye; mus.dogu@gmail.com
* Correspondence: krause-beate@ipfdd.de; Tel.: +49-351-4658-736

Abstract: In the present study, Poly(methyl methacrylate) (PMMA)/single-walled carbon nanotubes (SWCNT) composites were prepared by melt mixing to achieve suitable SWCNT dispersion and distribution and low electrical resistivity, whereby the SWCNT direct incorporation method was compared with masterbatch dilution. An electrical percolation threshold of 0.05–0.075 wt% was found, the lowest threshold value for melt-mixed PMMA/SWCNT composites reported so far. The influence of rotation speed and method of SWCNT incorporation into the PMMA matrix on the electrical properties and the SWCNT macro dispersion was investigated. It was found that increasing rotation speed improved macro dispersion and electrical conductivity. The results showed that electrically conductive composites with a low percolation threshold could be prepared by direct incorporation using high rotation speed. The masterbatch approach leads to higher resistivity values compared to the direct incorporation of SWCNTs. In addition, the thermal behavior and thermoelectric properties of PMMA/SWCNT composites were studied. The Seebeck coefficients vary from 35.8 µV/K to 53.4 µV/K for composites up to 5 wt% SWCNT.

Keywords: carbon nanotubes; melt-mixing; polymer composites; SWCNT; PMMA

Citation: Uçar, E.; Dogu, M.; Demirhan, E.; Krause, B. PMMA/SWCNT Composites with Very Low Electrical Percolation Threshold by Direct Incorporation and Masterbatch Dilution and Characterization of Electrical and Thermoelectrical Properties. *Nanomaterials* **2023**, *13*, 1431. https://doi.org/10.3390/nano13081431

Academic Editor: Muralidharan Paramsothy

Received: 24 March 2023
Revised: 18 April 2023
Accepted: 19 April 2023
Published: 21 April 2023

Copyright: © 2023 by the authors. Licensee MDPI, Basel, Switzerland. This article is an open access article distributed under the terms and conditions of the Creative Commons Attribution (CC BY) license (https://creativecommons.org/licenses/by/4.0/).

1. Introduction

Poly(methyl methacrylate) (PMMA) is a transparent, odorless polymer that stands out for its mechanical properties as well as optical properties [1]. It is also chemical and atmospheric corrosion resistant. With all these properties, PMMA is also a promising polymer for sensors, optics, coating and polishing materials, binders, high voltage, and electronic applications [1,2]. PMMA is also a very suitable candidate for passive electronic components and electromagnetic interference (EMI) materials if made electrically conductive [3]. Due to its compatibility with fillers and ability to be processed easily, it is used in polymer nanocomposites by being reinforced with carbon black, graphite, metal powder, and other inorganic components [1].

Carbon-based materials have outstanding properties and can be used in a wide variety of fields with their high surface area, unique carbon structures, and ability to form covalent bonds with different elements [4–9]. They can be used directly as well as fillers in polymer composites. The interest in polymer nanocomposites reinforced with carbon-based nanofiller is increasing due to their superior properties and their lightweight. Since carbon nanotubes (CNTs) have a high aspect ratio and unique mechanical properties, they stand out as fillers in polymer nanocomposites. CNTs are hydrophobic, electrically conductive, and have a large surface area compared to other carbon-based fillers such as graphite and fullerene [10]. In particular, single-walled carbon nanotubes (SWCNTs) have unique electrical and thermal conductivity, mechanical strength, and flexibility [11]. In view of all these properties, SWCNT and SWCNT-filled composites are potential materials for electronic applicatios such as supercapacitors, sensors, and EMI materials [12–16]. SWCNTs differ from other CNTs in terms of the number and diameter of concentric graphene layers. SWCNTs have a low diameter (0.5–2 nm), and thus a higher aspect ratio than multi-walled carbon nanotubes (MWCNTs), which can lead to better electrical conductivity at lower loading in SWCNT composites, although an improvement in electrical conductivity depends on multiple parameters. In other words, SWCNTs can be a good choice in cases where a low percolation threshold is desired [15,17].

On the other hand, SWCNTs tend to re-aggregate easily due to van der Waals interactions between nanotubes, which complicates their dispersion in the polymer matrix [18]. In addition, SWCNTs are more difficult to disperse in the matrix because they are usually more likely to appear in bundles than MWCNTs [17]. It is known that the formation of the electrical path in the polymer matrix depends on the nanotube dispersion and distribution. The more individual SWCNTs in the matrix, the better the conductive network is formed. Thus, higher electrical conductivity and a lower electrical percolation threshold are obtained [19]. Therefore, it is crucial to provide an optimum dispersion in improving the properties of polymer composites. Solvent-based methods [18,20–26] with surfactants and/or ultrasonication treatment, in situ polymerization [27,28], and functionalization or modification of SWCNT [18,22,23,26,29,30] are commonly used approaches to achieve the desired distribution of SWCNT in the polymer matrix [17].

The melt-mixing method is preferred in the preparation of polymer composites for large-scale industrial applications [31,32]. It does not cause pollution and environmental problems since no solvent is used in the melt mixing method [32]. Therefore, preparing polymer composites with improved properties using the melt-mixing method is one of the essential issues. The manufacturing conditions play a significant role [31,33–37]. For example, higher temperatures can reduce the melt viscosity and decrease the shear forces. In addition, a higher rotational speed also leads to higher shear forces, which is important for the dispersion and distribution of the fillers [31,35].

Another fact is that the filler type is also decisive for the final manufacturing conditions, e.g., the feeding position during melt extrusion, as described by Müller et al. [37]. MWCNTs that require high shear stresses for their dispersion (e.g., Baytubes® C150P) should be dosed in the main hopper, while for CNT powders that disperse well (e.g., NC7000), dosing into the melt at the side feeder is more advantageous.

There are limited studies on preparing PMMA/SWCNT composites with low percolation thresholds and high conductivity using melt-mixing as a simple approach. Fraser et al. [29] prepared PMMA by in situ polymerization in the presence of raw and purified SWCNTs. Then, they diluted this masterbatch with PMMA in a twin-screw extruder. Although this study stated that composites could be produced on a large scale, a conductivity value of only $1\text{--}5 \times 10^{-10}$ S/cm was obtained (0.09 wt% loading, both SWCNT types), which was below the electrical percolation threshold [29].

There are studies in the literature where PMMA is blended with different polymers to improve the nanofiller dispersion [30,38,39]. Bikshamaiah et al. [30] incorporated functionalized SWCNTs into polyamide 6 (PA6)/PMMA blends prepared with different polymer ratios. They focused in their study on the morphological and mechanical characteristics of the composites. The tensile modulus of the blends and nanocomposites increased by ~40% and ~54%, respectively, as the PMMA ratio increased from 20 up to 80 wt% for blends and from 19.5 up to 79.5 wt% for nanocomposites. They reported that the PA6/PMMA blends exhibited two different morphologies, namely dispersed PMMA particles or cocontinuous structures. In another study [39], MWCNTs were dispersed in PMMA and a PMMA/poly(vinylidene fluoride) (PVDF) blend using the melt-mixing method. It was demonstrated that PVDF improved the interactions between PMMA and MWCNTs by showing a compatibilizing effect. The electrical percolation threshold was between 0.5 and

1 wt% for both PMMA-CNT and PMMA/PVDF-CNT blend composites [39]. Guo et al. [38] dispersed CNTs in PMMA and PMMA/polystyrene (PS) blends using a twin screw microcompounder. The percolation thresholds were 1.25 wt% and 0.5 wt% in PMMA and PMMA/PS blends, respectively.

The masterbatch dilution is another method to ensure a good dispersion and homogeneous distribution of CNTs in a polymer matrix. In this method, a composite containing a high amount of filler material is prepared, and then this composite is diluted by a second mixing process with pure polymer. The masterbatch method is known to be a simple and effective method for incorporating CNTs into the polymer matrix [40]. Pötschke et al. [41] reported that the masterbatch approach is suitable for the dispersion of MWCNTs Nanocyl^{TM} NC7000 in a polypropylene (PP) matrix and leads to a better nanotube dispersion, albeit at the expense of slightly increased electrical resistivity, compared to the MWCNT direct incorporation. Annala et al. [42] incorporated MWCNTs Nanocyl^{TM} NC7000 into PS or PMMA matrix using a masterbatch prepared by direct and in situ polymerization. In PMMA/MWCNT composites prepared by direct incorporation, the samples were not conductive up to 4 wt% CNT loading. The electrical percolation threshold of the composites using the PMMA masterbatch was below 4 wt% [42].

Studies on melt-mixed composites of thermoplastics with SWCNTs have partially focused direct incorporation method [30,43–45]. To the best of our knowledge, there is no study in which the methods of SWCNT addition are compared and the parameters in the preparation of the masterbatch and their effects on the properties of SWCNT nanocomposites have been examined in detail.

In addition, thermoelectric investigations of the PMMA/SWCNT composites are also of interest. Previous studies have shown that various polymer-CNT composites can be used as thermoelectric materials. If commercial CNTs, which typically show p-type behavior, are incorporated into polymers, the composites usually show the p-type as well. This could be described for melt-mixed composites based on polypropylene (PP) [19,45,46], polycarbonate (PC) [44,47–49], poly(ether ether ketone) (PEEK) [49,50], PVDF [44,51], and poly(butylene terephthalate) (PBT) [44].

In this study, PMMA/SWCNT composites were prepared by direct incorporation and masterbatch dilution using the melt mixing method. Tuball^{TM} material, known as "graphene nanotubes", from OCSiAl company, was used as a filler in the study. Tuball^{TM} is a high-quality, low-cost material suitable for mass production [52]. Electrical, thermoelectrical, and morphological characterization of PMMA/SWCNT composites prepared by the direct incorporation method was performed. The effects of the rotation speed of direct incorporation and masterbatch preparation and dilution on the electrical and morphological properties of the composites were compared.

2. Materials and Methods

2.1. Materials

Pure PMMA (Plexiglas 8N Röhm GmbH, Darmstadt, Germany) and SWCNTs Tuball^{TM} (OCSiAl S.a.r.l., Leudelange, Luxembourg) were used as polymer matrix and nanofiller, respectively. Tuball^{TM} nanotubes has a carbon purity of 75%, an outside diameter of 1.6 nm, and lengths of more than 5 μm [17,52]. The density and melt volume-flow rate values of PMMA are 1190 kg/m^3 and 3 cm^3/10 min (230 °C and 3.8 kg), respectively. The SWCNT selection was based on a former study comparing different kinds of CNTs [44]. The Raman spectra of this SWCNT type are published in the supporting information [49]. The thermoelectric parameters of the SWCNT powder were described in [44]. This kind of SWCNTs is abbreviated as Tuball.

2.2. Composite Preparation

The melt-mixing of PMMA/SWCNT composites containing 0, 0.05, 0.075, 0.1, 0.2, 0.25, 0.5, 1, 3, and 5 wt% SWCNTs was performed in a 15 cm^3 conical twin-screw micro compounder Xplore 15 (Xplore Instruments BV, Sittard, The Netherlands) at 260 °C for

5 min with rotation speed of 250 rpm. Processing temperature and mixing time were kept constant throughout the entire study. Before melt mixing, PMMA and SWCNTs were dried overnight in a vacuum oven at 80 °C and 120 °C, respectively.

In order to examine the effect of rotation speed at the same processing temperature and mixing time (260 °C and 5 min), composites containing 0.1, 0.25, and 1 wt% SWCNTs were prepared by direct incorporation at rotation speeds varying between 50 and 250 rpm.

The masterbatch dilution method was used to study the effect of the nanofiller addition method on the electrical properties and nanotube macrodispersion. First, masterbatches with 5 wt% SWCNT content were prepared at 250 or 50 rpm rotation speed with the same procedure as specified in Section 2.2. Then, the masterbatches were diluted with PMMA at 250 and 50 rpm to prepare composites containing 0.05–1 wt% SWCNTs. Samples prepared by direct incorporation (direct) and masterbatch preparation (MB) and dilution (D) were named according to the corresponding rotation speeds. For example, the code of the sample prepared at 250 rpm and diluted at 50 rpm is PMMA/Tuball-MB250 + D50.

2.3. Shaping Process

The extruded strands were cut into pieces and compression-molded using a hot press PW40EH (Paul-Otto-Weber GmbH, Remshalden, Germany) with a compression molding time of 1 min at 260 °C. The plates were prepared with a diameter of 60 mm and a thickness of 0.5 mm.

2.4. Characterization

The SWCNT macrodispersion of the composites was investigated by transmission light microscopy using an integrated light microscope (Olympus BX 53M-RLA) equipped with an Olympus DP74 camera. The extruded strands were fixed by dipping them in a resin, and the cured resin with the embedded strand was cut using a Leica RM2265 instrument in thin slices with a thickness of 5 μm.

For characterisation of SWCNT nanodispersion, scanning electron microscopy (SEM) images of the composites were acquired using an ULTRA Plus (Carl Zeiss AG, Oberkochen, Germany) scanning electron microscope at 3 kV acceleration voltage using the SE2 detector. Cryo-fractured surfaces of strands were observed. All samples were sputtered with a 3-nm platin film.

The thermoelectric properties of strips (dimension $12 \times 5 \times 0.5$ mm) cut from the compression molded samples were determined using the device developed and built at the Leibniz Institute of Polymer Research Dresden (IPF) [53]. The measurements were carried out at 40 °C and at the following temperature differences: 32–40 °C, 36–40 °C, 44–40 °C, and 48–40 °C. The Seebeck coefficients were determined by measurements of two to four strips. The individual values were obtained by repeating measurements five times for each temperature difference. Electrical volume resistivity measurements were carried out at 40 °C on the same equipment using the same samples. The 4-wire technique was used in the measurements and the mean values of the resistivity values were calculated with ten measurements. The Keithley multimeter DMM2001 (Keithley Instruments, Cleveland, OH, USA) was used for resistance and thermovoltage measurements. More details on these measurements are given in references [54,55]. The Seebeck coefficient S was calculated from the quotient of the thermoelectric voltage U and the applied temperature difference dT. The power factor PF is calculated from the electrical volume conductivity σ and the squared Seebeck coefficient (PF = $S^2 \cdot \sigma$).

The electrical volume resistivity measurement of the compression-molded samples was performed using a Keithley 8009 Resistance test fixture coupled with a Keithley electrometer E6517A at room temperature. For samples with a resistance lower than 10^7 Ohm, strips were cut from these plates, and in-plane electrical resistivity was measured using a Keithley multimeter Model DMM2001 and a 4-point test fixture developed by IPF. Both surfaces of two different samples were measured for each composite sample. The averages of these values were taken.

The thermal behavior of the samples was measured using a Perkin Elmer DSC 4000 differential scanning calorimeter (DSC) under nitrogen atmosphere. The samples weighted 9.5 mg were heated from 25 to 250 °C at a heating rate of 10 K/min. The glass transition temperature (Tg) data were obtained from the second heating scan.

3. Results and Discussion

3.1. Composite Morphology

In the preparation of SWCNT-based polymer composites by the melt mixing process, the shear stresses to which the SWCNT primary material was exposed during the mixing process is of critical importance. Therefore, the shear stresses need to be optimized to achieve the desired dispersion and distribution without sacrificing the structural integrity of the SWCNTs [56]. With the increase of rotation speed, shear stresses or mixing energy input increased. To examine the effect of rotation speed, composites containing 0.1, 0.25, and 1 wt% SWCNT were prepared by direct incorporation at rotation speeds varying between 50 and 250 rpm.

The macrodispersion of 1 wt% SWCNTs in PMMA composites at the different rotation speeds was studied using transmission light microscopy (LM) on thin sections (Figure 1). The number and size of remaining agglomerates decreased with increasing rotation speed. The morphologically best SWCNT dispersion was obtained for the sample prepared at 250 rpm, indicated by the lowest number of agglomerates. It can be concluded that high shear stresses are favorable for good dispersion and homogeneous distribution of the SWCNTs.

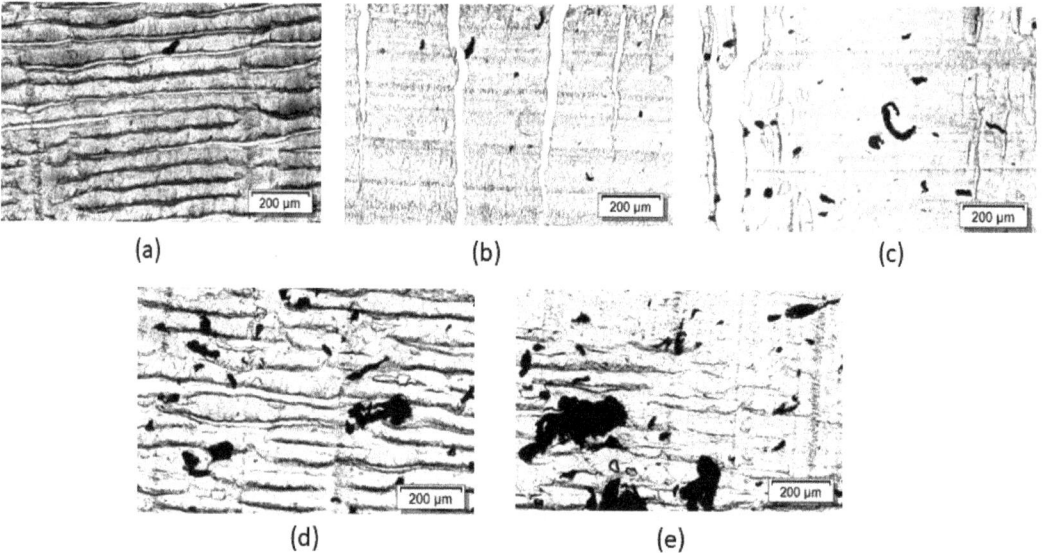

Figure 1. LM images of samples containing 1 wt% SWCNT prepared by direct incorporation at different rotation speeds: (**a**) 250 rpm, (**b**) 200 rpm, (**c**) 150 rpm, (**d**) 100 rpm, (**e**) 50 rpm (lines in the images are related to the brittleness of the samples and represents folds from the cutting process).

Composites containing 5 wt% SWCNTs were prepared by the direct incorporation method at 250 rpm and 50 rpm and regarded as a masterbatch, which was diluted to 0.05 to 1 wt% SWCNTs at 50 or 250 rpm. Figure 2 compares light microscope images of PMMA/1 wt% SWCNTs composites using the masterbatch approach at different rotation speeds in both steps. It can be seen that using 50 rpm in both steps caused a poor SWCNT dispersion with many large agglomerates (Figure 2a). The SWCNT dispersion is similar to that achieved with direct SWCNT incorporation (Figure 1e). If at least one mixing step took place at 250 rpm, significantly fewer agglomerates are seen in the composites (Figure 2b–d). Thereby, whether the higher rotation speed was applied in the masterbatch step or in the dilution step plays a subordinate role in the SWCNT dispersion.

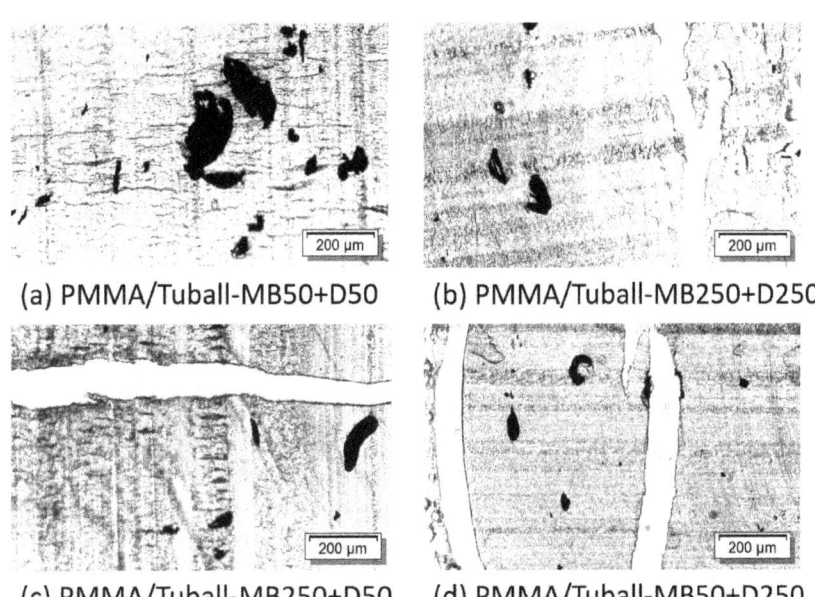

Figure 2. LM images of samples containing 1 wt% SWCNT prepared by masterbatch approach at different rotation speeds. 50 rpm/50 rpm (**a**), at 250 rpm/250 rpm (**b**), at 250 rpm/50 rpm (**c**), at 50 rpm/250 rpm (**d**) (rotation speed at masterbatch preparation/rotation speed at masterbatch dilution) (lines in the images are related to the brittleness of the samples and represents folds from the cutting process).

Figure 3 shows SEM images that can be used to assess SWCNT distribution at the nanoscale in the composites. It is remarkable that for the composite prepared by direct incorporation at 50 rpm (Figure 3a), the SWCNT are distributed significantly differently than in Figure 3b–d. Here, areas with SWCNT are visible and areas where no SWCNT can be seen (marked with dotted lines), indicating a very inhomogeneous SWCNT distribution. This observation correlates with the light microscopy image of this composite (Figure 1e), which shows very large agglomerates and thus a very inhomogeneous macroscopic SWCNT distribution. It can be concluded that when SWCNTs are directly incorporated into PMMA at the lowest rotation speed of 50 rpm, the SWCNTs are distributed very inhomogeneously in both the nanoscopic and macroscopic scales. For the composite prepared by direct incorporation at the highest speed of 250 rpm (Figure 3b) and the two composites produced by masterbatch dilutions (Figure 3c,d), uniform SWCNT distribution on a nanoscopic scale is seen in each case. The rotation speed seems to play a subordinate role in the masterbatch approach for the SWCNT distribution in nanoscale. However, larger and more numerous agglomerates were observed in the light microscopy images when only 50 rpm (Figure 2a) was used in the masterbatch approach instead of 250 rpm (Figure 2b).

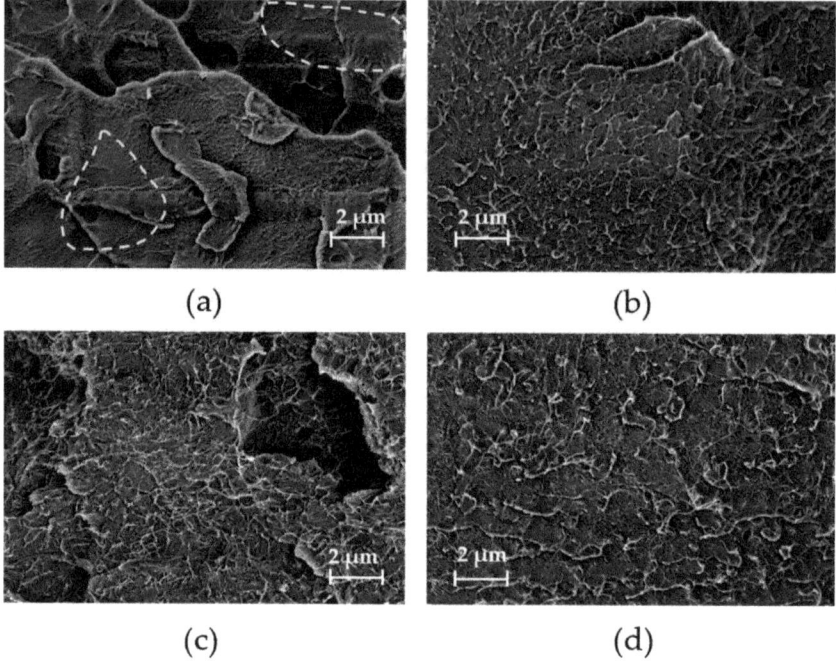

Figure 3. SEM images of cryofractures surface of composites containing 1 wt% SWCNT prepared by direct incorporation at different rotation speeds: (**a**) 50 rpm (dotted lines for highlighting), (**b**) 250 rpm, and masterbatch approach at different rotation speeds. 50 rpm/50 rpm (**c**), at 250 rpm/250 rpm (**d**) (rotation speed at masterbatch preparation/rotation speed at masterbatch dilution).

3.2. Electrical Properties

The electrical resistivity vs. the SWCNT content of the composites prepared by direct incorporation at 250 rpm is presented in Figure 4. A distinct decrease compared to the value of pure PMMA occurs at the addition of 0.05 wt% SWCNTs indicating the electrical percolation threshold to be in the range of 0.05 to 0.075 wt%. This is the lowest electrical percolation threshold value reported so far in literature for melt-mixed PMMA/SWCNT composites. The decrease in the electrical resistivity of the composites continued until the addition of 1 wt% SWCNT. At contents higher than 1 wt % SWCNT, the values levelled off at around 10 Ohm·cm.

To examine the effect of a shear stresses on the electrical resistivity of the composites, the rotation speed of the composites containing 0.1 wt%, 0.25 wt%, and 1 wt% SWCNTs was varied between 50 and 250 rpm. It appeared that the electrical resistivity decreased with the increasing rotation speed (Figure 5). The rotation speed effect is most pronounced at the filler content of 0.1 wt% SWCNTs. This can be explained by the fact that this is the concentration closest to the electrical percolation threshold, making the conductive network with the small number of contact points very sensitive to changes. Obtaining the lowest electrical resistivity value at the highest speed may be due to the high shear stresses making it easier to separate the SWCNT bundles. As seen in Figure 1, only a small amount of agglomerate is observed at high rotation speeds, whereas at 50 rpm, a high number of agglomerates are visible. These results of better dispersion and at the same time reduced resistivity at increasing rotation speed is what would be expected. However, this is in contrast to some other investigations on MWCNT composites. An example is the study by Krause et al. [31] on PA6/MWCNT (type NC7000) composites, where it was found that the percolation threshold increased from 2–2.5 wt% to 3–4 wt% when the rotation speed was increased from 50 to 150 rpm. A comparable trend was observed by Pötschke et al. [35] for composites of 0.5 wt% MWCNT NC7000 in poly(caprolactone) (PCL), where the electrical resistivity increased with increasing rotation speed (50–400 rpm) whereby at the same time the number of residual agglomerates in the polymer matrix decreased. The explanation of those effects, which on the first view are unexpected, was a possible MWCNT length shortening at higher rotation speeds [56]. However, in our case, for the SWCNTs investigated, high rotational speed of 250 rpm led to both a more homogeneous SWCNT dispersion and a lower resistivity, indicating that length shortening possibly may not have a significant effect.

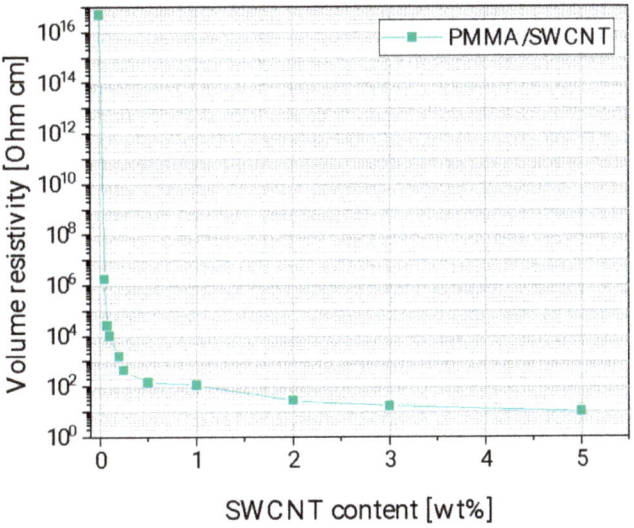

Figure 4. Electrical volume resistivity of PMMA/SWCNT composites prepared with direct incorporation method at 250 rpm.

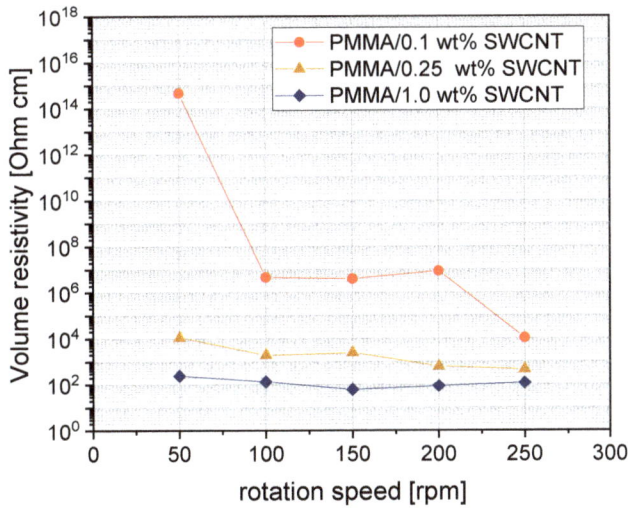

Figure 5. Electrical volume resistivity of PMMA/SWCNT composites prepared with the direct incorporation method at different rotation speeds.

The electrical resistivity results of the composites prepared by direct incorporation and masterbatch dilution are compared in Figure 6. The masterbatch (PMMA/5 wt% SWCNT) was prepared at 50 rpm or 250 rpm. Both masterbatches were diluted at 50 or 250 rpm in the dilution step.

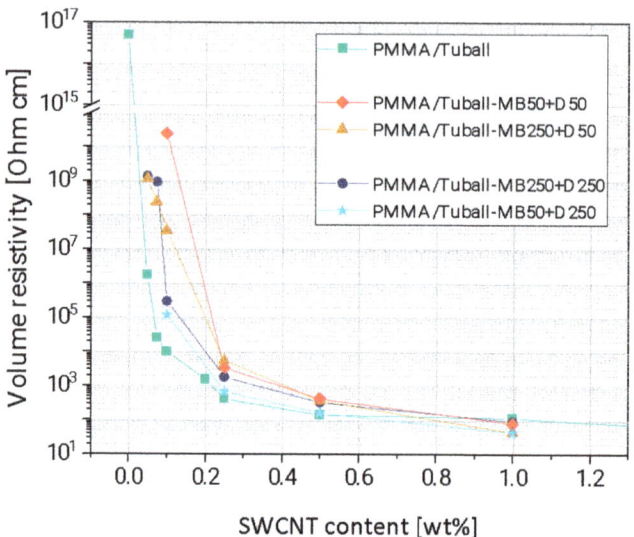

Figure 6. Electrical volume resistivity of PMMA/SWCNT composites prepared with the masterbatch diluting method at different rotation speeds compared with PMMA/SWCNT samples prepared at 250 rpm using direct incorporation.

The highest resistivity values were achieved at the rotation speed of 50 rpm in the masterbatch method and in the direct incorporation method. In particular, the electrical resistivity values of the composites are higher in composites prepared at a rotation speed of 50 rpm in both the masterbatch preparation and dilution step. The electrical resistivity of composites with 0.05 wt% and 0.075 SWCNT prepared at the masterbatch preparation speed of 250 rpm (regardless of high or low dilution speed) were three and four orders of magnitude higher, respectively, than those of composites prepared by direct incorporation at 250 rpm. Consequently, the electrical percolation threshold of the composites prepared with the masterbatch dilution method was higher compared to that using the direct incorporation. The best results (lowest resistivities) for almost all samples were obtained at a rotation speed of 250 rpm for both preparation approaches (MB and MBD). It was concluded that the rotation speed of 50 rpm was ineffective in masterbatch preparation and dilution as well as in direct incorporation. The electrical resistivities of the all composites prepared by the masterbatch dilution approach were up to a SWCNT content of 0.5 wt% higher than those prepared by the direct incorporation. At a SWCNTs content of 1 wt%, the resistivity values were in the same range and no significant influence of the preparation method can be observed. As a result, the most effective way to prepare composites with good dispersion and low electrical resistivity is the direct incorporation at 250 rpm.

3.3. Thermoelectrical Properties

The results of the thermoelectric properties of PMMA/SWCNT composites measured at 40 °C are given in Figure 7 and Table S1. The thermoelectric measurements were started with composites having a SWCNT concentration as low as 0.25 wt% and were carried out up to 5 wt%. All composites exhibit positive Seebeck coefficients due to the p-type characteristic of SWCNTs of the type Tuball™. The Seebeck coefficient, which increased with the addition of SWCNTs, achieved a maximum value of 53.4 µV/K at 2 wt% filler.

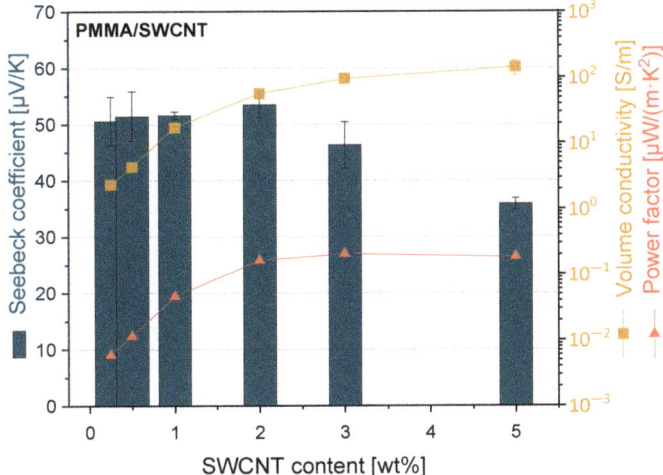

Figure 7. Thermoelectric properties of PMMA/SWCNT composites prepared by direct incorporation at a rotation speed of 250 rpm.

The power factor of the composites was calculated based on the Seebeck coefficient and electrical volume conductivity. The highest power factor value was 0.207 µW/m·K^2 at 3 wt% SWCNT loading. For PA6/PMMA/SWCNT composites (90/10/3), a maximum power factor of 0.135 µW/m·K^2 was previously determined at the same amount of filler content [43].

Comparing the Seebeck coefficient values of the composites with those of the pure SWCNT powder with a value of 39.6 µV/K [44] indicates that the Seebeck coefficient value increases after incorporation these SWCNTs into the PMMA polymer Matrix. This implies that there is a p-doping effect of the PMMA chains on the SWCNTs. A similar effect has already been described for polyether ether ketone (PEEK)/SWCNT Tuball [38] and polybutylene terephthalate (PBT)/SWCNT Tuball composites [44,54].

3.4. Thermal Behavior

The thermal behavior of PMMA/SWCNT composites was investigated by DSC analysis. The Tg value of pure PMMA, which was 113.7 °C, increased to 114.6 °C with the addition of 0.05 wt% SWCNTs. The Tg of the composites increased further with the addition of higher SWCNT amounts and reached 117.0 °C at a filler loading of 5 wt%. It is known that when an inorganic filler is incorporated into an organic polymer matrix, the chain mobility of polymers is affected by polymer–surface interactions [57]. Therefore, the increased Tg due to the addition of SWCNTs can be attributed to the fact that the mobility of PMMA chains is significantly reduced after mixing with SWCNTs. In other words, adding SWCNTs to the PMMA matrix reduced the free volume and chain mobility of PMMA. As a result, the Tg values increased with the increase of the amount of SWCNTs in the composites [22,57,58]. This increase is in agreement with other results. Badawi and Al Hosiny [58] found Tg values of 91.2, 92.7, and 99.5 °C for pure PMMA, 0.5 wt% SWCNTs, and 2 wt% SWCNT filled composites, respectively. In a study by Flory et al. [22], the Tg of pure PMMA increased from 99 °C to 116 °C with 1 wt% SWCNT loading.

The effect of the rotation speed and the masterbatch preparation method on the Tg of the preapared nanocomposites was also investigated. However, neither had a significant impact on the Tg of PMMA (Tables S2 and S3).

4. Conclusions

Electrically conductive PMMA/SWCNT nanocomposites were fabricated by melt mixing using direct incorporation and masterbatch dilution approaches. The volume resistivity of the composites varied between 1.81×10^6 and 9.92 Ohm·cm with the addition of 0.05–5 wt% SWCNTs to the insulating PMMA matrix. According to the volume resistivity results, the percolation threshold was in the range of 0.05–0.075 wt% SWCNTs. The investigation established that 250 rpm was the most effective rotation speed to get low electrical resistivity and good macrodispersion of the composites, regardless of whether the SWCNTs were incorporated directly or via the masterbatch approach. The number and size of agglomerations and the electrical resistivity decreased at higher rotation speeds. Masterbatches were prepared and diluted at low and high rotation speeds to examine the effect on electrical and morphological properties. Masterbatch samples prepared and/or diluted at 50 rpm were ineffective compared with those prepared by direct incorporation at 250 rpm in terms of dispersion and electrical properties. Similar results were obtained with samples prepared by direct incorporation at 250 rpm compared to samples prepared and diluted at 250 rpm (except for composites with low SWCNTs content). This result shows that, for the studied type of SWCNT, the selection of a suitable rotation speed is more important than the SWCNT addition method to produce good quality PMMA/SWCNT composites.

In the thermoelectric measurements on composites prepared by direct incorporation at 250 rpm, it was found that the Seebeck coefficient increased up to 2 wt% filler content and reached the maximum value of 53.4 µV/K. A high power factor, reaching values up to 0.207 µW/m·K^2, was recorded in the composite filled 3 wt% SWCNTs. The electrical conductivity increased with the addition of SWCNTs, and the maximum value was obtained at 5 wt% with 145 S/m.

Furthermore, DSC analysis showed that the addition of SWCNT increased the glass transition temperature of the PMMA by approximately 4 K.

In summary, the Tuball™ SWCNT material is an effective filler to obtain conductive composites with good dispersion and low electrical resistivity by melt-mixing. The thermoelectric measurements indicate that PMMA/SWCNT composites can be used as a thermoelectric material.

Supplementary Materials: The following supporting information can be downloaded at: https://www.mdpi.com/article/10.3390/nano13081431/s1, Table S1. Thermoelectric properties of PMMA/SWCNT composites; Table S2. Comparison of glass transition temperature Tg values of composites with the change of SWCNT (Tuball) amount; Table S3. Comparison of glass transition temperature Tg values of composites prepared at different rotation speeds.

Author Contributions: Conceptualization, B.K.; methodology, B.K.; validation, E.U.; formal analysis, E.U.; investigation, E.U.; writing—original draft preparation, E.U.; writing—review and editing, B.K., visualization, E.U.; supervision, B.K., M.D. and E.D.; project administration, B.K.; funding acquisition, E.U. All authors have read and agreed to the published version of the manuscript.

Funding: This research was funded by The Scientific and Technological Research Council of Türkiye (TÜBİTAK), grant numbers: 2214-A and 2244 (Project No:118C073).

Data Availability Statement: The data presented in this study are available on request from the corresponding author.

Acknowledgments: The authors would like to thank all collaborators of the IPF Functional Nanocomposites and Blends department for their support and Ulrike Jentzsch-Hutschenreuther (IPF) for the thermoelectric measurements. The authors thank the TÜBİTAK 2214-A International Research Fellowship Program for Ph.D. Students and 2244 Industrial Ph.D. Fellowship Program for the financial support of Ezgi Uçar.

Conflicts of Interest: The authors declare no conflict of interest. The funders had no role in the design of the study; in the collection, analyses, or interpretation of data; in the writing of the manuscript, or in the decision to publish the results.

References

1. Ali, U.; Karim, K.J.B.A.; Buang, N.A. A review of the properties and applications of poly (methyl methacrylate)(PMMA). *Polym. Eng. Rev.* **2015**, *55*, 678–705. [CrossRef]
2. Kuila, T.; Bose, S.; Khanra, P.; Kim, N.H.; Rhee, K.Y.; Lee, J.H. Characterization and properties of in situ emulsion polymerized poly(methyl methacrylate)/graphene nanocomposites. *Compos. Part. A Appl. Sci. Manuf.* **2011**, *42*, 1856–1861. [CrossRef]
3. Chen, B.; Cinke, M.; Li, J.; Meyyappan, M.; Chi, Z.; Harmon, J.P.; O'Rourke Muisener, P.A.; Clayton, L.; D'Angelo, J. Modifying the Electronic Character of Single-Walled Carbon Nanotubes Through Anisotropic Polymer Interaction: A Raman Study. *Adv. Funct. Mater.* **2005**, *15*, 1183–1187. [CrossRef]
4. Zhao, H.; Zhu, G.; Li, F.; Liu, Y.; Guo, M.; Zhou, L.; Liu, R.; Komarneni, S. 3D interconnected honeycomb-like ginkgo nut-derived porous carbon decorated with β-cyclodextrin for ultrasensitive detection of methyl parathion. *Sens. Actuators B Chem.* **2023**, *380*, 133309. [CrossRef]
5. Dos Santos, M.C.; Maynart, M.C.; Aveiro, L.R.; da Paz, E.C.; dos Santos Pinheiro, V. Carbon-based materials: Recent advances, challenges, and perspectives. In *Reference Module in Materials Science and Materials Engineering*; Elsevier: Amsterdam, The Netherlands, 2017.
6. Zhao, H.; Guo, M.; Li, F.; Zhou, Y.; Zhu, G.; Liu, Y.; Ran, Q.; Nie, F.; Dubovyk, V. Fabrication of gallic acid electrochemical sensor based on interconnected Super-P carbon black@ mesoporous silica nanocomposite modified glassy carbon electrode. *J. Mater. Res. Technol.* **2023**, *24*, 2100–2112. [CrossRef]
7. Zhao, H.; Chang, Y.; Liu, R.; Li, B.; Li, F.; Zhang, F.; Shi, M.; Zhou, L.; Li, X. Facile synthesis of Vulcan XC-72 nanoparticles-decorated halloysite nanotubes for the highly sensitive electrochemical determination of niclosamide. *Food Chem.* **2021**, *343*, 128484. [CrossRef] [PubMed]
8. Sabzehmeidani, M.M.; Mahnaee, S.; Ghaedi, M.; Heidari, H.; Roy, V.A. Carbon based materials: A review of adsorbents for inorganic and organic compounds. *Mater. Adv.* **2021**, *2*, 598–627. [CrossRef]
9. Tulliani, J.-M.; Inserra, B.; Ziegler, D. Carbon-based materials for humidity sensing: A short review. *Micromachines* **2019**, *10*, 232. [CrossRef]
10. Nurazzi, N.; Sabaruddin, F.; Harussani, M.; Kamarudin, S.; Rayung, M.; Asyraf, M.; Aisyah, H.; Norrrahim, M.; Ilyas, R.; Abdullah, N. Mechanical performance and applications of cnts reinforced polymer composites—A review. *Nanomaterials* **2021**, *11*, 2186. [CrossRef]
11. Kharlamova, M.V.; Kramberger, C. Applications of Filled Single-Walled Carbon Nanotubes: Progress, Challenges, and Perspectives. *Nanomaterials* **2021**, *11*, 2863. [CrossRef]
12. Husain, A.; Ahmad, S.; Shariq, M.U.; Khan, M.M.A. Ultra-sensitive, highly selective and completely reversible ammonia sensor based on polythiophene/SWCNT nanocomposite. *Materialia* **2020**, *10*, 100704. [CrossRef]
13. Sinha, R.; Roy, N.; Mandal, T.K. SWCNT/ZnO nanocomposite decorated with carbon dots for photoresponsive supercapacitor applications. *Chem. Eng. J.* **2022**, *431*, 133915. [CrossRef]
14. Byeon, J.-H.; Kim, J.-S.; Kang, H.-K.; Kang, S.; Kim, J.-Y. Acetone gas sensor based on SWCNT/Polypyrrole/Phenyllactic acid nanocomposite with high sensitivity and humidity stability. *Biosensors* **2022**, *12*, 354. [CrossRef]
15. Novikov, I.V.; Krasnikov, D.V.; Vorobei, A.M.; Zuev, Y.I.; Butt, H.A.; Fedorov, F.S.; Gusev, S.A.; Safonov, A.A.; Shulga, E.V.; Konev, S.D. Multifunctional elastic nanocomposites with extremely low concentrations of single-walled carbon nanotubes. *ACS Appl. Mater. Interfaces* **2022**, *14*, 18866–18876. [CrossRef]
16. Shahapurkar, K.; Gelaw, M.; Tirth, V.; Soudagar, M.E.M.; Shahapurkar, P.; Mujtaba, M.; MC, K.; Ahmed, G.M.S. Comprehensive review on polymer composites as electromagnetic interference shielding materials. *Polym. Polym. Compos.* **2022**, *30*, 09673911221102127. [CrossRef]
17. Krause, B.; Pötschke, P.; Ilin, E.; Predtechenskiy, M. Melt mixed SWCNT-polypropylene composites with very low electrical percolation. *Polymer* **2016**, *98*, 45–50. [CrossRef]
18. Kalakonda, P.; Banne, S. Thermomechanical properties of PMMA and modified SWCNT composites. *Nanotechnol. Sci. Appl.* **2017**, *10*, 45. [CrossRef]
19. Krause, B.; Bezugly, V.; Khavrus, V.; Ye, L.; Cuniberti, G.; Pötschke, P. Boron doping of SWCNTs as a way to enhance the thermoelectric properties of melt-mixed polypropylene/SWCNT composites. *Energies* **2020**, *13*, 394. [CrossRef]
20. Haggenmueller, R.; Gommans, H.H.; Rinzler, A.G.; Fischer, J.E.; Winey, K.I. Aligned single-wall carbon nanotubes in composites by melt processing methods. *Chem. Phys. Lett.* **2000**, *330*, 219–225. [CrossRef]
21. Al-Osaimi, J.; Al-Hosiny, N.; Abdallah, S.; Badawi, A. Characterization of optical, thermal and electrical properties of SWCNTs/PMMA nanocomposite films. *Iran. Polym. J.* **2014**, *23*, 437–443. [CrossRef]
22. Flory, A.L.; Ramanathan, T.; Brinson, L.C. Physical Aging of Single Wall Carbon Nanotube Polymer Nanocomposites: Effect of Functionalization of the Nanotube on the Enthalpy Relaxation. *Macromolecules* **2010**, *43*, 4247–4252. [CrossRef]
23. Liu, J.; Rasheed, A.; Minus, M.L.; Kumar, S. Processing and properties of carbon nanotube/poly(methyl methacrylate) composite films. *J. Appl. Polym. Sci.* **2009**, *112*, 142–156. [CrossRef]
24. Skákalová, V.; Dettlaff-Weglikowska, U.; Roth, S. Electrical and mechanical properties of nanocomposites of single wall carbon nanotubes with PMMA. *Synth. Met.* **2005**, *152*, 349–352. [CrossRef]
25. Pradhan, N.; Iannacchione, G. Thermal properties and glass transition in PMMA+ SWCNT composites. *J. Phys. D Appl. Phys.* **2010**, *43*, 305403. [CrossRef]

26. Ansón-Casaos, A.; Pascual, F.J.; Ruano, C.; Fernández-Huerta, N.; Fernández-Pato, I.; Otero, J.C.; Puértolas, J.A.; Martínez, M.T. Electrical conductivity and tensile properties of block-copolymer-wrapped single-walled carbon nanotube/poly (methyl methacrylate) composites. *J. Appl. Polym. Sci.* **2015**, *132*, 41547. [CrossRef]
27. Liu, M.; Zhu, T.; Li, Z.; Liu, Z. One-Step in Situ Synthesis of Poly(methyl methacrylate)-Grafted Single-Walled Carbon Nanotube Composites. *J. Phys. Chem. C* **2009**, *113*, 9670–9675. [CrossRef]
28. Clayton, L.M.; Sikder, A.K.; Kumar, A.; Cinke, M.; Meyyappan, M.; Gerasimov, T.G.; Harmon, J.P. Transparent Poly(methyl methacrylate)/Single-Walled Carbon Nanotube (PMMA/SWNT) Composite Films with Increased Dielectric Constants. *Adv. Funct. Mater.* **2005**, *15*, 101–106. [CrossRef]
29. Fraser, R.A.; Stoeffler, K.; Ashrafi, B.; Zhang, Y.; Simard, B. Large-Scale Production of PMMA/SWCNT Composites Based on SWCNT Modified with PMMA. *ACS Appl. Mater. Interfaces* **2012**, *4*, 1990–1997. [CrossRef]
30. Bikshamaiah, N.; Babu, N.M.; Kumar, D.S.; Ramesh, S.; Madhuri, D.; Sainath, A.V.S.; Madhukar, K. Carbon nanotube functional group-dependent compatibilization of polyamide 6 and poly(methyl methacrylate) nanocomposites. *Iran. Polym. J.* **2021**, *30*, 789–799. [CrossRef]
31. Krause, B.; Pötschke, P.; Häußler, L. Influence of small scale melt mixing conditions on electrical resistivity of carbon nanotube-polyamide composites. *Compos. Sci. Technol.* **2009**, *69*, 1505–1515. [CrossRef]
32. Ke, K.; Wang, Y.; Liu, X.-Q.; Cao, J.; Luo, Y.; Yang, W.; Xie, B.-H.; Yang, M.-B. A comparison of melt and solution mixing on the dispersion of carbon nanotubes in a poly(vinylidene fluoride) matrix. *Compos. B Eng.* **2012**, *43*, 1425–1432. [CrossRef]
33. Alig, I.; Pötschke, P.; Lellinger, D.; Skipa, T.; Pegel, S.; Kasaliwal, G.R.; Villmow, T. Establishment, morphology and properties of carbon nanotube networks in polymer melts. *Polymer* **2012**, *53*, 4–28. [CrossRef]
34. Alig, I.; Skipa, T.; Lellinger, D.; Pötschke, P. Destruction and formation of a carbon nanotube network in polymer melts: Rheology and conductivity spectroscopy. *Polymer* **2008**, *49*, 3524–3532. [CrossRef]
35. Pötschke, P.; Villmow, T.; Krause, B. Melt mixed PCL/MWCNT composites prepared at different rotation speeds: Characterization of rheological, thermal, and electrical properties, molecular weight, MWCNT macrodispersion, and MWCNT length distribution. *Polymer* **2013**, *54*, 3071–3078. [CrossRef]
36. Villmow, T.; Pötschke, P.; Pegel, S.; Häussler, L.; Kretzschmar, B. Influence of twin-screw extrusion conditions on the dispersion of multi-walled carbon nanotubes in a poly (lactic acid) matrix. *Polymer* **2008**, *49*, 3500–3509. [CrossRef]
37. Müller, M.T.; Krause, B.; Kretzschmar, B.; Pötschke, P. Influence of feeding conditions in twin-screw extrusion of PP/MWCNT composites on electrical and mechanical properties. *Compos. Sci. Technol.* **2011**, *71*, 1535–1542. [CrossRef]
38. Guo, J.; Briggs, N.; Crossley, S.; Grady, B.P. A new finding for carbon nanotubes in polymer blends: Reduction of nanotube breakage during melt mixing. *J. Thermoplast. Compos. Mater.* **2016**, *31*, 110–118. [CrossRef]
39. Yu, S.Z.; Juay, Y.K.; Young, M.S. Fabrication and characterization of carbon nanotube reinforced poly(methyl methacrylate) nanocomposites. *J. Nanosci. Nanotechnol.* **2008**, *8*, 1852–1857. [CrossRef]
40. Wang, Z.; Yang, X.; Wei, J.; Xu, M.; Tong, L.; Zhao, R.; Liu, X. Morphological, electrical, thermal and mechanical properties of phthalocyanine/multi-wall carbon nanotubes nanocomposites prepared by masterbatch dilution. *J. Polym. Res.* **2012**, *19*, 9969. [CrossRef]
41. Pötschke, P.; Mothes, F.; Krause, B.; Voit, B. Melt-mixed PP/MWCNT composites: Influence of CNT incorporation strategy and matrix viscosity on filler dispersion and electrical resistivity. *Polymers* **2019**, *11*, 189. [CrossRef]
42. Annala, M.; Lahelin, M.; Seppälä, J. Utilization of poly (methyl methacrylate)-carbon nanotube and polystyrene-carbon nanotube in situ polymerized composites as masterbatches for melt mixing. *Express Polym. Lett.* **2012**, *6*, 814–825. [CrossRef]
43. Krause, B.; Liguoro, A.; Pötschke, P. Blend Structure and n-Type Thermoelectric Performance of PA6/SAN and PA6/PMMA Blends Filled with Singlewalled Carbon Nanotubes. *Nanomaterials* **2021**, *11*, 1146. [CrossRef] [PubMed]
44. Krause, B.; Barbier, C.; Levente, J.; Klaus, M.; Pötschke, P. Screening of different carbon nanotubes in melt-mixed polymer composites with different polymer matrices for their thermoelectrical properties. *J. Compos. Sci.* **2019**, *3*, 106. [CrossRef]
45. Luo, J.; Cerretti, G.; Krause, B.; Zhang, L.; Otto, T.; Jenschke, W.; Ullrich, M.; Tremel, W.; Voit, B.; Pötschke, P. Polypropylene-based melt mixed composites with singlewalled carbon nanotubes for thermoelectric applications: Switching from p-type to n-type by the addition of polyethylene glycol. *Polymer* **2017**, *108*, 513–520. [CrossRef]
46. Jinji, B.; Beate, K.; Petra, P. Melt-mixed thermoplastic composites containing carbon nanotubes for thermoelectric applications. *AIMS Mater. Sci.* **2016**, *3*, 1107–1116.
47. Liebscher, M.; Gärtner, T.; Tzounis, L.; Mičušík, M.; Pötschke, P.; Stamm, M.; Heinrich, G.; Voit, B. Influence of the MWCNT surface functionalization on the thermoelectric properties of melt-mixed polycarbonate composites. *Compos. Sci. Technol.* **2014**, *101*, 133–138. [CrossRef]
48. Tzounis, L.; Gärtner, T.; Liebscher, M.; Pötschke, P.; Stamm, M.; Voit, B.; Heinrich, G. Influence of a cyclic butylene terephthalate oligomer on the processability and thermoelectric properties of polycarbonate/MWCNT nanocomposites. *Polymer* **2014**, *55*, 5381–5388. [CrossRef]
49. Konidakis, I.; Krause, B.; Park, G.-H.; Pulumati, N.; Reith, H.; Pötschke, P.; Stratakis, E. Probing the carrier dynamics of polymer composites with single and hybrid carbon nanotube fillers for improved thermoelectric performance. *ACS Appl. Energy Mater.* **2022**, *5*, 9770–9781. [CrossRef]

50. Gonçalves, J.; Lima, P.; Krause, B.; Pötschke, P.; Lafont, U.; Gomes, J.R.; Abreu, C.S.; Paiva, M.C.; Covas, J.A. Electrically conductive polyetheretherketone nanocomposite filaments: From production to fused deposition modeling. *Polymers* **2018**, *10*, 925. [CrossRef]
51. Hewitt, C.; Kaiser, A.; Roth, S.; Craps, M.; Czerw, R.; Carroll, D. Varying the concentration of single walled carbon nanotubes in thin film polymer composites, and its effect on thermoelectric power. *Appl. Phys. Lett.* **2011**, *98*, 183110. [CrossRef]
52. Predtechenskiy, M.R.; Khasin, A.A.; Bezrodny, A.E.; Bobrenok, O.F.; Dubov, D.Y.; Muradyan, V.E.; Saik, V.O.; Smirnov, S.N. New Perspectives in SWCNT Applications: Tuball SWCNTs. Part 1. Tuball by Itself–All You Need to Know about It. *Carbon. Trends* **2022**, *8*, 100175. [CrossRef]
53. Jenschke, W.; Ullrich, M.; Krause, B.; Pötschke, P. Messanlage zur Untersuchung des Seebeck-Effektes in Polymermaterialien. *tm-Technisches Messen.* **2020**, *87*, 495–503. [CrossRef]
54. Krause, B.; Pötschke, P. Polyethylene Glycol as Additive to Achieve N-Conductive Melt-Mixed Polymer/Carbon Nanotube Composites for Thermoelectric Application. *Nanomaterials* **2022**, *12*, 3812. [CrossRef]
55. Gnanaseelan, M.; Chen, Y.; Luo, J.; Krause, B.; Pionteck, J.; Pötschke, P.; Qi, H. Cellulose-carbon nanotube composite aerogels as novel thermoelectric materials. *Compos. Sci. Technol.* **2018**, *163*, 133–140. [CrossRef]
56. Khan, W.; Sharna, R.; Saini, P. Carbon Nanotube-Based Polymer Composites: Synthesis, Properties and Applications. In *Carbon Nanotubes*; Mohamed Reda, B., Inas Hazzaa, H., Eds.; IntechOpen: Rijeka, Croatia, 2016; p. 15.
57. Zhi, C.Y.; Bando, Y.; Wang, W.L.; Tang, C.C.; Kuwahara, H.; Golberg, D. Mechanical and Thermal Properties of Polymethyl Methacrylate-BN Nanotube Composites. *J. Nanomater.* **2008**, *2008*, 642036. [CrossRef]
58. Badawi, A.; Al Hosiny, N. Dynamic mechanical analysis of single walled carbon nanotubes/polymethyl methacrylate nanocomposite films. *Chin. Phys. B* **2015**, *24*, 105101. [CrossRef]

Disclaimer/Publisher's Note: The statements, opinions and data contained in all publications are solely those of the individual author(s) and contributor(s) and not of MDPI and/or the editor(s). MDPI and/or the editor(s) disclaim responsibility for any injury to people or property resulting from any ideas, methods, instructions or products referred to in the content.

Article

Carbon Nanotube Migration in Melt-Compounded PEO/PE Blends and Its Impact on Electrical and Rheological Properties

Calin Constantin Lencar, Shashank Ramakrishnan and Uttandaraman Sundararaj *

Department of Chemical and Petroleum Engineering, University of Calgary, 2500 University Drive NW, Calgary, AB T2N 1N4, Canada
* Correspondence: ut@ucalgary.ca

Abstract: In this work, the effects of MWCNT concentration and mixing time on the migration of multi-walled carbon nanotubes (MWCNTs) within polyethylene oxide (PEO)/polyethylene (PE) blends are studied. Two-step mixing used to pre-localize MWCNTs within the PE phase and subsequently to observe their migration into the thermodynamically favored PEO phase. SEM micrographs show that many MWCNTs migrated into PEO. PEO/PE 40:60 polymer blend nanocomposites with 3 vol% MWCNTs mixed for short durations exhibited exceptional electromagnetic interference shielding effectiveness (EMI SE) and electrical conductivity (14.1 dB and 22.1 S/m, respectively), with properties dropping significantly at higher mixing times, suggesting the disruption of percolated MWCNT networks within the PE phase. PE grafted with maleic anhydride (PEMA) was introduced as a compatibilizer to arrest the migration of MWCNTs by creating a barrier at the PEO/PE interface. For the compatibilized system, EMI SE and electrical conductivity measurements showed a peak in electrical properties at 5 min of mixing (15.6 dB and 68.7 S/m), higher than those found for uncompatibilized systems. These improvements suggest that compatibilization can be effective at halting MWCNT migration. Although utilizing differences in thermodynamic affinity to draw MWCNTs toward the polymer/polymer interface of polymer blend systems can be an effective way to achieve interfacial localization, an excessively low viscosity of the destination phase may play a major role in reducing the entrapment of MWCNTs at the interface.

Keywords: polymer nanocomposite; electrical conductivity; mixing; polymer blends; electromagnetic interference shielding; carbon nanotube; filler migration; polyethylene

Citation: Lencar, C.C.; Ramakrishnan, S.; Sundararaj, U. Carbon Nanotube Migration in Melt-Compounded PEO/PE Blends and Its Impact on Electrical and Rheological Properties. *Nanomaterials* **2022**, *12*, 3772. https://doi.org/10.3390/nano12213772

Academic Editor: Ilaria Armentano

Received: 1 October 2022
Accepted: 19 October 2022
Published: 26 October 2022

Publisher's Note: MDPI stays neutral with regard to jurisdictional claims in published maps and institutional affiliations.

Copyright: © 2022 by the authors. Licensee MDPI, Basel, Switzerland. This article is an open access article distributed under the terms and conditions of the Creative Commons Attribution (CC BY) license (https://creativecommons.org/licenses/by/4.0/).

1. Introduction

Electronic devices have made the world more connected than ever. Smart phones, laptops, and other such devices have made the internet more accessible, and internet of things have connected everything in our daily lives to our devices. The utility of smart devices has fueled huge growth in their global demand. This growing demand has led to an explosion in wireless data traffic [1]. Unfortunately, all electronic devices (especially wireless devices) emit electromagnetic (EM) waves as part of their regular operation. These EM waves can interfere with other critical devices, including medical equipment, navigational equipment, and communications devices [2–4]. This interference can lead to faulty operation or even total failure of the affected equipment. Furthermore, there are health implications due to high exposure to EM waves, and serious chronic health issues may arise in humans [5,6]. Due to the hazards of electromagnetic interference (EMI), the demand for proper shielding measures has grown significantly. Historically, metals have been utilized to shield against EMI, but they are expensive, difficult to shape, and susceptible to corrosion [7,8]. Furthermore, metals protect against incident EM waves by reflecting them back into the environment, and these reflected waves can still interfere with other devices [9]. Polymer, ceramic, and metal-based nanocomposites offer numerous advantages, including their light weight, resistance to corrosion, and high tuneability. Metal matrix nanocomposites utilize the exceptional electrical properties offered by metals and seek to combat the low mechanical yield of metals via nano-reinforcement [10]. Other metal-based nanocomposites, such as those prepared by Ji et al., utilize open-celled foams containing metal nanowires grafted on carbon nanotubes (CNTs) to shield against incident EMI [11]. Ceramic nanocomposites containing nanoparticles such as CNTs negate the typically high brittleness seen in ceramics and offer unique opportunities for multifaceted applications due to the high electrical properties of CNTs [12]. Polymer nanocomposites (PNCs) containing multi-walled carbon nanotubes (MWCNTs) are especially interesting, due to their high impact properties and processability [13]. Additionally, PNCs containing MWCNTs primarily attenuate incident EMI via absorption and this can significantly reduce EMI smog [14–16]. MWCNTs possess superb conductivity, tensile strength, and a high aspect ratio, which makes them ideal for imparting electrical properties to polymers through interconnected conductive networks [17,18]. Single walled carbon nanotubes (SWCNTs) can also be utilized, but their application is often inhibited by high synthesis costs. Several works have achieved exceptional electrical properties and thermal stability of SWCNTs via polybenzoxazines and oxadiazole-linked conjugated microporous polymers, but these processes have limited scalability for industrial applications [19,20].

Previous works have shown that polymer blend nanocomposites (PBNs) containing MWCNTs can be used to develop inexpensive conductive materials for advanced applications, including in the medical and aerospace sectors. Sumita et al. [21] were the first to show that by localizing carbon black (CB) within one phase of HDPE/PP and PP/PMMA PBN systems, the effective local concentration of CB in one phase could be increased, thus reducing the total quantity of conductive filler required to form a percolated network within the blend, in a phenomenon often dubbed "double percolation". Double percolation has also been studied in various PBN systems containing MWCNTs [22–24]. In these systems, it is important that the polymer blend has a co-continuous morphology, so that the percolation of nanofiller within one of the phases leads to the formation of a continuous nanofiller network. Recent work has shown that the concept of double percolation of MWCNTs within polymer blends can be taken further, by locating them at the polymer/polymer interface. Zhang et al. [25] used amine functionalized MWCNTs in blends of PA6/PVDF to achieve interfacial localization, resulting significantly reduced percolation thresholds. Wu et al. [26] made use of carboxylic-functionalized MWCNTs within blends of poly (ε-caprolactone)/polylactide (PCL/PLA) with similar effect. Solution mixing techniques have also been adopted to achieve interfacial localization of MWCNTs within polymer blends [27,28]. Unfortunately, the use of solution mixing strategies or functionalized MWCNTs increases the cost of preparing PBN materials, hindering the scalability of these systems for commercial purposes.

Understanding how MWCNTs move within a given polymer blend is an important first step to preparing PBN systems with MWCNTs localized at the polymer/polymer interface. MWCNTs should initially be in the phase with lower thermodynamic affinity to encourage their migration toward the interface with their preferred phase. Young's equation is often used to predict the thermodynamic preference of MWCNTs within binary polymer blends [21]:

$$\omega_{A/B} = \frac{\sigma_{A/MWCNT} - \sigma_{MWCNT/B}}{\sigma_{A/B}}$$

where

$\omega_{A/B}$—wettability of MWCNTs within a blend of polymer A and B,
$\sigma_{A/MWCNT}$—surface energy between MWCNTs and polymer A,
$\sigma_{MWCNT/B}$—surface energy between the MWCNTs and polymer B, and
$\sigma_{A/B}$—surface energy between polymers A and B.

When $\omega_{A/B} > 1$, MWCNTs will prefer polymer B. When $\omega_{A/B} < -1$, MWCNTs will prefer polymer A. Finally, when $-1 < \omega_{A/B} < 1$, MWCNTs will tend to settle at the polymer A/polymer B interface.

In our previous work [29], poly (vinylidene difluoride) (PVDF)/polyethylene (PE) blends of varying blend ratios containing 2 vol% MWCNTs were studied to observe the effect of blend morphology and mixing time on the migration behavior of MWCNTs during melt mixing. MWCNTs were initially localized within PE and migrated towards PVDF during subsequent melt-mixing. Although MWCNTs thermodynamically favor PVDF (based on Young's equation), the higher viscosity of PVDF relative to PE was expected to retard the migration of MWCNTs from PE into PVDF when MWCNTs reached the interface. SEM images both confirmed that MWCNTs migrated toward the PVDF/PE interface and became trapped there. A modified version of Göldel et al.'s [30] "Slim-Fast" mechanism was used to conceptualize the migration behavior of MWCNTs within the PVDF/PE blend. Short, straight MWCNTs are more likely to penetrate the interface while coiled MWCNTs and MWCNT agglomerates are more likely to become trapped at the polymer/polymer interface.

The objective of the current work is to study the phase migration of MWCNTs within co-continuous blends of polyethylene oxide (PEO) and PE, with the aim of producing MWCNT-based PBN materials with exceptional electrical properties at low MWCNT concentrations. The concentration of MWCNTs within the PE phase at the start of mixing with PEO was varied to study the difference in migration behavior of individual and agglomerated MWCNTs. MWCNTs were selected for their superb electrical and mechanical properties, high aspect ratio, and low-cost relative to nanoparticles, such as single-walled carbon nanotubes (SWCNTs) or graphene [31]. PEO was chosen because it has good affinity for MWCNTs compared to PE, it has low viscosity, and its water solubility allows it to be easily extracted from blends with PE [32]. The same PE was used in this work as in our previous work with PVDF [29] to provide a direct comparison between the two systems. A low-cost polymer with exceptional impact properties and high processability, PE is an excellent choice for commercial PBN systems. The purpose of using a PE with a relatively high viscosity versus PEO was to study the relative significance of nanoparticle geometry (i.e., individual MWCNTs vs. agglomerated MWCNTs) and polymer viscosity on MWCNT phase migration. Better understanding of the migration behavior of MWCNTs allows us to better design high performance MWCNT-based materials for commercial applications.

2. Materials and Methods

2.1. Materials and Sample Preparation

PBN samples were prepared using PEO (Polyox™ WSR N10) supplied by DuPont (Wilmington, DE, USA), PE (Lumicene™ M3581 uv) supplied by Total SA (Houston, TX, USA), PE grafted with maleic anhydride (PEMA) (OREVAC® 18340) supplied by Arkema S.A. (Colombes, France), and MWCNTs (NC7000) supplied by Nanocyl S.A. (Sambreville, Belgium). Based on the specifications provided by the manufacturer, NC7000 has an average diameter of 9.5 nm, an average length of 1.5 μm, and aspect ratio of 158, 90% purity and an electrical conductivity of approximately 10^6 S/m. PEO and PE were vacuum dried at 60 °C for 24 h before use.

PBN samples with a PEO/PE ratio of 40:60 were prepared at MWCNT concentrations of 0.5, 1.5, 2, and 3 vol% using two-step mixing. MWCNT powder and PE powder were dry mixed and added to the mixing cup of an Alberta Polymer Asymmetric Mini-mixer (APAM) (University of Calgary, Calgary, AB, Canada) and left to melt for 2 min without rotation [33]. The mixture was then melt-compounded at 200 rpm for 5 min to create a well-mixed PE/MWCNT composite. Mixing was then halted, and PEO powder was introduced to the mixing cup, and left to melt for 2 min without rotation. Finally, the mixture was melt-compounded at 200 rpm for an additional 1, 5, and 10 min. All melting and mixing steps in the APAM were done at a constant temperature of 150 °C. Samples were rapidly removed at the end of the final blending step and chunks of the sample were quenched in liquid nitrogen to freeze the sample morphology. The rest of the sample was molded into circular discs (diameter = 25 mm, thickness = 0.45 mm) using a Carver compression molder (model 3912) (Carver Inc., Wabash, IN, USA) at 150 °C and 35 MPa for

10 min. A minimum of four specimens were prepared for each sample to measure electrical conductivity, EMI shielding and rheological properties. An outline of the procedure for preparing PEO/PE/MWCNT samples can be found in Figure 1 below.

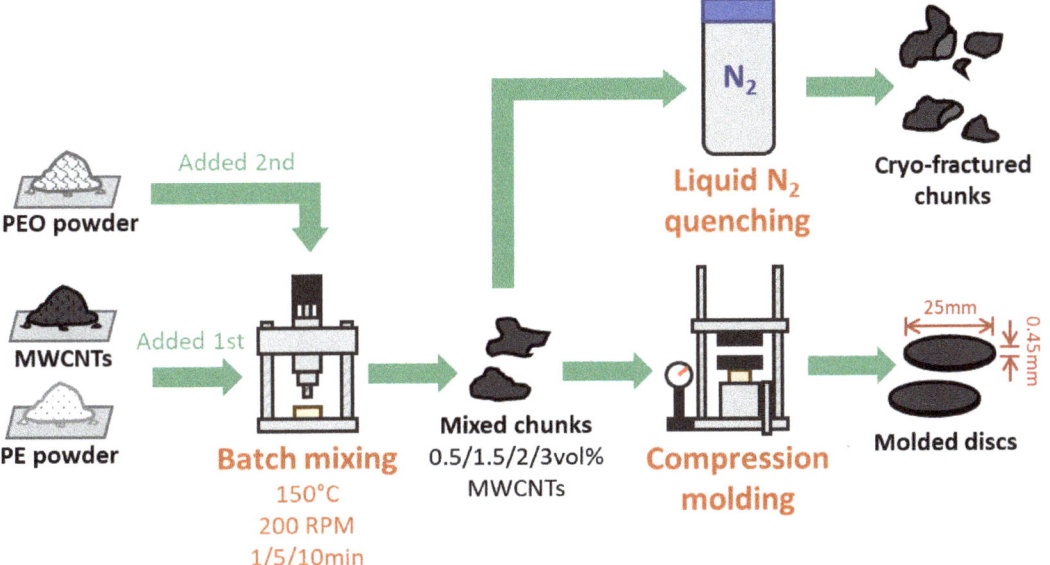

Figure 1. Schematic representation of procedure used to prepare the PEO/PE/MWCNT nanocomposite samples studied in this work.

2.2. Sample Characterization

Cryo-fractured samples were mounted and imaged under a low vacuum using a Quanta FEG 250 VP-FESEM (variable pressure field emission SEM) (FEI Company, Hillsboro, OR, USA). A large field detector (LFD) was used to take secondary electron images to observe the detailed sample topography (including MWCNTs).

EMI shielding measurements were performed in the X-band frequency range (8.2–12.4 GHz) using a vector network analyzer (ENA Model E5071C) (Agilent Technologies, Santa Clara, CA, USA), with a connected WR-90 rectangular waveguide. The X-band is a radar frequency usually used in civil and military applications, aircraft and sea craft detection and monitoring [34]. Although the X-band exists in a higher frequency range than the frequencies typically used by wireless smart devices, several works have shown that EMI shielding effectiveness increases with decreasing frequency [7]. This suggests that EMI shielding materials that perform well in the X-band will perform even better at lower frequencies. EMI SE values were derived from scattering parameters (see Supplementary Materials) based on measured data [7]. DC electrical conductivity of the samples was measured via a Loresta GP (model MCP-T610) resistivity meter (Mitsubishi Chemical Co., Tokyo, Japan), attached to an ESP probe. Measurements were performed on 4 specimens for each sample, with the average values of EMI SE and conductivity reported within this work.

Rheological tests were performed using an Anton-Paar rheometer (MCR 302) (Anton-Paar GmbH, Graz, Austria) with a 25 mm diameter parallel plate and a gap size of 0.45 mm. Linear frequency sweeps in the range of 600–0.1 rad/s were performed at a constant strain of 0.1% for 3 specimens. Frequency sweeps were followed by strain sweeps in the range of 0.1–1000% strain to confirm the linear viscoelastic region (LVR) of the specimens. All tests were conducted at a constant temperature of 150 °C.

3. Results

3.1. Theoretical Surface Energy Models

Prior to preparing the PEO/PE/MWCNT blend nanocomposites, a theoretical model based on a modified Young's equation adapted by Sumita et al. [21] was used to predict the surface energies of the blend components within the system. The surface energy calculations were performed using surface energy values for the individual components reported in literature, obtained at 25 °C [35,36]. Details on the equations used to calculate the surface energies and wettability values can be found in the Supplemental Materials.

Based on the surface energy and wettability values in Table 1, MWCNTs will prefer to localize within PEO over PE. Consequently, MWCNTs were pre-localized within PE prior to subsequent melt-compounding with PEO to see if MWCNT would migrate from PE to the thermodynamically preferred PEO phase. For more details on the data used to calculate the parameter in Table 1, refer to Table S1 of the Supplementary Materials.

Table 1. Surface Energies and Wettability of MWCNTs within PEO and PE at 150 °C.

Parameter	Geometric Mean [1]	Harmonic Mean [2]
$\sigma_{PEO/PE}$ [mJ/m^2]	9.31	9.39
$\sigma_{PEO/MWCNT}$ [mJ/m^2]	5.80	9.32
$\sigma_{MWCNT/PE}$ [mJ/m^2]	27.65	28.39
Wettability	−2.35	−2.03

[1,2] Wettability parameters were calculated using data from Wu [35] and Owens [36].

3.2. Imaging Results

Samples were initially studied using light microscopy (LM) imaging to better understand the dispersion and localization of MWCNTs within the PEO/PE blends, and the results can be found in Figure S1 of the Supplementary Materials. The detailed morphology of the prepared 40:60 blends systems were studied via SEM to better understand the morphology of the system, and to study the localization of MWCNTs within the blend system, especially at the PEO/PE interface. SEM images of pure PE/PEO blends without MWCNTs can be seen in Figure S2 of the Supplementary Materials. Figure 2 shows SEM micrographs of PEO/PE 40:60 blends containing 0.5 vol% MWCNTs mixed for 1 min and 10 min of mixing. At 1 min of mixing Figure 2a), the PE phase appears to have a grainy structure, which is due to the MWCNTs within the PE phase. In contrast, the PEO phase is not present within the micrograph, having fallen off the fracture surface due to poor adhesion to the PE phase. This poor adhesion suggests that MWCNTs have not yet migrated to the PEO/PE interface, which would improve interfacial adhesion. At higher magnification (Figure 2(a1)), individual MWCNTs appear as hairs along the surface of the PE phase.

At 10 min of mixing (Figure 2b), PEO domains (which appear far smoother than the PE domains) can be seen along the sample surface. The PEO appears as large co-continuous domains with PE, and also as small droplets that are imbedded within PE. Droplets of PEO are likely sticking to the PE phase due to the presence of MWCNTs at the blend interface, i.e., MWCNT improves the interfacial adhesion via bridging. At higher magnification (Figure 2(b1)), individual MWCNTs can be seen along the PE surface, and PE ridges that are concentrated with MWCNTS can be seen surrounding PEO droplets. Thus, it is apparent that MWCNTs are migrating towards the PEO/PE interface and likely penetrating PEO.

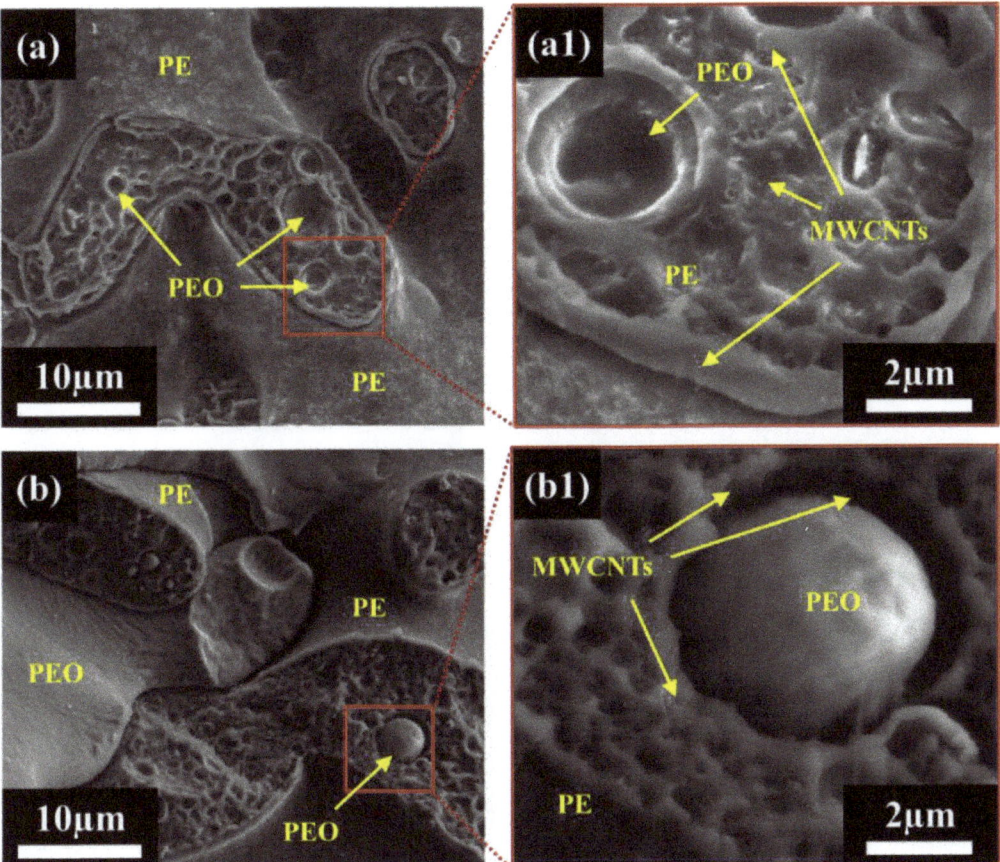

Figure 2. Low field detector (LFD) SEM images of PEO/PE 40:60 with 0.5 vol% MWCNTs mixed for (**a**) 1 min and (**b**) 10 min. Images (**a1**) and (**b1**) show PEO droplets at higher magnification.

Figure 3 shows SEM micrographs of PEO/PE 40:60 blends containing 3 vol% MWCNTs mixed for 1 min and 10 min. At 1 min of mixing (Figure 3a), the PE phase appears the have a very hair-like texture, due to the very high concentration of MWCNTs present therein. Similar to the 0.5 vol% blend at 1 min of mixing (Figure 3a or Figure 3(a1)), the PEO phase is seldom actually present since much of it has likely fallen from the fracture surface due to poor interfacial adhesion with PE. At higher magnifications (Figure 3(a1)), individual MWCNTs can clearly be seen along the surface of the PE phase, and smaller droplets of PEO that have been filled with MWCNTs can also be seen. At 10 min of mixing (Figure 3b), many domains of PEO (i.e., smaller PEO droplets and larger continuous PEO domains) can be observed. Once again, this is likely due to MWCNTs bridging the PEO/PE interface as they migrate into PEO. At higher magnification (Figure 3(b1)), dense networks of MWCNTs can be seen surrounding PEO droplets, and many MWCNTs can be seen straddling the PEO/PE, and even fully embedded within PEO.

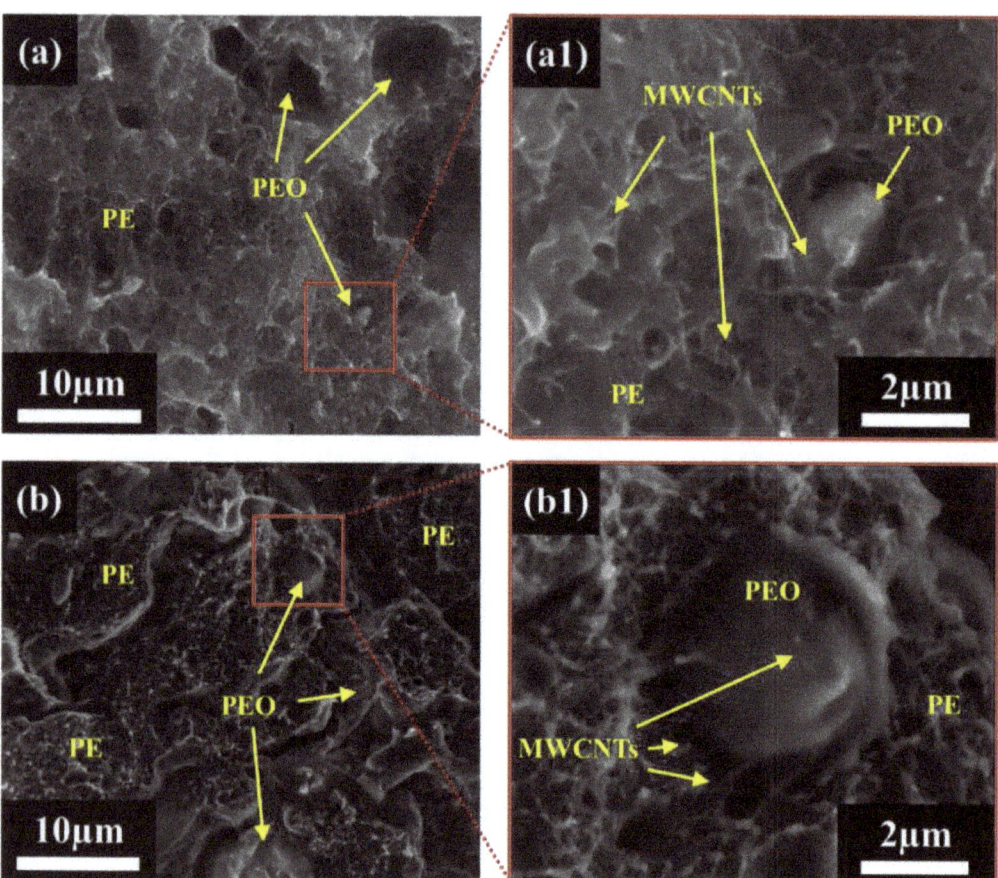

Figure 3. Low field detector (LFD) SEM images of PEO/PE 40:60 with 3 vol% MWCNTs mixed for (**a**) 1 min and (**b**) 10 min. Images (**a1**) and (**b1**) show PEO droplets at higher magnification.

3.3. Electrical Properties: DC Conductivity, EMI Shielding Effectiveness and Permittivity

Figure 4 shows the effect of mixing time and MWCNT concentration on the final observed DC electrical conductivity (σ_{DC}) and EMI SE of PEO/PE 40:60 blends containing MWCNTs. σ_{DC} and EMI SE values for PEO and PE nanocomposites containing 3 vol% MWNCTs were also prepared to serve as a baseline, and the results can be found in Figure S6 of the Supplemental Materials. PEO/PE 40:60 blends at all MWCNT concentrations (Figure 4a,c,d) showed a decrease in the values of EMI SE and σ_{DC} with increasing mixing time. The overall decreasing trend can likely be attributed to the migration of MWCNTs from the PE phase into PEO, resulting in a breakdown of the conductive network. With MWCNTs localizing in both phases, double percolation no longer exists, i.e., because of the dilution of MWCNT concentration in PE phase, there is a reduction in the extent of MWCNT network formation. The high initial electrical properties can likely be attributed to MWCNTs forming a percolated network within PE exclusively, because of the firs mixing step before PEO was added. The most significant mechanism of EMI attenuation in all prepared 40:60 samples was absorption, which is common for PBNs [7]. PBNs tend to absorb incident EMI because the relatively miniscule quantity of MWCNTs at the surface of the samples allows EM waves to easily penetrate the bulk of the sample. Once within the sample, the electric and magnetic fields generated by the incident EM wave interact with the electric and magnetic dipoles within the embedded MWCNTs, leading to dissipation through ohmic losses.

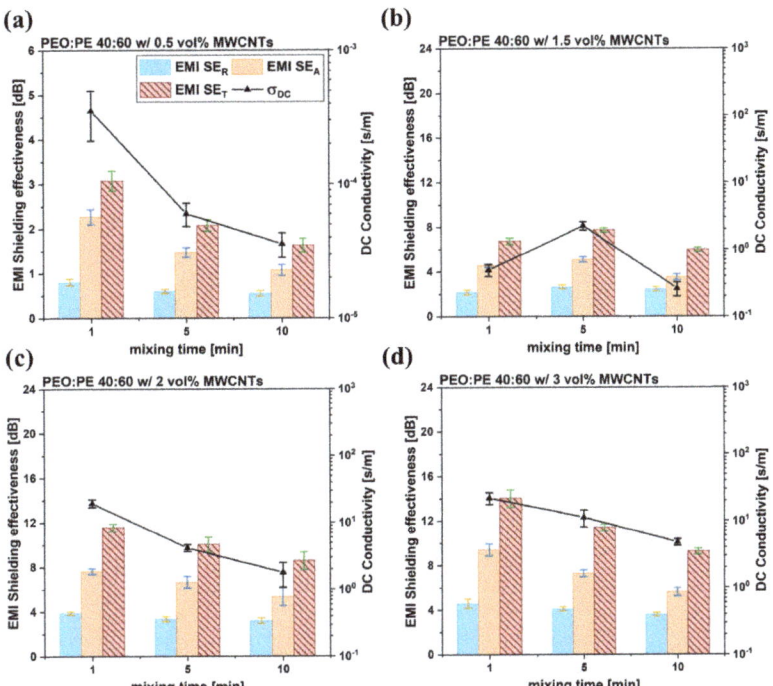

Figure 4. EMI SE values and DC electrical conductivity data for PEO/PE 40:60 blends containing MWCNTs at concentrations of (**a**) 0.5 vol%, (**b**) 1.5 vol%, (**c**) 2 vol%, and (**d**) 3 vol%.

The dielectric properties of the PEO/PE blends made in this work can also be studied. Complex electrical permittivity is typically described as:

$$\varepsilon^* = \varepsilon' + i\varepsilon'' \text{ or } \varepsilon^* = \sqrt{(\varepsilon')^2 + (\varepsilon'')^2}$$

where ε' is the real permittivity and ε'' is the imaginary permittivity. The real permittivity of a material describes its ability to store electrical energy when subjected to an electric field, and the imaginary permittivity describes its ability to dissipate energy. In the case of PBN systems containing MWCNTs, ε' is caused by interfacial polarization between the polymer and MWCNTs, and ε'' is caused by charges dissipating through interconnected MWCNT networks [7]. Plots of ε' and ε'' for PEO/PE blends containing MWCNTs can be found in Figure S5 of the Supplementary Materials. Additionally, real and complex permittivity plots of PEO and PE containing 3 vol% MWCNTs can be found in Figure S7 of the Supplementary Materials. Figure 5 shows the ε^* data for all prepared PEO/PE samples. When studying electrical permittivity in terms of ε^*, the similarities to the EMI SE and σ_{DC} values seen in Figure 4 become apparent. Looking at the ε^* data in Figure 5, all curves closely match the results seen in Figure 4.

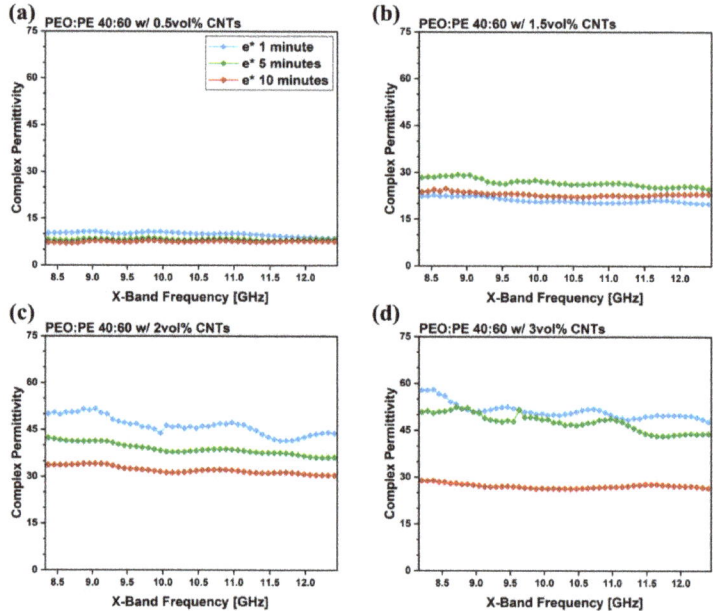

Figure 5. Complex permittivity data within the X-band for PEO/PE 40:60 blends containing MWCNTs at concentrations of (**a**) 0.5 vol%, (**b**) 1.5 vol%, (**c**) 2 vol%, and (**d**) 3 vol%.

3.4. Rheological Properties: Linear Frequency Sweeps and Strain Sweeps

Frequency sweeps were performed on the mixed PBNs at a constant strain of 0.1% to ensure the samples were being studied in the linear viscoelastic (LVE) region, reducing the risk of disrupting the MWCNT networks formed within the PEO/PE blend samples. The results of the frequency sweeps are plotted in Figure 6. In addition to the rheology curves for PEO/PE blends plotted in Figure 6, frequency sweep plots for PEO and PE containing 3 vol% MWCNTs were also prepared to serve as a baseline and can be found in Figure S8 of Supplemental Materials. Frequency sweeps of pure PEO and PE, as well as their blends are available in Figure S3 and S4 of the Supplementary Materials. At MWCNT concentrations below the percolation threshold (0.5 and 1.5 vol% MWCNTs, shown in Figure 6a,b), the peak in viscoelastic moduli for low angular frequencies appears at 5 min of mixing. Previous works studying the rheological properties of MWCNT-filled polymer nanocomposites have suggested that a higher plateau of the storage and loss modulus (G' and G") at low angular frequencies suggests a more pronounced network structure of MWCNTs within the PBN system. Since MWCNTs are not effectively percolating within these systems, the peak in rheological properties at 5 min is likely due to MWCNTs dispersing into both the PEO and PE phase.

In the case of PEO/PE blend nanocomposites above the percolation threshold (2 vol% and 3 vol% MWCNTs, shown in Figure 6c,d), the trend in G' and G" is not as consistent. This is likely due to competing effects between the disruption of the pre-existing MWCNT network within the PE phase, and the migration of MWCNTs into the PEO phase. For the system containing 2 vol% MWCNTs (Figure 6c), G' and G" rise slightly with increasing mixing time, suggesting that the increase in properties caused by MWCNTs migrating into PEO outweighs the loss in properties caused by the disruption of the initial MWCNT network within PE. For 3 vol% MWCNT nanocomposites (Figure 6d), G' and G" decrease over time. The loss in properties caused by the disruption of the highly robust MWCNT network initially present in PE has more significance than the increase in properties resulting from dispersing MWCNTs uniformly into PEO.

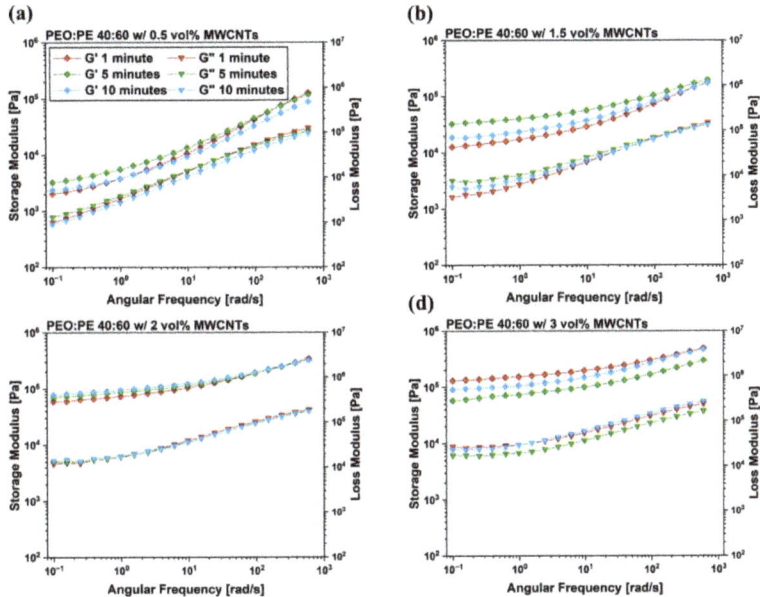

Figure 6. Frequency sweep results for PEO/PE 40:60 blends containing MWCNTs at concentrations of (**a**) 0.5 vol%, (**b**) 1.5 vol%, (**c**) 2 vol%, and (**d**) 3 vol%.

Linear frequency sweeps were followed-up with strain sweeps to link the rupture of the microstructures within the blend systems to the extent of MWCNTs networks formed therein. Figure 7 shows the relationship between the strain amplitude and the observed rheological properties of the prepared 40:60 blends at varying MWCNT concentrations and mixing times. In addition to the strain sweep data prepared for the 40:60 blends, plots for PEO and PE containing 3 vol% were also prepared to serve as a baseline, which are plotted in Figure S8 of the Supplementary Materials. Furthermore, strain sweeps of pure PEO and PE, as well as their blends can be found in Figures S3 and S4 of the Supplementary Materials. As strain amplitude within the test exceeds a critical value, G' values drop rapidly, and the samples exhibit a crossover point (i.e., the strain amplitude at which the storage modulus becomes larger than the loss modulus) [37]. This decrease in G' is typically associated with the destruction of the existing MWCNT network [38]. Consequently, crossover points occurring at higher strain amplitudes are indicative of a more substantial MWCNT network within the system. The change in crossover points observed with increasing mixing time closely follows the trends observed in the EMI SE and σ_{DC} values seen in Figure 4. 40:60 blends with 0.5 vol%, 2 vol% and 3 vol% MWCNTs (Figure 7a,c,d) all show the most delayed crossover point occurring at 1 min of mixing, due to the robust MWCNT network present with PE at this time.

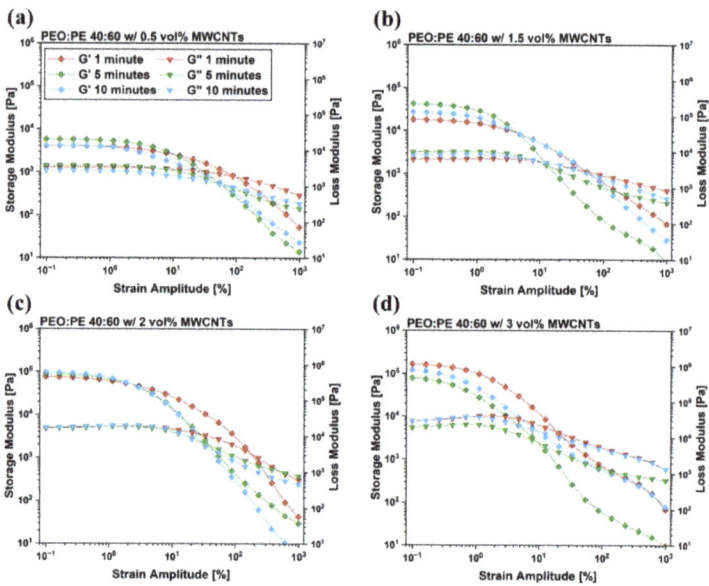

Figure 7. Strain sweep results for PEO/PE 40:60 blends containing MWCNTs at concentrations of (**a**) 0.5 vol%, (**b**) 1.5 vol%, (**c**) 2 vol%, and (**d**) 3 vol%.

3.5. Compatibilization Effects—Changes in Morphological, Electrical, and Rheological Properties with the Addition of PEMA Compatibilizer

Studies into the effect of MWCNT concentration and mixing time on the morphological, electrical, and rheological properties of PEO/PE blends indicated substantial migration of MWCNTs from the PE phase into the PEO phase over time, as suggested by SEM and electrical conductivity/EMI SE measurements. PEO has a significantly lower viscosity than PE, which likely played a role in the migration of MWCNTs from PE to PEO. A PEMA compatibilizer was introduced to improve the interfacial adhesion between PE and PEO and to arrest the migration of MWCNTs at the blend interface. By forming crosslinks between the maleic anhydride groups of the PEMA molecules and the hydroxyl end groups of the PEO molecules, a more rigid interface, with superior adhesion can be created [39,40]. This interface serves to trap the MWCNTs as they migrate into PEO, arresting their movement, and preserving the desirable electrical and rheological properties observed in the PEO/PE 40:60 blends at 1 min of mixing.

SEM imaging was performed to study the blend morphology, and to identify the localization of MWCNTs within the prepared samples. Figure 8 shows SEM images of 40:54:6 blend samples containing 3 vol% MWCNTs. At 1 min of mixing (Figure 8a), PEO and PE phases can both be seen on the fracture surface, suggesting improved interfacial adhesion between the phases compared to the uncompatibilized blends. Furthermore, the PEO phases appear very rippled, which can also be indicative of compatibilization [41]. At higher magnification (Figure 8(a1)), numerous MWCNTs can clearly be seen along the PE surface, whereas MWCNTs appear near PEO only along the interface with PE. At 5 min of mixing (Figure 8b), the blend morphology appears far more refined, with smaller co-continuous domains of PE and PEO intermixing. At higher magnification (Figure 8(b1)), MWCNTs can once again be seen coating the PE surface and MWCNTs are visible along the interface with the PEO phase, which aligns with the high electrical and rheological properties seen at this point. At 10 min of mixing (Figure 8c), the PEO/PE blend morphology appears as before, suggesting that the limit of droplet breakup has been reached (coalescence is greatly inhibited by the presence of PEMA compatibilizer).

At higher magnification (Figure 8(c1)), MWCNTs can be seen clearly in both PE and in PEO, suggesting that the migration of MWCNTs into PEO is extensive. This explains the greatly diminished electrical properties seen at 10 min of mixing, when compared to 5 min of mixing, since MWCNTs is now spread over both polymer phases.

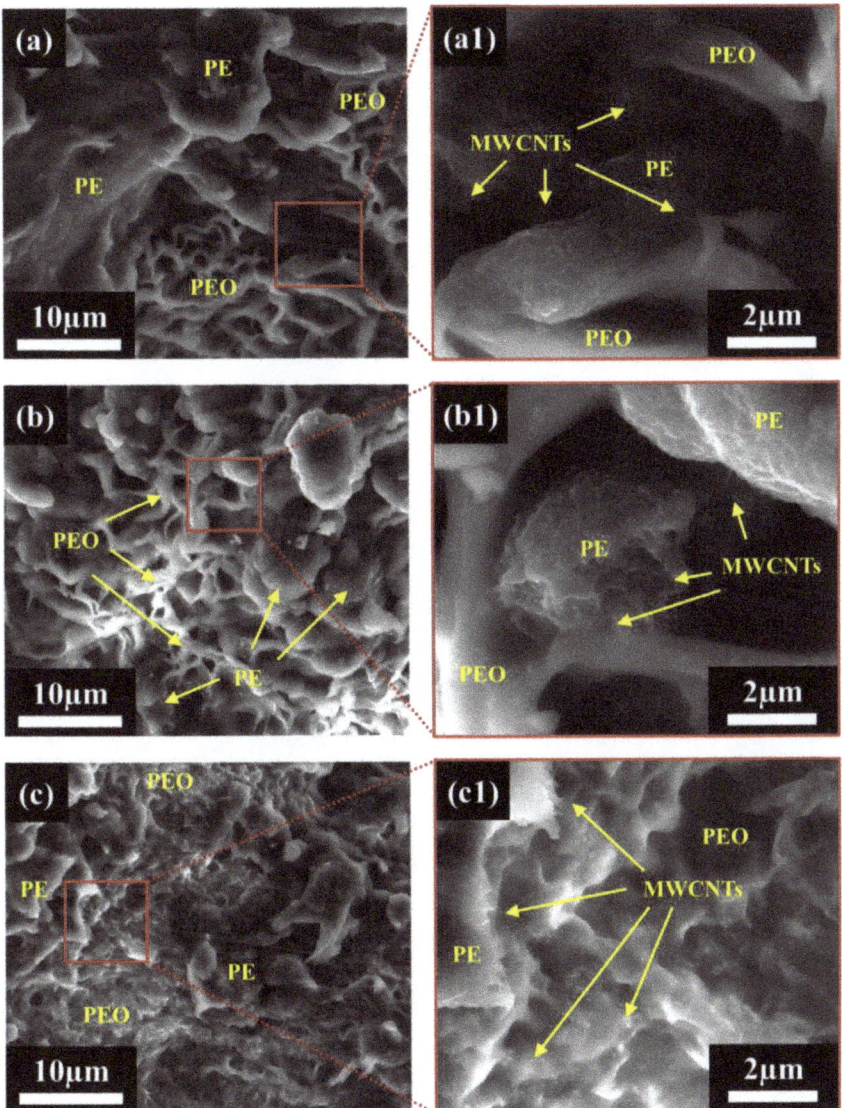

Figure 8. Low field detector (LFD) SEM images of PEO/PE/PEMA 40:54:6 with 3 vol% MWCNTs mixed for (**a**) 1 min, (**b**) 5 min, and (**c**) 10 min. Images (**a1**), (**b1**), and (**c1**) show PEO/PE/PEMA blend morphology at higher magnification.

Figure 9 shows the EMI SE and σDC values for PEO/PE/PEMA 40:54:6 blends containing 3 vol% MWCNTs. Because PEMA is more polar than PE, MWCNT may tend to move near the PEMA molecules. The electrical properties at 1 min of mixing are quite low, possibly due to MWCNTs concentrating in and around the PEMA phase within PE at the onset of mixing, leading to a loss of continuity in the MWCNT network. At 5 min of mixing however, there is a huge rise in both EMI SE and σDC. These values are even higher than the 40:60 blends with 3 vol% MWCNTs at 1 min mixing (15.6 dB and 68.7 S/m compared to 14.1 dB and 22.1 S/m), marking a significant improvement in the MWCNT network for the compatibilized blend. This improvement is likely due to MWCNTs being concentrated at the interface of the blend at 5 min, as evidenced by SEM images. The subsequent drop in properties at 10 min of mixing is likely due to MWCNTs finally crossing into PEO.

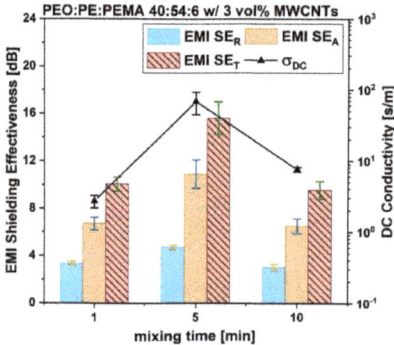

Figure 9. Low field detector (LFD) SEM images of PEO/PE/PEMA 40:54:6 with 3 vol% MWCNTs mixed for 1 min, 5 min, and 10 min.

Figure 10 shows linear frequency and strain sweeps for 40:54:6 blends containing 3 vol% MWCNTs. As expected, 1 min of mixing yielded the lowest G' and G" values at low angular frequencies (Figure 10a), likely due to no MWCNTs being present in PEO, and an inability of MWCNTs to form networks within the PE phase. At 5 min of mixing, the viscoelastic moduli rise sharply. This is due to the high degree of MWCNTs networks present within the blend, likely concentrated along the PEO/PE interface at this point. Furthermore, the critical strain in Figure 10b for 40:54:6 blends mixed for 5 min occurs at a substantially higher strain than at other times, suggesting a robust MWCNT network at this point. Although 1 min and 10 min samples yielded similar electrical properties, the values of G' and G" are substantially higher at 10 min than at 1 min. This difference between the electrical and rheological properties is due to MWCNTs being more uniformly dispersed throughout both PE and PEO phases at 10 min, whereas at 1 min, MWCNTs are confined to the PE phase.

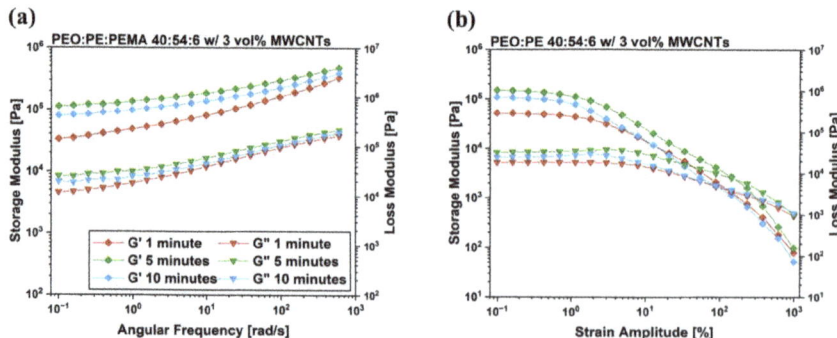

Figure 10. (a) Frequency sweep and (b) strain sweep results for PEO/PE 40:54:6 blends containing 3 vol% MWCNTs.

4. Discussion
4.1. Characteristics of MWCNT Migration in PEO/PE Blends

SEM imaging confirmed that more MWCNT phase migration from PE into PEO occurs with increasing mixing time. Figures 2 and 3 show MWCNTs in the PE phase at all mixing times, with MWCNTs only being visible near the PEO/PE interface or within PEO at higher mixing times. The speed and extent to which this occurs also to rise with increasing MWCNT concentration. The faster migration of MWCNTs at 3 vol% is because there is a higher probability of MWCNTs encountering the PEO/PE interface during mixing at higher concentrations, and consequently MWCNT cross the interface and enter PEO.

The migration of MWCNTs into PEO is also supported by electrical properties of the PBNs with increasing mixing time (Figures 4 and 5). At 1 min of mixing, MWCNTs are still within the PE phase, and form efficient interconnected networks due to the double percolation phenomenon. As the mixing time is increased, many MWCNTs reach the PEO/PE interface and migrate into the PEO phase, eventually becoming uniformly dispersed throughout the blend system. This leads to loss of double percolation, and a corresponding reduction in conductive properties. The rheological properties (Figures 6 and 7) are similarly impacted. There are competing effects between having a uniform dispersion of MWCNTs throughout the blend system and having interconnected networks of MWCNTs that are entangled with one another. Below the percolation threshold (e.g., 0.5 vol% and 1.5 vol% MWCNTs), MWCNTs do not form networks or entangling with each other, therefore the improvement in rheological properties only occurs with MWCNTs dispersing uniformly throughout the blend. At MWCNT concentrations above the percolation threshold, the rheological properties at low mixing times may be higher due to the presence of entangled MWCNT networks. As the mixing time is increased, these MWCNT networks are disrupted as MWCNTs migrate into PEO and disperse throughout the blend. At 2 vol% (Figure 6c or Figure 7c), there is little change in properties over time because as the MWCNT network is destroyed, there is a gain in properties due to more uniform MWCNT dispersion; that is, the two effects cancel each other. In the case of 3 vol% (Figure 6d or Figure 7d), the pre-existing MWCNT network is far more significant, and its rupture over time leads to a decrease in properties with time.

The incorporation of PEMA compatibilizer slowed the migration of MWCNTs, and there was a delayed onset of peak electrical and rheological properties, which occurred at 5 min rather than 1 min of mixing for the uncompatibilized blend (Figure 9). SEM images showed that the blend morphology became finer over time, compared to uncompatibilized blends, suggesting that PEMA reached the PEO/PE interface, and inhibited the coalescence of the PEO phase. Thus, it is important to note that the addition of MWCNT may not only change the properties but also the type of morphology formed. However, the drop in electrical properties at 10 min of mixing, suggests that MWCNTs are ultimately able to cross the reinforced interface, despite the crosslinking of PE and PEO chains.

4.2. Predicting MWCNT Migration in Immiscible Polymer Blends

Based on our previous work studying the phase migration of MWCNTs in PVDF/PE blends, and the present work studying the migration of MWCNTs in PEO/PE blends, it is clear that there are several competing effects governing the final location of MWCNTs within polymer blend nanocomposites. In the case of the PVDF/PE system presented in our previous work [29], the high viscosity of PVDF (the phase with higher thermodynamic affinity for MWCNTs) helped to entrap MWCNTs at the blend interface. Although some individual MWCNTs were able to penetrate the PVDF/PE interface, and fully migrate into PEO, the majority of MWCNT agglomerates remained trapped on the PE side of the PVDF/PE interface. A modified "Slim-Fast Mechanism" originally presented by Göldel et al. [30] was proposed to explain the impact of MWCNT geometry on the ability of MWCNTs to migrate across the PVDF/PE interface; i.e., lone, straight MWCNTs penetrate the interface easily while MWCNT agglomerates do not and become locked at the interface. As MWCNT agglomerates jam at the PVDF/PE interface, they act as barriers for subsequent

MWCNTs, leading to a cascading effect of most MWCNT become trapped at the interface, regardless of their geometry. In contrast, for the PEO/PE blend system studied in the present work, MWCNTs penetrate the PEO/PE interface and fully migrate into the low viscosity PEO and the migration continues as mixing time is increased.

The differences in outcomes of MWCNT migration can be explained by looking at the surface energy data (Table 2) and the complex viscosity data (Table 3). Although the values of wettability are similar for the PVDF/PE and PEO/PE systems, the surface energies between individual component pairs varies significantly. In the case of the PVDF/PE system, the interfacial energies between either polymer and MWCNT is higher than the interfacial energies between the two polymers. This suggests that it is energetically unfavorable for incoming MWCNTs to penetrate the PVDF/PE interface, so MWCNTs failed to migrate from PE into PVDF, despite having a higher thermodynamic affinity for PVDF. In contrast, for the PEO/PE system, PEO has a higher thermodynamic affinity for MWCNT than PEO has for PE (i.e., $\sigma_{A/MWCNT} < \sigma_{A/B}$). This means that it is energetically favorable for MWCNTs to migrate fully into PEO.

Table 2. Surface Energy and Wettability Data for PVDF/PE and PEO/PE Blend Systems Containing MWCNTs (calculated based on geometric mean) [29].

PBN System [A/B]	Temperature [°C]	$\sigma_{A/B}$ [mJ/m^2]	$\sigma_{A/MWCNT}$ [mJ/m^2]	$\sigma_{B/MWCNT}$ [mJ/m^2]	ω
PVDF/PE	200	7.00	11.74	27.18	−2.21
PEO/PE [1]	150	9.31	5.80	27.65	−2.35

[1] Current work.

Table 3. Complex Viscosity Data of PVDF/PE and PEO/PE Systems [29].

PBN System [A/B]	Temperature [°C]	μ_A [Pa·s]	μ_B [Pa·s]	$\frac{\mu_A}{\mu_B}$
PVDF/PE	200	3.14·10^2	4.70·10^2	0.67
PEO/PE [1]	150	4.66·10^2	1.62·10^2	2.88

[1] Current work.

The complex viscosity data for the PVDF/PE blends shows that PVDF (i.e., the destination phase) had a significantly higher viscosity than PE throughout mixing, suggesting that the migration of MWCNTs would be kinetically hindered once they reached the PVDF side of the PVDF/PE interface. In contrast, the complex viscosity data for the PEO/PE blends show that PEO has a significantly lower viscosity than PE, which will facilitate the complete migration of MWCNTs once MWCNTs reach and penetrate the PEO/PE interface during mixing. While the same grade of PE is used as the initial phase in both systems, the viscosity of PE in the PEO/PE system is higher due to the lower processing temperature. Based on the literature, it is expected that this higher viscosity would impede the movement of MWCNTs towards the interface, but this does not appear to be the case since many MWCNTs rapidly migrated into PEO. This suggests that the viscosity of the pre-localized phase does not play as large of a role as the viscosity of the destination phase or of the specific interfacial surface energies between component pairs of the blend systems (e.g., $\sigma_{A/B}$, $\sigma_{A/MWCNT}$, etc.). When selecting a polymer blend in which interfacial localization of nanofillers is desired, it is crucial to select a blend in which the destination phase has a higher viscosity than the pre-localized phase. It is also important to select a blend not only based on the theoretical wettability values obtained via Young's equation [21] but also based on the interfacial surface energies of the individual component pairs. For example, if the two polymer phases have a higher affinity for each other than with the nanofiller, it will be energetically unfavorable for the nanofiller to penetrate the interface.

5. Conclusions

The effects of MWCNT concentration, mixing time, and compatibilizer addition on the migration of MWCNTs from the polyethylene (PE) phase to a polyethylene oxide (PEO) phase of a 60:40 PEO/PE blend and the subsequent impact on electrical properties and rheological properties were investigated. Two-step mixing was used to pre-localize MWCNTs in the less thermodynamically favored PE phase and observe their migration into the thermodynamically favored PEO phase. SEM micrographs showed that MWCNTs migrate into the PEO phase as the mixing time increases at all concentrations of MWCNTs studied. This migration is also supported by EMI SE and DC conductivity measurements, which showed significant reductions in electrical properties over time, suggesting a disruption of conductive networks as MWCNTs migrate into PEO.

PEO/PE 40:60 samples containing 3 vol% MWCNTs showed a high conductivity of 22.1 S/m, respectively, suggesting effective MWCNT networks were present at the onset of mixing. To arrest the migration of MWCNTs into PEO, a PE-graft-maleic anhydride (PEMA) compatibilizer was added to the PEO/PE blend. SEM images confirm an improvement in the formation of MWCNT networks along the PEO/PE interface at 5 min of mixing for the compabilized PBN. Furthermore, major improvements in electrical conductivity (68.7 S/m) were observed. Comparisons to the PVDF/PE system studied in our previous work suggest that the viscosity of the destination phase, as well as the interfacial surface energies of the blend components play significant roles in determining whether MWCNTs will successfully migrate across polymer/polymer interfaces or whether they will become trapped at the interface. The migration behavior was shown to significantly influence the electrical and rheological properties of PBNs.

Supplementary Materials: The following supporting materials can be downloaded at https://www.mdpi.com/article/10.3390/nano12213772/s1. Figure S1. LM images for PEO/PE 40:60 blends containing MWCNTs, including—blends containing 0.5vol% MWCNTs mixed for (a) 1min, (b) 5min, and (c) 10min; blends containing 1.5vol% MWCNTs mixed for (d) 1 min, (e) I.5min, and (f) 10min; blends containing 2vol% MWCNTs mixed for (g) 1min, (h) 5min, and (i) 10min; blends containing 3vol% MWCNTs mixed for (j) 1min, (k) 5min, and (l) 10min. PE composites containing MWCNTs at concentrations of (m) 0.5vol%, (n) 1.5vol%, (o) 2vol%, and (p) 3vol% are also included; Figure S2. SEM images of neat (a) 40:60 PEO/PE blends, (b) 40:57:3 PEO/PE/PEMA blends, and (c) 40:54:6 PEO/PE/PEMA blends, all mixed for 10 minutes; Figure S3. (a) Frequency sweep data and (b) strain sweep data for pure PEO, PE and PEMA; Figure S4. (a) Frequency sweep data and (b) strain sweep data for PEO/PE 40:60 blends containing varying concentrations of PEMA compatibilizer; Figure S5. Real and imaginary permittivity data within the X-band for PEO/PE 40:60 blends containing MWCNTs at concentrations of (a) 0.5vol%, (b) 1.5vol%, (c) 2vol%, and (d) 3vol%; Figure S6. EMI SE and DC conductivity data for (a) PE and (b) PEO nanocomposites containing 3vol% MWCNTs; Figure S7. Electrical permittivity data across the X-band for PE and PEO nanocomposites containing 3vol% MWCNTs, including—(a) (b) real and imaginary permittivity, and (c) (d) complex permittivity; Figure S8. (a,b) Frequency sweep data and (c,d) strain sweep data for PE and PEO nanocomposite samples containing 3vol% MWCNTs;. Surface energy values for each component within the PVDF/PE system at 150 °C [29,30].

Author Contributions: Conceptualization, C.C.L. and U.S.; methodology, C.C.L. and S.R.; formal analysis, C.C.L., S.R. and U.S.; investigation, C.C.L.; data curation, C.C.L. and S.R; writing—original draft preparation, C.C.L. and S.R.; writing—review and editing, S.R. and U.S.; visualization, C.C.L.; supervision, U.S.; funding acquisition, U.S. All authors have read and agreed to the published version of the manuscript.

Funding: This research was supported by the Natural Sciences and Engineering Research Council of Canada (NSERC) Discovery Grant 05503-2020.

Data Availability Statement: The data presented in this work are available upon request from the corresponding author.

Acknowledgments: We would like to thank NSERC for funding this work.

Conflicts of Interest: The authors declare no conflict of interest.

References

1. Jonsson, P.; Moller, R.; Carson, S.; Davies, S.; Lundvall, A. *Ericsson Mobility Report*; Ericsson: Stockholm, Sweden, 2022.
2. Niehaus, M.; Tebbenjohanns, J. Electromagnetic interference in patients with implanted pacemakers or cardioverter-defibrillators. *Heart* **2001**, *86*, 246–248. [CrossRef] [PubMed]
3. Song, P.; Liu, B.; Qiu, H.; Shi, X.; Cao, D.; Gu, J. MXenes for polymer matrix electromagnetic interference shielding composites: A review. *Compos. Commun.* **2021**, *24*, 100653. [CrossRef]
4. Nazir, A.; Yu, H.; Wang, L.; Haroon, M.; Ullah, R.S.; Fahad, S.; Naveed, K.-u.-R.; Elshaarani, T.; Khan, A.; Usman, M. Recent progress in the modification of carbon materials and their application in composites for electromagnetic interference shielding. *J. Mater. Sci.* **2018**, *53*, 8699–8719. [CrossRef]
5. Rahimpour, S.; Kiyani, M.; Hodges, S.E.; Turner, D.A. Deep brain stimulation and electromagnetic interference. *Clin. Neurol. Neurosurg.* **2021**, *203*, 106577. [CrossRef] [PubMed]
6. Lapinsky, S.E.; Easty, A.C. Electromagnetic interference in critical care. *J. Crit. Care* **2006**, *21*, 267–270. [CrossRef]
7. Pawar, S.P.; Biswas, S.; Kar, G.P.; Bose, S. High frequency millimetre wave absorbers derived from polymeric nanocomposites. *Polymer (Guildf)* **2015**, *84*, 398–415. [CrossRef]
8. Huang, J.-C. EMI Shielding Plastics: A Review. *Adv. Polym. Technol.* **1995**, *14*, 137–150. [CrossRef]
9. Kittur, J.; Desai, B.; Chaudhari, R.; Loharkar, P.K. A comparative study of EMI shielding effectiveness of metals, metal coatings and carbon-based materials. *IOP Conf. Ser. Mater. Sci. Eng.* **2020**, *810*, 12019. [CrossRef]
10. Pan, S.; Wang, T.; Jin, K.; Cai, X. Understanding and designing metal matrix nanocomposites with high electrical conductivity: A review. *J. Mater. Sci.* **2022**, *57*, 6487–6523. [CrossRef]
11. Ji, K.; Zhao, H.; Zhang, J.; Chen, J.; Dai, Z. Fabrication and electromagnetic interference shielding performance of open-cell foam of a Cu–Ni alloy integrated with CNTs. *Appl. Surf. Sci.* **2014**, *311*, 351–356. [CrossRef]
12. Cho, J.; Boccaccini, A.R.; Shaffer, M.S.P. Ceramic matrix composites containing carbon nanotubes. *J. Mater. Sci.* **2009**, *44*, 1934–1951. [CrossRef]
13. Chung, D.D.L. Materials for electromagnetic interference shielding. *J. Mater. Eng. Perform.* **2000**, *9*, 350–354. [CrossRef]
14. Otero-Navas, I.; Arjmand, M.; Sundararaj, U. Carbon nanotube induced double percolation in polymer blends: Morphology, rheology and broadband dielectric properties. *Polymer (Guildf)* **2017**, *114*, 122–134. [CrossRef]
15. Arjmand, M.; Apperley, T.; Okoneiwski, M.; Sundararaj, U. Comparative study of electromagnetic interference shielding properties of injection molded versus compression molded multi-walled carbon nanotube/polystyrene composites. *Carbon* **2012**, *50*, 5126–5134. [CrossRef]
16. Arjmand, M.; Mahmoodi, M.; Gelves, G.A.; Park, S.; Sundararaj, U. Electrical and electromagnetic interference shielding properties of flow-induced oriented carbon nanotubes in polycarbonate. *Carbon* **2011**, *49*, 3430–3440. [CrossRef]
17. Kanoun, O.; Bouhamed, A.; Ramalingame, R.; Bautista-Quijano, J.R.; Rajendran, D.; Al-Hamry, A. Review on Conductive Polymer/CNTs Nanocomposites Based Flexible and Stretchable Strain and Pressure Sensors. *Sensors* **2021**, *21*, 341. [CrossRef]
18. Han, Z.; Fina, A. Thermal conductivity of carbon nanotubes and their polymer nanocomposites: A review. *Prog. Polym. Sci* **2011**, *36*, 914–944. [CrossRef]
19. Mohamed, M.G.; Kuo, S.W. Functional Silica and Carbon Nanocomposites Based on Polybenzoxazines. *Macromol. Chem. Phys.* **2019**, *220*, 1800306. [CrossRef]
20. Mohamed, M.G.; Samy, M.M.; Mansoure, T.H.; Sharma, S.U.; Tsai, M.S.; Chen, J.-H.; Lee, J.-T.; Kuo, S.-W. Dispersions of 1,3,4-Oxadiazole-Linked Conjugated Microporous Polymers with Carbon Nanotubes as a High-Performance Electrode for Supercapacitors. *ACS Appl. Energy Mater.* **2022**, *5*, 3677–3688. [CrossRef]
21. Sumita, M.; Sakata, K.; Asai, S.; Miyasaka, K.; Nakagawa, H. Dispersion of fillers and the electrical conductivity of polymer blends filled with carbon black. *Polym. Bull.* **1991**, *25*, 265–271. [CrossRef]
22. Al-Saleh, M.H.; Al-Anid, H.K.; Hussain, Y.A. Electrical double percolation and carbon nanotubes distribution in solution processed immiscible polymer blend. *Synth. Met.* **2013**, *175*, 75–80. [CrossRef]
23. Pötschke, P.; Bhattacharyya, A.R.; Janke, A. Carbon nanotube-filled polycarbonate composites produced by melt mixing and their use in blends with polyethylene. *Carbon* **2004**, *42*, 965–969. [CrossRef]
24. Bose, S.; Bhattacharyya, A.R.; Bondre, A.P.; Kulkarni, A.R.; Pötschke, P. Rheology, electrical conductivity, and the phase behavior of cocontinuous PA6/ABS blends with MWNT: Correlating the aspect ratio of MWNT with the percolation threshold. *J. Polym. Sci. Part B: Polym. Phys.* **2008**, *46*, 1619–1631. [CrossRef]
25. Zhang, Z.; Cao, M.; Chen, P.; Yang, B.; Wu, B.; Miao, J.; Xia, R.; Qian, J. Improvement of the thermal/electrical conductivity of PA6/PVDF blends via selective MWCNTs-NH2 distribution at the interface. *Mater. Des.* **2019**, *177*, 107835. [CrossRef]
26. Wu, D.; Zhang, Y.; Zhang, M.; Yu, W. Selective Localization of Multiwalled Carbon Nanotubes in Poly(ε-caprolactone)/Polylactide Blend. *Biomacromolecules* **2009**, *10*, 417–424. [CrossRef] [PubMed]
27. Sultana, S.M.N.; Pawar, S.P.; Sundararaj, U. Effect of Processing Techniques on EMI SE of Immiscible PS/PMMA Blends Containing MWCNT: Enhanced Intertube and Interphase Scattering. *Ind. Eng. Chem. Res.* **2019**, *58*, 11576–11584. [CrossRef]

28. Ravindren, R.; Mondal, S.; Nath, K.; Das, N.C. Investigation of electrical conductivity and electromagnetic interference shielding effectiveness of preferentially distributed conductive filler in highly flexible polymer blends nanocomposites. *Compos. Part A: Appl. Sci. Manuf.* **2019**, *118*, 75–89. [CrossRef]
29. Lencar, C.; Ramakrishnan, S.; Erfanian, E.; Sundararaj, U. The Role of Phase Migration of Carbon Nanotubes in Melt-Mixed PVDF/PE Polymer Blends for High Conductivity and EMI Shielding Applications. *Molecules* **2022**, *27*, 933. [CrossRef]
30. Göldel, A.; Marmur, A.; Kasaliwal, G.R.; Pötschke, P.; Heinrich, G. Shape-dependent localization of carbon nanotubes and carbon black in an immiscible polymer blend during melt mixing. *Macromolecules* **2011**, *44*, 6094–6102. [CrossRef]
31. Maheswaran, R.; Shanmugavel, B.P. A Critical Review of the Role of Carbon Nanotubes in the Progress of Next-Generation Electronic Applications. *J. Electron. Mater.* **2022**, *51*, 2786–2800. [CrossRef]
32. Arya, A.; Sharma, A.L. Insights into the use of polyethylene oxide in energy storage/conversion devices: A critical review. *J. Phys. D: Appl. Phys.* **2017**, *50*, 443002. [CrossRef]
33. Breuer, O.; Sundararaj, U.; Toogood, R.W. The design and performance of a new miniature mixer for specialty polymer blends and nanocomposites. *Polym. Eng. Sci.* **2004**, *44*, 868–879. [CrossRef]
34. Pawar, S.P.; Marathe, D.A.; Pattabhi, K.; Bose, S. Electromagnetic interference shielding through MWNT grafted Fe3O4 nanoparticles in PC/SAN blends. *J. Mater. Chem. A* **2015**, *3*, 656–669. [CrossRef]
35. Wu, S. *Polymer Interface and Adhesion*; Marcel Dekker Inc.: New York, NY, USA, 1982.
36. Owens, D.K.; Wendt, R.C. Estimation of the surface free energy of polymers. *J. Appl. Polym. Sci.* **1969**, *13*, 1741–1747. [CrossRef]
37. Kamkar, M.; Sultana, S.M.N.; Pawar, S.P.; Eshraghian, A.; Erfanian, E.; Sundararaj, U. The key role of processing in tuning nonlinear viscoelastic properties and microwave absorption in CNT-based polymer nanocomposites. *Mater. Today Commun.* **2020**, *24*, 101010. [CrossRef]
38. Kamkar, M.; Aliabadian, E.; Zeraati, A.S.; Sundararaj, U. Application of nonlinear rheology to assess the effect of secondary nanofiller on network structure of hybrid polymer nanocomposites. *Phys. Fluids* **2018**, *30*, 23102. [CrossRef]
39. Trifkovic, M.; Hedegaard, A.T.; Sheikhzadeh, M.; Huang, S.; Macosko, C.W. Stabilization of PE/PEO Cocontinuous Blends by Interfacial Nanoclays. *Macromolecules* **2015**, *48*, 4631–4644. [CrossRef]
40. Trifkovic, M.; Hedegaard, A.; Huston, K.; Sheikhzadeh, M.; Macosko, C.W. Porous Films via PE/PEO Cocontinuous Blends. *Macromolecules* **2012**, *45*, 6036–6044. [CrossRef]
41. Vuksanović, M.M.; Heinemann, R.J. Chapter 10-Micro and nanoscale morphology characterization of compatibilized polymer blends by microscopy. In *Compatibilization of Polymer Blends*; Ajitha, A.R., Sabu, T., Eds.; Elsevier: Amsterdam, The Netherlands, 2020; pp. 299–330. [CrossRef]

Review

Mechanical Performance and Applications of CNTs Reinforced Polymer Composites—A Review

N. M. Nurazzi [1,2], F. A. Sabaruddin [1], M. M. Harussani [3], S. H. Kamarudin [4], M. Rayung [5], M. R. M. Asyraf [6], H. A. Aisyah [1,7,*], M. N. F. Norrrahim [8,*], R. A. Ilyas [9,10,*], N. Abdullah [2,*], E. S. Zainudin [1,7], S. M. Sapuan [1,3] and A. Khalina [1,*]

[1] Institute of Tropical Forestry and Forest Products (INTROP), Universiti Putra Malaysia (UPM), Serdang 43400, Malaysia; mohd.nurazzi@gmail.com (N.M.N.); atiyah88@gmail.com (F.A.S.); edisyam@upm.edu.my (E.S.Z.); sapuan@upm.edu.my (S.M.S.)
[2] Centre for Defence Foundation Studies, Universiti Pertahanan Nasional Malaysia (UPNM), Kem Perdana Sungai Besi, Kuala Lumpur 57000, Malaysia
[3] Advanced Engineering Materials and Composites (AEMC), Department of Mechanical and Manufacturing Engineering, Universiti Putra Malaysia (UPM), Serdang 43400, Malaysia; mmharussani17@gmail.com
[4] Faculty of Applied Sciences, School of Industrial Technology, Universiti Teknologi MARA (UiTM), Shah Alam 40450, Malaysia; sitihasnahkam@uitm.edu.my
[5] Faculty of Science, Universiti Putra Malaysia (UPM), Serdang 43400, Malaysia; marwahrayung@yahoo.com
[6] Department of Aerospace Engineering, Universiti Putra Malaysia (UPM), Serdang 43400, Malaysia; asyrafriz96@gmail.com
[7] Department of Mechanical and Manufacturing Engineering, Faculty of Engineering, Universiti Putra Malaysia (UPM), Serdang 43400, Malaysia
[8] Research Centre for Chemical Defence, Universiti Pertahanan Nasional Malaysia (UPNM), Kem Perdana Sungai Besi, Kuala Lumpur 57000, Malaysia
[9] Faculty of Engineering, School of Chemical and Energy Engineering, Universiti Teknologi Malaysia (UTM), Skudai 81310, Malaysia
[10] Centre for Advanced Composite Materials (CACM), Universiti Teknologi Malaysia (UTM), Skudai 81310, Malaysia
* Correspondence: aisyah.humaira@upm.edu.my (H.A.A.); faiznorrrahim@gmail.com (M.N.F.N.); ahmadilyas@utm.my (R.A.I.); norli.abdullah@upnm.edu.my (N.A.); khalina@upm.edu.my (A.K.)

Abstract: Developments in the synthesis and scalable manufacturing of carbon nanomaterials like carbon nanotubes (CNTs) have been widely used in the polymer material industry over the last few decades, resulting in a series of fascinating multifunctional composites used in fields ranging from portable electronic devices, entertainment and sports to the military, aerospace, and automotive sectors. CNTs offer good thermal and electrical properties, as well as a low density and a high Young's modulus, making them suitable nanofillers for polymer composites. As mechanical reinforcements for structural applications CNTs are unique due to their nano-dimensions and size, as well as their incredible strength. Although a large number of studies have been conducted on these novel materials, there have only been a few reviews published on their mechanical performance in polymer composites. As a result, in this review we have covered some of the key application factors as well as the mechanical properties of CNTs-reinforced polymer composites. Finally, the potential uses of CNTs hybridised with polymer composites reinforced with natural fibres such as kenaf fibre, oil palm empty fruit bunch (OPEFB) fibre, bamboo fibre, and sugar palm fibre have been highlighted.

Keywords: CNTs; MWCNTs; SWCNTs; polymer composite; mechanical performance

1. Introduction

CNTs are cylindrical molecules made up of hexagonally arranged hybridised carbon atoms. Carbon nanotubes are formed from micrometre-scale graphene sheets folded into nanoscale cylinders and topped with spherical fullerenes. Due to the presence of delocalised electrons in the z-axis, CNTs have distinct electrical properties. CNTs are classified according to their wall thickness into single-wall carbon nanotubes (SWCNTs) and multiwall carbon nanotubes (MWCNTs). MWCNTs are multilayered rolled graphene sheets, whereas SWCNTs are nanocylinders constructed from a single graphene sheet. The van der Waals force between CNTs and the weak interplanar interactions of graphene sheets (highly polarised π-electron clouds in CNTs) firmly bind CNTs in nature. As a result, the aggregation and solvent chemistry of CNTs nanomaterials regulate their size, shape, and surface area [1,2]. CNTs are now only used as reinforcements in polymer matrices. Nanotubes have outstanding mechanical and physical properties, making them ideal building blocks for high-performance multifibres and composites [3]. Because of their superior mechanical properties and high aspect ratio, CNTs have long been considered a desirable filler for polymer composites [4]. As shown in Table 1, CNT-reinforced polymer composites were first used commercially in the 1980s and have since risen in popularity as a high-performance material in the aerospace, automotive, sports, biomedical, and electronics industries due to their high specific stiffness, strength-to-weight ratio, low thermal expansion coefficient, and high thermal conductivity. Aside from that, CNTs are broadly used as sensing materials in chemical and biosensor applications [5,6].

Table 1. Shows several examples of CNTs reinforced polymer composites made in the 1980s, organised by fabrication method.

Year	CNTs	Matrix	Fabrication Method	Ref.
1998	MWCNTs	Epoxy	Solution casting–curing	[7]
1999	CNTs	PVA	Solution casting	[8]
2002	MWCNTs	Epoxy	CVD–injection molding	[9]
2002	MWCNTs	PS	Spin-casting	[10]
2003	SWCNTs	Alumina	Spark-plasma sintering	[11]
2003	MWCNTs	Epoxy	Solution-casting	[12]
2004	MWCNTs	P(MMA-co-EMA)	Solution-mixing	[13]
2004	MWCNTs	Nylon 6	Melt compounding	[14]
2005	MWCNTs	PA	In situ polymerization	[15]
2006	MWCNT–NH$_2$	Nylon 6	Solution-casting–melt compounding	[16]
2007	MWCNTs	Aluminium	Isostatic pressing–hot extrusion techniques	[17]
2007	SWCNTs	PVC	Film casting	[18]
2007	MWCNTs	PVC	Film casting	[18]
2008	MWCNTs	PMMA	CVD–solvent casting	[19]
2008	MWCNTs	PS	CVD–solvent casting	[19]
2010	MWCNTs	Epoxy	Ultrasonication technique–sputtering	[20]
2010	DWCNTs	Magnesia	In situ polymerization–spark-plasma-sintering	[21]
2010	MWCNTs	PP	Melt mixing–in situ polymerization	[22]
2011	MWCNTs	Epoxy	Chemical functionalization–cast molding	[3]
2013	Dense-CNTs	PP	CVD	[23]
2014	MWCNTs	PVC	Film casting	[24]
2015	Amino-MWCNTs	Epoxy	Direct stirring–resin infusion molding	[25]
2015	MWCNTs	HDPE	Melt-mixing–compression molding	[26]
2016	SWCNTs	Chitosan	Solution-casting	[27]
2016	CNTs	Epoxy	Press cured method	[28]
2017	MWCNTs	Epoxy	EPD	[29]
2018	MWCNTs	PMMA	Chemical functionalization–micro compounding–injection molding	[30]
2019	MWCNTs	Epoxy	Non-destructive synthesis technique	[31]
2020	MWCNTs	Epoxy	Solution-casting–hand lay-up–resin infusion	[32]
2020	MWCNTs	PVC	CVD–ultrasonic dispersion–extrusion	[33]
2020	MWCNTs	PVC	CVD–ultrasonic dispersion–extrusion	[33]
2021	MWCNTs	Epoxy	Resin castings (injection-molding)	[34]

CVD—chemical vapor deposition, EPD—electrophoretic deposition, ESD—electrospray deposition and CF—chemical functionalization, GF—glass fibre, NBCNT—nitrogen-doped bamboo-shaped CNT, PP—polypropylene, DWCNT—double-walled CNT, PMMA—Poly (methyl methacrylate), PVA—poly (vinyl alcohol), PVC—polyvinyl chloride, P(MMA-co-EMA)—copolymer of methyl and ethyl methacrylate.

Even so, agglomeration and restricted dispersion in the polymer matrix, as well as the van der Waals force between CNTs and weak interplanar interactions of graphene sheets (highly polarised π-electron clouds in CNTs) firmly bind CNTs in nature, making production of advanced composites with CNTs as reinforcement difficult. As a consequence, the size, shape, and surface area of CNT nanomaterials are controlled by aggregation and solvent chemistry. Thus, in the use of carbon-based nanomaterials, overcoming aggregation is critical. When compared to other carbon compounds such as graphite and fullerene, CNTs are hydrophobic and electrically conductive by nature, and they have a large surface area. The large surface area of CNTs results in a high viscosity of the nanotube/epoxy combination when fabricating composites with a high nanotube loading level, making nanotube dispersion difficult. As a consequence, controlled particle size distribution and dispersion are important factors in composite material production. Because the fillers are small, the composites have a high interfacial area [20]. The schematic diagram of SWCNTs and MWCNTs from a rolled graphene sheet is shown in Figure 1 [35,36].

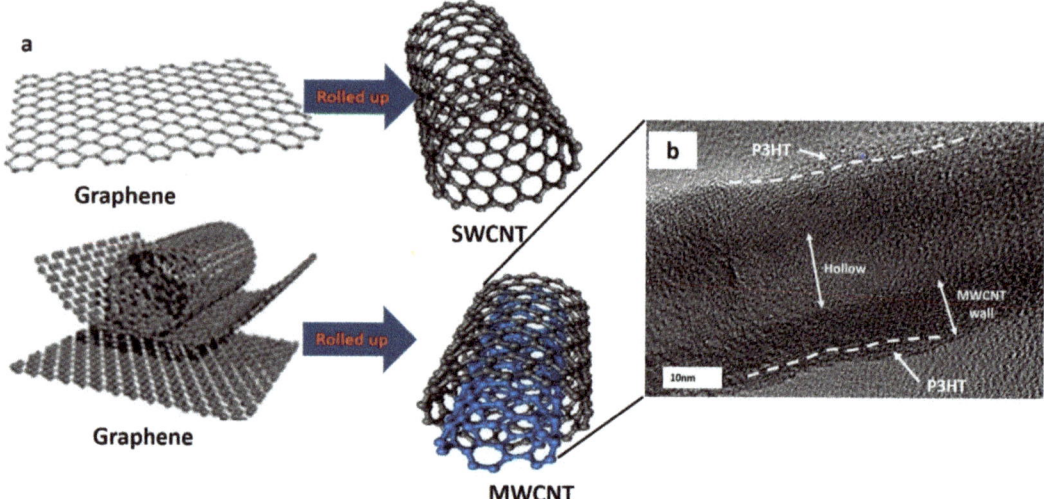

Figure 1. (a) Schematic diagram of SWCNT and MWCNT (reproduced from [35]) and (b) the MWCNT wrapped with poly(3-hexylthiophene). Reproduced from [37].

It can be concluded that so far the performance on CNTs in reinforcing polymer matrices has proved inadequate, which several researchers have attributed to two main issues: (1) difficulties in distributing CNTs in polymers, and (2) insufficient bonding of nanotubes with the polymer interface. A substantial amount of research has been conducted on the chemical functionalisation of CNTs to achieve homogeneous dispersion of CNTs in the polymer matrix and high interfacial adhesion between CNTs and polymer matrix [3,16,25]. The results revealed that functionalisation of CNTS surface-enhanced both the adsorption energy, mechanical and electrical characteristics. This happens towards the carbon layer's margins but can also appear further from the edges if the incorporation sites are related to vacancies. These vacancies and edges might act as adsorption sites, explaining the unusual structures of doped CNTs [33]. Previous studies have shown that controlling CNTs contact improves CNTs-polymer matrix interactions. The degree of interaction between the filler and the polymer modify the mobility of the polymer chain, the degree of curing and the crystallinity of the polymer. The successful integration of interfacial adhesion between CNTs and the relevant polymer matrix could result in significant structural benefits for a variety of applications. As a result, in this brief overview, the mechanical performance and factors influencing the mechanical performance of CNTs reinforced polymer composites and CNTs-reinforced polymer composites for structural applications and their prospects have been discussed.

2. Mechanical Characteristics of CNTs

Dispersion and distribution are key characteristics in the manufacturing of composites. CNTs with good dispersion and homogeneous distribution are favourable for the creation of linked networks [38]. Nevertheless, depending on the type of polymer matrix, a certain degree of agglomeration and a carefully tailored non-homogeneous distribution may lead to segregated structures with excellent mechanical properties.

Overney et al. [39] computed the rigidity of short SWCNTs using ab initio local density calculations to obtain the parameters in a Keating potential. Another study led by Wang and Zhang [40] found out that the effective thickness of SWCNTs should be smaller than 0.142 nm. In this case, Young's modulus of SWCNTs composites can be attained between 0.65 TPa and 5.5 TPa. Subsequent initial mechanical measurements on MWCNTs created by the arc discharge technique were made. Poncharal et al. [41] generated electromechanical resonant vibrations with moduli ranging from 0.7 TPa to 1.3 TPa. Wong et al. [42] investigated the mechanical characteristics of MWCNTs and found Young's modulus average value of 1.28 TPa. More importantly, they were able to conduct the initial strength tests, getting an average bending strength of 14 GPa. Table 2 lists the mechanical characteristics of CNTs and other examples of reinforcing materials [36,43–47].

Table 2. Mechanical properties of CNTs with other example of reinforcing materials.

Reinforcement Materials	Young's Modulus (TPa)	Tensile Strength (GPa)
SWCNTs	0.65 to 5.5	126
MWCNTs	0.2 to 1.0	>63
Monolayer Graphene	1.0	130
Stainless steel	0.186 to 0.214	0.38 to 1.55
Kevlar	0.06 to 0.18	3.6 to 3.8
Diamond	1.22	>60
Aluminium	71	0.65
Glass fibres	72	3
Carbon fibres	300	3
Silicon carbide fibres	450	10
Sugar palm fibre	0.0049	0.00016
Kenaf fibre	0.053	0.00025
Bamboo fibre	0.0011 to 0.0017	0.00014 to 0.00023

3. Factors Influencing the Mechanical Performance of CNTs Reinforced Polymer Composites

The bonding and strength at the interface, in addition to mechanical load transmission from the matrix to the nanotubes surface, all have a major impact on the performance of CNTs-reinforced polymer composites. The mechanism of interfacial load transmission from matrix to nanotubes may be classified into two types: weak van der Waals forces between the polymer matrix and CNTs as reinforcement [48]. Furthermore, one of the most important factors influencing the performance of CNTs-reinforced polymer composites is the dispersion of CNTs in the polymer matrix via physical, functionalisation of CNT surfaces, and their structures [49].

Microcracks can develop during the curing process or due to the wettability of CNTs and matrix interfaces. Microcracking can occur in high modulus resin systems. This is particularly the case at high processing temperatures and low service temperatures, where there is a substantial disparity in thermal expansion between the polymer matrix and the CNT reinforcements. Hence, the use of CNTs as a toughening reinforcement to a polymer resin matrix increases microcracking prevention while compromising performance at elevated temperatures [50]. As a result of the hydrophobic surface regions of the matching micelles surrounding the nanotubes, constraints such as the agglomeration of CNTs arise frequently. Therefore, a detailed understanding of the factors that influence the mechanical properties of CNTs-reinforced polymer composites has been a major consideration.

Aside from the previously mentioned issues of dispersion and agglomeration, the aspect ratio of CNTs is an important factor in the longitudinal elastic modulus. CNTs have a high aspect ratio in general, but their ultimate performance in a polymer composite is dependent on the type of polymer matrix used. Arash et al. [51] investigated the influences of CNT aspect ratio on Young's modulus and yield strength of CNTs/polymethyl methacrylate (PMMA). The results revealed an increase in Young's modulus of PMMA polymer reinforced by CNTs, as well as an increase in the CNT aspect ratio. The diameter of the (5, 5) CNTs reinforcements was 0.68 nm, and the length-to-diameter ratio (L/d) ranged from 7.23 to infinity (∞). The stress transfer between the CNTs and the polymer was then enhanced by increasing the aspect ratio of the CNTs. Finally, the CNTs reinforced polymer composites have high strength and stiffness values. Coleman et al. [52] stated the higher the aspect ratio of CNTs, the higher the stress transfers from the polymer matrix to the dispersed CNTs. This is because the CNTs, which have a high aspect ratio, may lead to an adequate load transmission via interfacial shear stress. As a result, the full strength of CNTs can be utilised. Figure 2 shows the effect of different types of nano-scale particle distribution caused by the exceptionally large surface area of the nanocomposites.

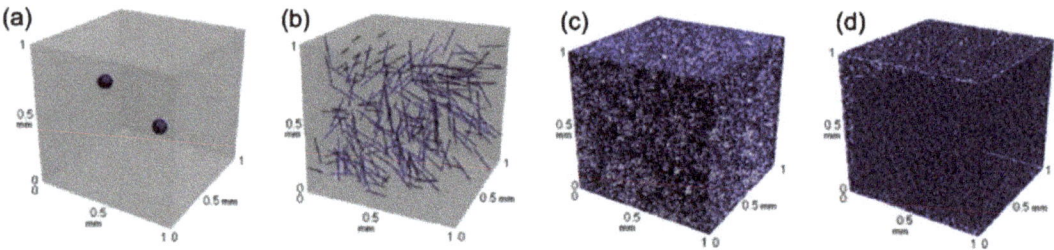

Figure 2. Micro and nano scale distribution of; (**a**) Al_2O_3 particles, (**b**) carbon fibers, (**c**) graphene nanoplatelets (GNPs), and (**d**) CNTs. Reproduced with permission from [38]. Copyright Elsevier, 2010.

A significant amount of research has been directed toward the fabrication of CNT-reinforced polymer composites for functional and structural applications [53–55]. Referring to Ma et al. [38], however, the potential for using CNTs as reinforcements has been greatly restricted due to difficulties associated with entangled CNT dispersion during processing and poor interfacial interaction between CNTs and polymer matrix (Figure 3). The limits to dispersing CNTs differ from those of other conventional fillers such as spherical particles and carbon fibres because CNTs have nanometer-scale properties with aspect ratios greater than 1000, resulting in an exceptionally large surface area. Figure 2 depicts a schematic representation of the 3D distribution of micro-and nanoscale fillers in a polymer matrix, which demonstrates the strong influence of particle size and geometry on the varied distribution behaviour of particles in the matrix. The distribution of micro-scale fillers is homogenous throughout the matrix, as shown in Figure 2a,b, and a differentiation between individual particles in a matrix can be easily created. When CNTs are filled into the same volume of matrix system, however, it is difficult to disperse individual particles equally, as shown in Figure 2c,d. Besides that, a large surface area of nano CNTs means a large interface or interphase area present between the filler and the matrix.

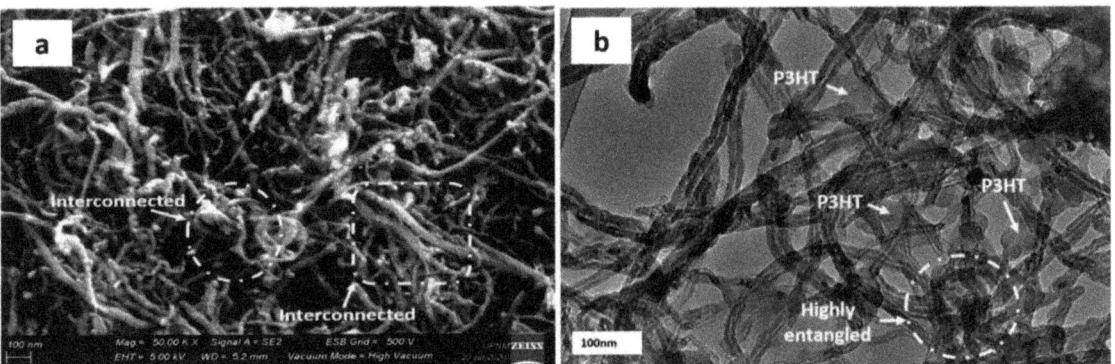

Figure 3. Entangled nature of MWCNT (**a**) under FESEM and (**b**) under HRTEM. Reproduced from [37].

The "interface" in composites is a surface formed by a common boundary of reinforcing fillers and matrix that is in contact and maintains the interfacial bonding in between for load transfer mechanism occurs [56]. The "interphase" is defined as the region with altered chemistry, polymer chain mobility, degree of cure and the crystallinity index that are unique from those of the filler or the matrix. The interphase size of CNTs polymer–matrix composites has been reported to be as large as about 500 nm according to the size and dimension of fillers [57]. Even if the interfacial region is only a few nanometres thick, this would lead to tremendous issues with uniform dispersion and distribution that finally deteriorate the mechanical stability and performance.

Related to MWCNTs and based on Paramsothy [58] in regards to the dispersion of nanotubes, adhesion (contact) at the nanotube–polymer matrix interface, and alignment of nanotubes with the polystyrene (PS) composites, dispersion refers to how individual nanotubes were spread out within the PS matrix after solvent casting and before stretching. It was observed that the dispersion of individual CNTs in composite films of 5 wt.% CNTs content was good but poor (due to the occurrence of CNT clumps) in films of higher (10 wt.% and 30 wt.%) CNTs content. Paramsothy also mentioned that agglomerations or clumps of CNTs are caused by two reasons. Before solvent casting, the purified CNT/PS/toluene suspension was treated with ultrasound (sonicated) for 30 min for homogenisation purposes. The purified CNTs used to form the suspension were mainly in the form of clumps. The first reason was that the ultrasound treatment was only capable of partially separating individual CNTs from the purified CNTs clumps in the suspension. The CNT clumps were made up of individual CNTs that were interlocked with one another. It was possible that the ultrasound treatment did not provide enough energy to overcome completely the interlocking between individual CNTs forming the purified CNT clump. Ultrasonic treatment of the purified CNT/PS/toluene suspension (during its preparation for solvent casting) also resulted in uniform distribution of individual CNTs.

The second reason for the observation of CNT agglomerations in the composite film was that reformation of CNT clumps from individual CNTs (in suspension) was possible in the absence of ultrasound. This was due to the high binding energy between individual CNTs, which resulted from van der Waal's interactions between the CNTs. The van der Waals interaction among the CNTs was sufficient to physically attract them to one another. The resulting high binding energy among the CNTs was high enough to keep them physically close to one other. Ruoff et al. [59] showed that the van der Waal's interaction between individual CNTs is sufficient to cause substantial deformation (destruction of the cylindrical symmetry of the CNT) when the CNTs are aligned and adjacent, and that the binding energy between a C60 molecule and a graphite plane is high at 1 eV. Also, no defloculent was used during the fabrication of the composite film. With insufficient dissolved polymer physically separating the individual CNTs and no use of any defloculent and ultrasound treatment, it was possible that the van der Waals interaction among the individual CNTs was sufficient to physically attract the individual CNTs to one another and that the resulting binding energy among the individual CNTs physically attracted to one another was high enough to keep them physically close to one another, during solvent casting of composite films of higher CNTs content.

In conclusion, two types of interfaces can be formed in CNT-reinforced polymer composites [58]. In the first type of interface (Type 1 interface), wetting of the CNT by the polymer matrix is good, but the adhesion of the CNT to the polymer matrix is weak. This results in the CNT getting pulled out of the polymer matrix before it can experience fragmentation during composite fracture. In the second type of interface (Type 2 interface), wetting of the CNT by the polymer matrix is good. However, the adhesion of the CNT to the polymer matrix is also good. The Type 2 interface can be further sub-categorised into two forms, Type 2a and Type 2b interfaces. In the Type 2a interface, the good adhesion of the CNT to the polymer matrix results in CNT fragmentation during composite fracture. The polymer matrix is not too ductile such that the interface it shares with the CNT is not held in place during CNT pull-out. In the Type 2b interface, the good adhesion of the CNT to the polymer matrix results in a matrix fracture around the CNT during composite fracture, instead. Following the matrix fracture, the polymer coats the CNT as it is pulled out of the matrix. The polymer matrix is too ductile even after work-hardening such that the interface it shares with the CNT is not held in place during CNT pull-out.

4. Mechanical Performance of CNTs Reinforced Polymer Composites

The remarkable success of polymer nanocomposites with the incorporation of CNTs to impart superior performance, particularly in mechanical properties, has been widely reported [4]. Among all the factors that contribute to the excellent properties of the nanocomposites, the individual morphological features of CNTs contribute significantly to determining the performance of the nanocomposites [38]. Their mechanical properties are based on the sp^2 strength of the C-C bonds of the nanotubes, which is stronger than sp^3 found in a diamond. This characteristic then makes CNTs good candidates for reinforcement in polymer composites [4]. Meanwhile, the novel properties of CNTs include lightweight, distinct optical characteristics, high aspect ratios and surface area, high mechanical strength, and high thermal and electrical conductivity help to impart excellent properties to the polymer nanocomposites they are incorporated into and make them suitable for a wide range of applications [60].

The mechanical properties of the individual CNTs have also become one of the most vital features that contribute to the outstanding mechanical properties of polymer nanocomposites. Theoretically, CNTs have a Young's modulus of roughly at 1 TPa, which is approximately five times greater than that of steel, and their tensile strength is in the vicinity of 11 GPa to 100 GPa, which is nearly 100 times higher than that of steel. Because of these characteristics, they are the strongest materials ever invented by mankind [4,61–65]. Similar yet more detailed values have been reported by Vankataraman et al. [59] in their review indicating that the tensile strength of MWCNTs is in the range of 11 GPa to 63 GPa, whereas the elastic modulus for the individual MWCNTs is around 1 TPa. Meanwhile, the tensile strength of SWCNTs is in the vicinity of 22 GPa, whilst Young's modulus was directly measured and determined to be in the range of 0.79 TPa to 3.6 TPa. The compressive strength of the MWCNTs, on the other hand, was estimated to be in the range of 1 GPa to 150 GPa.

The utilisation of CNTs in polymer nanocomposites relies on their very small size with a high aspect ratio that contributes to the high stiffness and strength of the resulting nanocomposites [4]. Despite their small size, CNTs can also have different dimensions, diameters and lengths that determine the dispersion properties which affect the properties of the nanocomposites. The van der Waals interactions between CNTs also cause agglomeration, resulting in poor dispersion properties. Poor dispersion of CNTs can deteriorate the overall performance of the nanocomposites, especially the mechanical and electrical characteristics. In contrast, homogenous dispersions enable uniform load distributions, thus reducing the load concentration and improving the mechanical properties of the nanocomposites [1,4,66]. Besides, the mechanical properties of the CNTs reinforced polymer composites are also greatly influenced by the type of bonding between the two components, the strength of the interface and the mechanical load transfer from the surrounding to the CNTs filler.

The mechanical characteristics of the CNT-reinforced polymer composites can be further improved via various functionalisation techniques, including physical and chemical functionalisation, to enhance the dispersion capability and improve the CNTs interface. As a result, the interfacial bonding between the CNTs reinforcement and matrix components in the composite system will be improved. Chemical modification, for example, aids in improving the dispersion and solubility of the CNTs in solvents or polymers, thus improving the interaction and reactivity with the matrix via hydrogen bonding [67]. This treatment usually involves the use of strong acids to remove the end caps as well as reduce the length of the CNTs. Oxygenated groups like carboxylic acids, carbonyl and hydroxyl groups were added in the acid treatment to the tube ends and defect sites of the CNTs. These oxygen-containing groups can be further treated with other groups like amides, thiols, etc. [68–71]. As mentioned by Norizan et al. [1] the mechanical properties of the CNT-reinforced polymer composites can be enhanced by incorporating chemical-functionalised CNTs into the polymer matrix that enables covalent bonding between SWCNTs and MWCNTs. Chemical functionalisation can improve the CNTs and polymer matrix interface, which imparts enhancement to the interfacial strength, thus improving the load transfer mechanism to the CNTs [72].

To date, a variety of polymers have been used to be incorporated with CNTs, including liquid crystalline, water-soluble, thermoplastics, and polymer [66]. The CNTs loading was usually reported to be under 10 wt.% to avoid the agglomeration, which resulted in poor processability and weak properties of the resulting polymer composites [4]. CNTs- reinforced thermoplastics have been commonly reported in the past years based on their positive attributes like high strength, high modulus and low density. Thermoplastic composites offer advantages over thermoset composites in terms of damage tolerance, faster component manufacturing times, indefinite shelf life, better recyclability and an improved work environment [73,74]. Like the aforementioned stress transfer criteria that are required for mechanical improvement, the interfacial adhesion between CNTs and the thermoplastics matrix is unfortunately weak as there is no or little chemical bonding at the CNTs-reinforced thermoplastics interface [75]. To date, various chemical modifications and advanced types of thermoplastics were applied to improve the mechanical properties of the composites. A study by Sattar et al. [65] reported on the mechanical behaviour of PU-reinforced MWCNTs nanocomposites indicating that the most challenging issue with MWCNTs in the matrix is increasing the dispersion of the filler to enhance the load transfer capacity of the composite to the nanotube network. The authors compiled the findings about thermoplastic PU-reinforced MWCNTs and discovered that increasing the nanotube concentration from 0 wt.% to 17.7 wt.% produced a non-monotonic trend, with 9.3 wt.% exhibiting the optimum tensile strength nearly 2.4 times higher than that of neat PU polymer. Meanwhile, Young's modulus and tensile strength of the sample with the amide-functionalised MWCNTs sample considerably improved with no loss in elongation at break [65,76,77]. In a separate discussion, further improvement in interlaminar shear strength (ILSS) and impact toughness was reported by Liu et al. [73] in mechanical properties of thermoplastic-reinforced composites using hybrid CNTs and commercial carbon fibres in the form of multiscale composites.

Other than thermoplastics, CNTs are added to other polymers to improve the mechanical properties of engineering polymers such as epoxy resins. Among all the types of epoxies available, amine-cured epoxies are considered for the polymer matrix due to their superior engineering performance [78,79]. For example, Uthaman et al. [80] found the addition of CNTs into the epoxy imparted an optimum increment in percentage by 52.9% (flexural strength) and 25.5% (flexural modulus), 29.5% (tensile strength) and 48.1%

(tensile) with only 1.5 wt.% addition of CNTs. However, the mechanical properties of the CNT- reinforced epoxy nanocomposites decreased by the addition of 2.0 wt.% of CNTs. In contrast, the mechanical properties of the epoxy-reinforced CNTs were also observed to increase even at high loadings (20 wt.%) CNTs. This finding has been proven by Herceg et al. [81], whereby the addition of the highest loading of CNTs provides a maximum measured Young's modulus of 5.4 GPa and yield strength of 90 MPa. Although the nanocomposites produced had some porosity (2 vol.%), the modulus and the strength were shown to increase. Better improvement can be achieved with the addition of treated CNTs. For example, Lopes et al. [82] utilised oxidised in thermoset polyurethane elastomer (PU). In that study, addition of only 0.5 wt.% of MWCNT-ox was able to increase the elastic modulus of the PU nanocomposites by 47% with better dispersion as compared to non-oxidised MWCNT.

A comparison study also has been done by Zahid et al. [83] between thermoplastic PU and epoxy thermoset-based composites enhanced with MWCNTs. With the addition of 0.5% MWCNTs, the ILSS showed an improvement of 24.37% in epoxy-based composites and 10.05% in thermoplastic PU composites. Even though the ILSS showed thermoplastic-based composites having lower values compared to thermoset based composites, the thermoplastic PU composites impart inelastic deformation without any trace of brittle fractures. In contrast, the CNTs reinforced epoxy composites showed inelastic deformation followed by brittle fracturing. The brittleness properties, on the other hand, decrease with a higher concentration of MWCNTs due to the crack bridging effect of the CNTs. Table 3 shows the comparison of CNTs and other carbon-based reinforcement materials in polymer composites on mechanical strength.

The amount of CNTs plays a vital role in the mechanical properties of nanocomposites. Yazik et al. [84] investigated the effect of MWCNTs on the mechanical properties of shape memory epoxy (SMEP) nanocomposites. Accordingly, it can be seen that the increment in the tensile properties of nanocomposites could be achieved with the addition of low filler content of CNTs, which is around 0.5 wt.%. Notably, the improvement in tensile strength can be attributed to the high surface area of nanofillers that provide more efficient stress transfer, thus strengthening the materials.

Table 3. Mechanical properties of various carbon-based as reinforcement materials in polymer composites.

Reinforcement Materials	Matrix	Mechanical Strength					Ref.
		Tensile Strength (MPa)	Flexural Strength (MPa)	Impact Strength (J/m)	Elastic Modulus (GPa)	Hardness (GPa)	
CB	PVC	35 (−34%)	-	-	-	-	[85]
CB	PP	25 (−47%)	-	-	0.25 (−23%)	-	[86]
CB	PP	60 (100%)	68 (70%)	56 (65%)	4.2 (68%)	-	[87]
CB	Epoxy	58 (190%)	90 (125%)	-	2.6 (200%)	-	[88]
CB	Unsaturated polyester	40 (−14%)	72 (−25%)	-	1.3 (80%)	0.17 (17%)	[89]
CB	NBR/EPDM	16.7	-	-	-	-	[90]
Carbon fabric	Epoxy	580	-	-	67.5	-	[91]
MLG	PVC	19 (17%)	-	-	6 (1%)	-	[92]
Graphene	PVC	55 (130%)	-	-	2 (58%)	-	[93]
Graphite	PS	29 (16%)	-	21 (−28%)		-	[94]
Graphite	POBDS	NA	42.5 (0%)	-	-	-	[95]
Graphene oxide	PMMA	180 (−18%)	-	-	8 (−33%)	-	[96]
Graphene sheets	PS	40 (60%)	-	-	2.25 (50%)	-	[97]
Graphite	Epoxy	41 (21%)	-	-	3.3 (10%)	-	[98]
Graphene	PVC	140 (8%)	-	-	5.3 (10%)	-	[24]
MWCNTs	PVC	NA	-	-	NA	-	[24]
MWCNTs	Epoxy	-	105 (110%)	-	-	-	[20]
MWCNTs	Epoxy	52.4	-	-	3.23	-	[3]

Table 3. Cont.

Reinforcement Materials	Matrix	Mechanical Strength					Ref.
		Tensile Strength (MPa)	Flexural Strength (MPa)	Impact Strength (J/m)	Elastic Modulus (GPa)	Hardness (GPa)	
MWCNTs	Epoxy	85.6 (13%)	121.6 (0.7%)	23.4 (60%)	2.9 (10%)	-	[34]
MWCNTs	Epoxy	720 (16%))	-	-	54 (4%)	-	[31]
CNTs	Epoxy	-	-	-	9 (−18%)	-	[28]
NBCNTs	PVC	29.5 (−5%)	-	-	0.35 (0%)	-	[33]
MWCNTs	PVC	28 (−9%)	-	-	0.3 (−14%)	-	[33]
MWCNTs	P(MMA-co-EMA)	74 (57%)	-	-	2.3 (130%)	-	[13]
MWCNTs	PMMA	25 (0%)	-	-	2 (33%)	-	[19]
MWCNTs	PS	16 (0%)	-	-	1.5 (36%)	-	[19]
MWCNTs	PS	30.6 (36%)	-	-	3.4 (122%)	-	[10]
CNTs	PP	24 (71%)	34 (35%)	155 (34%)	-	-	[23]
CNTs	Epoxy	1300 (24%)	1078 (10%)	-	-	-	[29]
Amino-CNTs	Epoxy	370 (37%)	225 (80%)	-	8 (33.3%)	-	[25]
MWCNTs	Epoxy	535.4 (4%)	-	-	-	-	[32]
MWCNTs	HDPE	-	-	-	4.7 (47%)	0.1 (15%)	[26]
MWCNTs	PP	35 (25%)	-	4 (54%)	0.8 (23%)	-	[22]
MWCNTs	PA	65.9 (8.2%)	-	-	-	-	[15]
MWCNTs	PMMA	60 (20%)	-	1.3 (−36%)	-	-	[30]
DWCNTs	Magnesia	-	-	-	-	12.2	[21]
CNTs	Epoxy	-	-	-	3.7 (19%)	-	[7]
MWCNTs	Epoxy	6 (500%)	-	-	0.5 (290%)	-	[9]
MWCNTs	Nylon 6	40.3 (124%)	-	-	0.9 (115%)	-	[14]
MWCNTs	Nylon 6	59.3 (70%)	-	-	3.6 (90%)	100 (67%)	[16]
SWCNTs	Alumina	-	-	-	-	16.1 (−21%)	[11]
SWCNTs	Chitosan	-	-	-	8 (25%)	-	[27]
CNTs	Aluminium	520 (33%)	-	-	103 (41%)	1.3 (30%)	[17]

CB—carbon black, MLG—multi-layer graphene, CF—carbon fibre, SPS—sugar palm starch, NBR/EPDM—acrylonitrile-butadiene/ethylene-propylene-diene rubber blends, PDMS—polydimethylsiloxane, PS—polystyrene, POBDS—poly (4,4′-oxybis (benzene) disulfide), PA—polyamide 6, NA—non-applicable.

When the CNTs content was increased to 1.5 wt.%, the tensile strength value dropped due to agglomeration that occurred at higher filler content. They found out that the higher MWCNTs content caused poor interfacial adhesion between the polymer and the MWCNTs, which caused aggregations and lumping of the nanofillers [99,100]. This led to a stress concentration area and disrupted the wetting of the nanofillers by epoxy, thus preventing the stress transfer of epoxy to nanomaterials. In addition, the flexural strength of the nanocomposite was also improved significantly by 176% with the addition of 1 wt.% of MWCNTs into the SMEP matrix compared to neat SMEP. The presence of higher dispersion of CNTs inside the SMEP matrix inhibits the mobility of the polymer chain under flexural load [101]. Moreover, the uniform dispersion of CNTs filler provided a uniform distribution of stress, hence, reduced the sites of stress concentrations in the SMEP matrix.

Zakaria et al. [102] analysed the influence of SWCNTs and single-layer graphene (SLG) as reinforcing nanofillers on the mechanical properties of epoxy nanocomposites. Different filler loadings of SWCNTs and SLG (0 wt.%, 0.1 wt.% and 0.5 wt.%) were used in this experimental work. The results showed an improvement in the mechanical performance of epoxy nanocomposites with both SWCNTs and SLG fillers compared with the undoped epoxy matrix. The composites' tensile strength and modulus increased by 14% and 21%, respectively, when 0.5 wt.% SWCNTs were added, which was attributed to several factors, including cross-linking interactions that enhanced the polymer to nanofiller interactions. Interestingly, the SWCNTs/epoxy nanocomposites showed higher tensile strength and modulus as compared with SLG/epoxy nanocomposites. The tensile strength of SWCNTs-based nanocomposites was higher than that of the SLG-based nanocomposite, as the SWCNTs filler has a high filler length and aspect ratio. SWCNTs-based nanocomposite with 0.5 wt.% SWCNT displayed the highest tensile strength and tensile modulus of 49.07 MPa and 1.70 GPa, respectively, as compared with SLG-based nanocomposite with 0.5 wt.% SLG with 48.01 MPa and 1.62 GPa, respectively. An increment of about 2% and 5% in tensile strength and modulus value of SWCNT-based nanocomposite is higher than that of SLG-based nanocomposite. The enhancement is easily explained by the properties of SWCNTs, which have higher dispersion and different shapes of filler than SLG. In the case of SWCNTs, some of the wire-like structures of the SWCNTs show twists and kinks which could prevent the detachment of the SWCNTs from the epoxy matrix. Meanwhile, for SLG, the crumpled and wrinkled thin film of the SLG structure seemed to detach more easily from the epoxy matrix compared with the SWCNTs structure. Therefore, SWCNTs was able to be dispersed more effectively in the epoxy matrix than SLG. Furthermore, the weak interaction of SLG-based nanocomposites than SWCNT nanocomposites could be because of van der Waals forces acting between the adjacent SLG, resulting in lower tensile strength and modulus value of SLG nanocomposites than SWCNT nanocomposites [103].

Sapiai et al. [104] reported on the mechanical properties of functionalised CNTs added to kenaf-reinforced epoxy composites. The tensile, flexural, and impact properties of the kenaf/epoxy composite were strengthened by 43.30%, 21.10%, and 130%, respectively, when 1 wt.% acid-silane-treated CNTs (ACNTs) were included. The mechanical study indicates that the composite with 1 wt.% acid silane-treated CNTs loading exhibited the best value mechanical performance. With increasing ACNTs filler contents of 0.5 wt.%, 0.75 wt.% and 1.0 wt.%, the ACNTs/kenaf/epoxy composites demonstrate increments of 0.08%, 0.76% and 8.66% in flexural strength compared to the unfilled kenaf composites. It was concluded that acid and silane treatment on CNT surfaces increased the flexural strength and modulus because the acid and silane treatment process aided in functionalising the CNTs surfaces. This is because the existence of the –COOH and Si–OH groups had improved CNTs surfaces by enhancing dispersibility and reducing agglomeration of CNTs in the epoxy matrix. Moreover, the impact strength continued to increase for 0.5%, 0.75% and 1.0% of CNTs kenaf/epoxy composites where the increments observed were about 84.12%, 86.51% and 130%, respectively. The ability of CNTs to absorb more impact energy compared to the epoxy matrix contributes to the remarkable improvement in value in impact strength. Therefore, the toughness of the material could be further improved with more energy absorbed by the material.

A comparison of bamboo/CNT reinforced epoxy hybrid composite and alkali-treated bamboo epoxy composite was conducted by Kushwaha et al. [105]. The functional groups which are formed on the CNTs surface had improved the interfacial bonding between the CNTs and the surrounding matrix. CNTs addition results in an improvement in the interfacial bonding by giving rise to additional sites of mechanical interlocking that facilitate load transfer. The formation of covalent bonds between the CNTs and epoxy resin facilitates load transfer between the CNTs and epoxy matrix and contributes to the improvement in the mechanical properties of the composites. Remarkably, there was a significant increase in impact strength by 84.5% due to the flexibility of the interface molecular chain, resulting in comparatively greater energy absorption.

5. Potential Applications of CNTs

Carbon-based nanofillers reinforced polymer composites have gained popularity for a variety of applications due to their superior properties [38]. The varied applications of these polymer nanocomposites rely on the superior properties possessed by the CNTs themselves. Furthermore, the good compatibility of CNTs with polymer matrices has increased the potential of these materials for being used in a variety of advanced applications, such as electronics, automotive, textiles, aerospace, sports equipment, sensors, energy storage devices, filters [4,106–110].

Polymer nanocomposites reinforced with CNTs have also been reported as an excellent choice for the fabrication of ballistic armour materials, owing to their outstanding stiffness and strength, large fracture resistance, light density, and high energy absorption, which increases their potential for use in body armour [111]. When a bullet hits body armour, the material's fibres absorb and distribute the impact energy to subsequent layers so that the bullet does not penetrate through the body armour. However, blunt force trauma or non-penetrating injuries may still be caused by dissipation forces. The collision and resultant trauma will cause severe damage and injure critical organs, even when the bullet is stopped by the body armour. Thus, a high degree of elastic storage energy should be used as the ideal material for body armour which causes the bullet to be rebuffed or deflected. According to Benzait et al. [111], polymer-reinforced CNTs are an excellent choice for ballistic armour materials due to their remarkable stiffness and strength, low density, large fracture resistance and high energy absorption. The findings reported by Hanif et al. [112] on the influence of CNTS inclusion on the fracture and ballistic resistance in twaron/epoxy composite panels support this statement. The study revealed that with only 1 wt.% addition of MWCNTs, they were able to significantly improve fracture toughness and ballistic resistance with increased impact energy absorption value. Another study conducted by Mylvaganam and Zhang [113] found the highest ballistic resistance capacity of a CNTs is when the bullet hits its centre and a larger tube withstands a higher bullet speed. They also fabricated a body armour made of six layers of 100 μm nanotube yarn with a thickness of 600 μmin that could bounce off a bullet with the muzzle energy of 320 J. A study led by Han and Elliott [114] conducted a study on classical molecular dynamics simulations of model polymer/CNT composites constructed by embedding a single wall (10, 10) CNT into two different amorphous polymers matrices. They found out that it is possible to use CNTs to mechanically reinforce an appropriate polymer matrix, especially in the longitudinal direction of the nanotube. Other literature reports on dynamic molecular simulation studies conducted of CNTs-reinforced polymer composites are those of Zhang and Shen [115], Chang [116], Ni et al. [117], Shen et al. [118], Fan et al. [119] and Lin et al. [120]. Figure 4 displays the molecular dynamics model of a CNT subjected to ballistic impact.

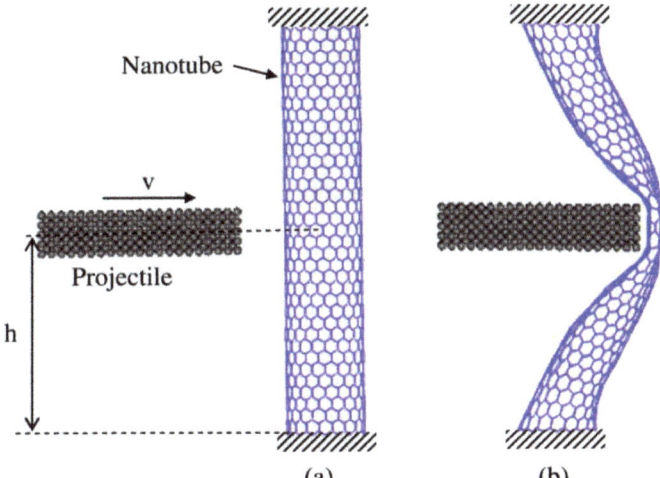

Figure 4. The molecular dynamics model of a CNT subjected to ballistic impact. (**a**) Initial model, (**b**) a deformed (18, 0) nanotube at its maximum energy absorption. Reproduced with permission from [113]. Copyright IOP Publishing, 2007.

Recently, the development of CNTs-based nanocomposites for biomedical applications has been reported, particularly in tissue engineering and drug delivery [60]. The unique graphitic structure and the superior performance of CNTs for their mechanical, electrical, optical and biological characteristics have allowed them to be used in biomedical field applications like gene/drug delivery and tissue engineering. According to Huang et al.'s review paper [121], researchers have documented the use of CNTs as substrates for neuronal tissue engineering because CNTs can assist neuron attachment, allow the generation of longer and more elaborate neuritis, as well as promote cell differentiation. CNTs- based polymer composites are also employed in the formation of bone scaffolding materials. Tanaka et al. [122], for instance, employed the 3D block structure of CNTs to study their efficacy as scaffold materials for bone repair. They found the CNTs scaffolds for cell adhesion as compared to PET reinforced collagen scaffolds with good osteogenesis behaviour, as shown in Figure 5. Other than that, CNTs have also been considered to serve as drug and gene delivery carriers. Their easy surface functionalisation has prompted their use to deliver different genes, including plasmid DNA (PDNA), micro-RNA, and small infecting RNA as gene delivery vectors for various diseases for instance, cancers [123].

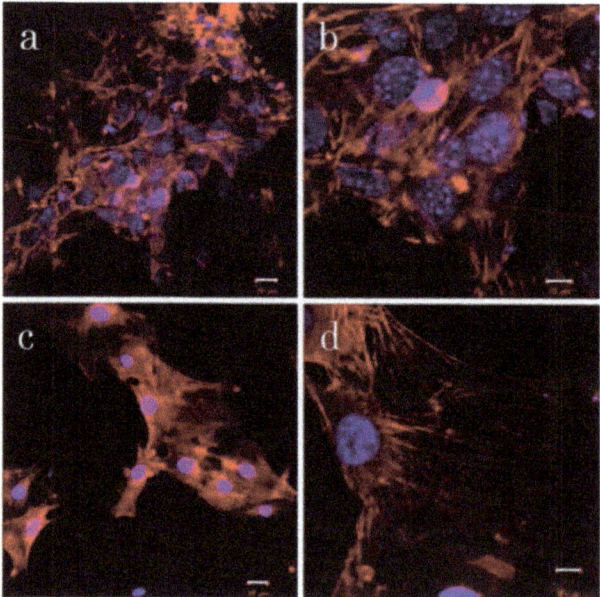

Figure 5. Fluorescence photomicrographs of cell cultures on (**a**,**b**) PET reinforced collagen sheets and (**c**,**d**) MWCNTs blocks. Reproduced from [122].

CNTs are ideal materials for gas sensors due to their inherent characteristics such as high porosity and high specific surface area [124]. The main concern with the burning of fossil fuels is toxic gas emissions. The identification of these gases is crucial for saving the environment and humans from the dangers posed by the gases generated by the combustion of fossil fuels. In consideration of gas sensor applications, the physical and chemical characteristics of CNTs were discussed critically in many works [125–130]. Some metallic nanoparticles such as Pd, Pt, Au, Ag, Rh, Pb, and Sn have catalytic properties and allow for the specific binding of gas molecules. Variations in the barrier potential of CNTs-metal contact or CNTs-CNTs junctions cause changes in CNT resistance in defect free CNTs. The gases released during the combustion of fossil fuels, such as CO_2 [131–133], CO [134–137], SO_2 [138–147], NO_2 [136,148–155], and NO [156,157], adsorb on the CNT surface either physically or chemically. Figure 6 shows the bonding behaviour and charge transfer between CNTs and the molecules of C–O. The H atom of functionalised O-H modified CNTs bonds to the electronegative oxygen of carbon monoxide. During the purification procedures, OH groups are attached to CNTs to remove the contaminants.

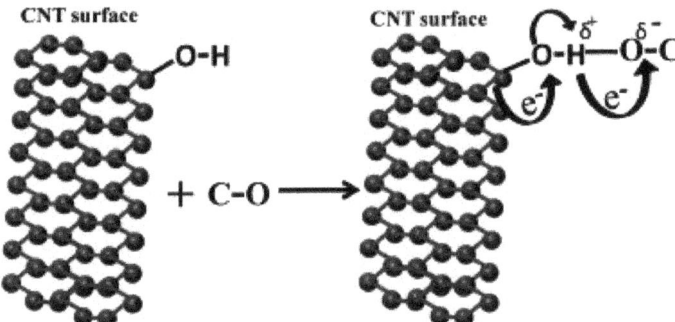

Figure 6. Adsorption of CO gas molecule on the hydroxyl modified CNTs. Reproduced from [158].

Several other advanced applications of CNTs-based composites have also been reported in the automotive, aerospace, marine and sporting goods industries. The potential of these materials to be applied in the aforementioned advanced applications can be improved by hybridising the CNTs with other materials, including natural fibres [45,110,159–163]. For example, CNT-polymer composites have been applied to the production of vehicles with the goal of reducing the weight of the body parts, which allows the vehicle to have lower fuel consumption and minimise global warming effects by reducing carbon dioxide emissions. Yang et al. [164] discovered that a 25% reduction in vehicle weight can reduce up to 250 million barrels of crude per year. Therefore, many car manufacturers have employed CNTs-based composites in vehicle parts, including trunk lids, car seats, dashboard coverings, and roofs.

The CNTs-based polymer composite applications in the automobile industry include advancements in current technology such as in body components, electrical systems, and engine parts. The addition of CNTs reinforced with fibreglass in epoxy composites could increase the strength and impact energy by 60% and 30%, respectively. This would subsequently contribute to a reduction in fuel consumption and greenhouse gas emissions by 16% and 26% [165]. The next use of CNTs is for a bendable or flexible battery that is produced by applying an ink-coated sheet of paper or plastic to a CNTs/Ag nanowire-infused substrate. This battery is adaptable to many vehicle applications because of its potential use in portable and wearable electronics. CNTs offer many potential benefits due to their advantages, including high electrical conductivity, the unique structure of 1D nanoscale, suitable surface chemical properties, high degree of graphitisation, and superior electrical performance, which may play a key role in the development of high-performance flexible batteries [166]. CNTs are also found in vehicles tires, which are surrounded by a matrix of polybutadiene and styrene-butadiene rubber (SBR), which are employed as colouring and reinforcing agents during tire production. Andrews et al. [167] used CNTs thin-film transistors (CNTs-TFT) as a tool for sensing environmental pressure on the tire. Shao et al. [168] found that CNTs-filled in passenger tire tread compounds have been shown to offer better handling and traction properties, making them ideal for racing and sports vehicle tires.

Extensive research on the potential of CNTs in the aerospace industry has been conducted to produce composite materials as very high strength and durability aircraft components. The incorporation of CNTs into complicated aircraft designs creates lightweight, minimal cost materials for engines and components, as well as reduced waste in the production processes [169]. The vibration damping factors of the polymer nanocomposites with CNTs sheet reinforcement were found to be significantly reduced, with an enhancement in mechanical, electrical, and thermal characteristics of the MWCNTs composite for structural aerospace applications [170]. Venkatesan et al. [171] observed reductions in coefficient of friction of wear properties in glass fibre hybrid CNT-based composites as a result of the combination of polymer resins in an aluminium–titanium–magnesium matrix, which represent an alternative for passive thermal coverings. A study by Kwon et al. [172] successfully fabricated well-dispersed CNT-based aluminium matrix composites using ball milling and hot pressing processes. In this work, they discovered that the hardness of the CNT-Al composites was significantly enhanced about seven times compared to pure aluminium. Another interesting study in the aerospace application was performed by Laurenzi et al. [173] emphasising the effect of varying loadings of SWCNT and GO nanoplatelets on the equivalent dose received by the nanocomposites in various radiation fields in space, as well as numerical analysis that showed how atoms in nanomaterial formations were arranged. It was noted that CNTs and GNPs suspended in an epoxy matrix decreased the impact damage produced by micrometeoroid orbital debris (MMOD), and the loading and radiation shielding were improved with the addition of GO fillers. Thus, CNTs and GNPs were used to make sensors for aerospace applications. It also recorded a reduction of about 18% in weight and 2.4% in neutron production of radiation shielding spacesuit applications produced from improved MWCNT embedded in PMMA matrix [174]. Furthermore, the electromagnetic interference (EMI) shielding efficacy of MWCNT/polypropylene composites increased as CNTs content and shielding plate thickness increased, demonstrating the efficiency of the CNTs nanocomposites as a heat-absorbing media in the aerospace industry [175].

6. Environmental, Health, and Safety Concerns in Utilisation of CNTs

The toxicity, health and safety concerns of CNTs are influenced by several factors, such as aspect ratio, length, surface area, degree of aggregation, purity, and concentration or loading [176]. According to Donaldson et al. [177], repeated exposure of CNTs over a long period may contribute to some common diseases associated with asbestos exposure that has a high mortality burden, triggering global pandemics in the 20th century. Chronic inflammation, formation of granuloma, and fibrosis are among those common anticipated diseases from CNTs persistence [178].

6.1. Aspect Ratio

The fact that CNTs have smaller aspect ratios than other reinforcing fillers like carbon fibres, carbon blacks, and clay, means they have better compatibility with the polymer matrix, due to the formation of uniformity of CNTs in the composite's matrix. Other than uniformity concerns, an international standard regarding the allowance of inhalation of respirable fibre into the lungs has been highlighted by the World Health Organisation (WHO). Only CNTs with a length greater than 5 μm and a diameter of less than 3 μm with a minimum aspect ratio of 3:1 are accepted to be inhaled into the lungs. Otherwise, the large aspect ratio of CNTs affects their behaviour in which they are more difficult to be engulfed and cleared off from the site of deposition of targeted organs due to their propensity to aggregate and form bundle structures of CNTs [179]. Consequently, prolonged exposure to bundle pathogenic CNTs causes bronchogenic carcinoma, mesothelioma, asbestosis, pleural fibrosis, and pleural plaque which cause the pleural pathologies in the end [177].

The standard of occupational exposure limit values (OELs) has been established as legislation applicable to handling nanomaterials to ensure health and safety protection during exposure to CNTs to the environment. Table 4 presents the information OELs for nanomaterials [180]. In detailed findings, the United States of America-National Institute for Occupational Health and Safety (NIOSH) recommends OELs for CNTs to be in the range of 1 to 50 μg/m^3 as an 8 hours' time weighted average-TWA μg/m^3 [181].

Table 4. OELs for nanomaterial handling.

Category	Benchmark Exposure Level
Fibrous, a high aspect ratio insoluble nanomaterial	0.01 fibres/mL
Any nanomaterial that is already classified in its molecular or in its larger particle form a as carcinogenic, mutagenic, reproductive, and sensitizing (CMRS) toxin	0.1 × OEL
Insoluble or poorly soluble nanomaterials not in the fibrous or CMRS categories	0.066 × OEL
Soluble nanomaterials not in the fibrous or CMRS categories	0.5 × OEL

6.2. Length

The relationships between the lengths of CNTs like MWCNTs in connection to pulmonary fibrosis have been investigated [44]. The result shows that long MWCNTs have higher detrimental pulmonary effects than MWCNTs. However, long CNTs cannot pass through the stomata and are retained, thus causing inflammation diseases. According to Poland et al. [182], the effect of short CNTs (<15 μm) through direct instillation of fibre into the pleural cavity of mice was investigated as compared to long CNTs with a length of 5 μm to 20 μm. With the long type of CNTs (>15 μm), significant inflammation leading to various cell damages could happen due to the disability of the long CNTs to be effectively engulfed by gathering macrophages, resulting in frustrated phagocytosis.

6.3. Surface Area

CNTs surface area has been pointed out as another critical aspect as a factor of toxicity. Kim et al. [183] investigated the toxicity of a nanomaterial to be highly affected by its physical properties, such as size distribution and surface area reactivity of particles. In bronchoalveolar lavage fluid (BALF) cell analysis, MWCNTs are found to induce more severe acute inflammatory cell recruitment than acid-treated multiwalled carbon nanotubes (tMWCNTs). This is due to the reduction in the size of the nanoscale increasing the surface area ratio of the materials. As a result, the potential for damage has increased, but this was not possible while they were in larger forms [184].

Considering the higher surface area and lower density of CNTs characteristics, these toxicant particles provide a higher contact area with biological structures, including gas exchanges across alveolar walls in which the total surface area of the alveoli may exceed 100 m^2. As demonstrated for high aspect ratio materials, this high surface area of CNTs often leads to pronounced biological activity [185]. In comparison with MWCNT, the toxicity of SWCNT was found to be 8.5-fold more fibrogenic than MWCNTs per microgram of dose, causing inflammation in the lungs, resulting in respiratory failure. Dong and Ma (2014) have shown that the lighter and larger surface area of SWCNT than MWCNT are the two factors contributing to the higher level of toxicity of SWCNT on an equal weight basis [179].

Volume per specific surface area is among the complementary criterion for exposure assessment and identification of potential risk [186]. Therefore, the surface of CNT requires modification to alter its toxic responses. With respect to that, the modification of the surface of CNT has been accomplished through the use of acid treatment. This technique is an effective modification by oxidising CNT to introduce carboxyl and hydroxyl groups on the surface of CNTs, resulting in changes in bioactivity and interaction with other molecules [179]. Carrero et al. [187] revealed that nitrogen-doped MWCNTs showed significantly reduced toxicity as well as better tolerance in exposed mice than pristine MWCNTs. In another study conducted by Taylor et al. [188], a thin film of aluminium oxide (Al_2O_3) coated MWCNTs induced lower fibrosis in mice as compared with pristine MWCNTs exposures.

6.4. Concentration

A compilation of several sets of literature of cell viability to interaction with different types and concentrations of functionalised SWCNTs (f-SWCNTs) and functionalised MWCNTs (f-MWCNTs) is presented in Table 5. Based on this review, the observation from tests on T-lymphocytes by Bottini et al. [189] found out that a safe dosage value of CNTs is around 40 µg/mL. Further, Bianco et al. [190] discovered death in 50% of HeLa (Henrietta Lacks) cells in culture after 6 h of incubation with increasing doses of f-SWCNTs and f-MWCNTs at a concentration of 5 mg/mL to 10 mg/mL. CNTs concentration ranging from six orders of magnitude (from 5 mg CNT/mL to 10 mg) could imply toxicity and resistance within the biological system [191]. Another related study was conducted on the negligible toxicity in the main organs (liver, lung and spleen) of exposed mice after intravenous exposure to CNTs of increasing concentration through constant malondialdehyde (MDA) levels for three months [192]. Results from the long-term accumulation and toxicity of intravenously injected SWCNTs indicate that slight inflammation and inflammatory cell infiltration occurred in the lungs. However, serum immunological indicators (CH 50 level and TNF-α level) remain unchanged and no apoptosis was found in the main organs.

Table 5. Compilation literature studies of toxicity cellular and tissue of different concentration and types of CNT.

Types of CNTs	Concentration	Biological System	Toxicity	Ref.
Plasmid DNA-SWCNTs and Plasmid DNA-MWCNTs	10 mg/mL	f-CNTs: HeLa cell lines in vitro	50% survival of HeLa cells	[193]
Pristine SWCNTs	7.5 µg/mL water	SWCNT: Mesothelioma cell line MSTO-211H in vitro	10% decrease in cell proliferation and activity	[194]
RNA-polymer SWCNTs conjugate	1 mg/mL	MCF-7 breast cancer cells in vitro	No significant cell damage	[195]
Pristine MWCNTs	40 µg/mL	Human T lymphocytes in vitro	No toxicity on human T lymphocytes	[189]
Ammonium chloride-SWCNTs, and poly(ethylene glycol)-SWCNTs	10 µg/mL water	Macrophages, B and T lymphocytes from BALB/c mice spleen and lymph nodes in vitro	5% decrease in viability of B lymphocytes, but no adverse effects on T lymphocytes and macrophages	[196]
125I-SWCNT-OH	1.5 µg/mouse	Intraperitoneal, intravenous, subcutaneous, in male KM mice in vivo	Accumulate in bone, but good biocompatibility	[197]
Streptavidin-SWCNT	0.025 mg/mL	HL60 and Jurkat cells in vitro	No adverse effects	[198]
SWCNTs dispersed in DMEM with 5% (vol/vol) fetal bovine serum	100 µg/mL	Human epithelial-like HeLa cells in vitro	No effect on growth rate	[199]
0.5 DMSO pristine SWCNTs	25 µg/mL	Human embryo kidney (HEK 293) cells in vitro	G1 cell arrest and apoptosis	[200]

Patlolla et al. [201] investigated hepatotoxicity and oxidative stress in male Swiss-Webster mice exposed to functionalised MWCNTs (f-MWCNTs) at different dosages. The investigation aims to assess the effects, after intraperitoneal (ip) injection, of f-MWCNT on various hepatotoxicity and oxidative stress biomarkers. The mice were dosed at 0.25 mg/kg/day, 0.5 mg/kg/day, and 0.75 mg/kg/day for 5 days. The results show a short-term and high toxicity in mice exposed to f-MWCNTs were recorded and ROS induction, increase in the level of LHP, serum biochemical changes, and damage to the liver tissue were observed. The result indicates that the f-MWCNT induces hepatotoxicity. The authors also suggested that the high toxicity of f-MWCNTs does not imply that they should be banned for biomedical applications, but rather improving the dispersion and excretion of MWCNTs by further chemical modification is essential for safe occupational and environmental exposure to nanomaterials.

7. Conclusions and Future Perspectives

In this review, the mechanical performance of CNTs-reinforced polymer composites has been discussed. In essence, CNTs have excellent chemical and physical properties that make them ideal and promising reinforcements in polymer composites. Based on existing studies, it has been acknowledged that the mechanical properties of the CNTs polymer composites are influenced by the interactions between the nanofillers and the polymer matrices. The challenge is mainly the tendency of the CNTs to agglomerate, resulting in poor dispersion properties, which can deteriorate the whole performance of the composite structures. Researchers have come up with various methods for distributing and orienting the CNTs. Further, it has been found that dispersing a small amount of filler in the polymer matrix enhances the properties of the composites. Though many excellent CNT composites have been achieved, constant progress is needed to obtain composites with the best performance. Several aspects, such as the number of CNTs used, size of fillers, spatial distribution and orientation, suitable surface modifications on CNTs surface, and methods of fabrications, affect the mechanical properties of the composites. It is crucial to find an optimum balance between these parameters. Therefore, addressing all the concerns raised will be fascinating to study in the forthcoming investigation into utilising the potential of CNTs in polymer composites.

Author Contributions: Conceptualization, N.M.N.; validation, S.M.S., A.K. and E.S.Z.; writing—original draft preparation, N.M.N., S.H.K., M.M.H., M.R.M.A., M.N.F.N., M.R., R.A.I., H.A.A., F.A.S.; supervision, N.A., S.M.S., A.K. and E.S.Z.; project administration, H.A.A.; funding acquisition, A.K. All authors have read and agreed to the published version of the manuscript.

Funding: This research was funded by Higher Education Center of Excellence (HICoE), Ministry of Higher Education, Malaysia (Grant number 6369109).

Institutional Review Board Statement: Not applicable.

Informed Consent Statement: Not applicable.

Data Availability Statement: Not applicable.

Acknowledgments: The authors gratefully acknowledge the technical and financial support from the Universiti Putra Malaysia (UPM).

Conflicts of Interest: The authors declare no conflict of interest.

References

1. Norizan, M.N.; Moklis, M.H.; Demon, S.Z.N.; Halim, N.A.; Samsuri, A.; Mohamad, I.S.; Knight, V.F.; Abdullah, N. Carbon nanotubes: Functionalisation and their application in chemical sensors. *RSC Adv.* **2020**, *10*, 43704–43732. [CrossRef]
2. Lee, J. Carbon Nanotube-Based Membranes for Water Purification. In *Nanoscale Materials in Water Purification*; Elsevier: Amsterdam, The Netherlands, 2019; pp. 309–331.
3. Zhang, J.; Jiang, D. Interconnected multi-walled carbon nanotubes reinforced polymer-matrix composites. *Compos. Sci. Technol.* **2011**, *71*, 466–470. [CrossRef]
4. Nurazzi, N.M.; Asyraf, M.R.M.; Khalina, A.; Abdullah, N.; Sabaruddin, F.A.; Kamarudin, S.H.; Ahmad, S.; Mahat, A.M.; Lee, C.L.; Aisyah, H.A.; et al. Fabrication, functionalization, and application of carbon nanotube-reinforced polymer composite: An overview. *Polymers* **2021**, *13*, 1047. [CrossRef]
5. Norizan, M.N.; Zulaikha, N.D.S.; Norhana, A.B.; Syakir, M.I.; Norli, A. Carbon nanotubes-based sensor for ammonia gas detection–an overview. *Polimery* **2021**, *66*, 175–186. [CrossRef]
6. Nurazzi, N.M.; Harussani, M.M.; Siti Zulaikha, N.D.; Norhana, A.H.; Imran Syakir, M.; Norli, A. Composites based on conductive polymer with carbon nanotubes in DMMP gas sensors—An overview. *Polimery* **2021**, *66*, 85–97. [CrossRef]
7. Schadler, L.S.; Giannaris, S.C.; Ajayan, P.M. Load transfer in carbon nanotube epoxy composites. *Appl. Phys. Lett.* **1998**, *73*, 3842–3844. [CrossRef]
8. Shaffer, M.S.P.; Windle, A.H. Fabrication and characterization of carbon nanotube/poly (vinyl alcohol) composites. *Adv. Mater.* **1999**, *11*, 937–941. [CrossRef]

9. Allaoui, A.; Bai, S.; Cheng, H.-M.; Bai, J.B. Mechanical and electrical properties of a MWNT/epoxy composite. *Compos. Sci. Technol.* **2002**, *62*, 1993–1998. [CrossRef]
10. Safadi, B.; Andrews, R.; Grulke, E.A. Multiwalled carbon nanotube polymer composites: Synthesis and characterization of thin films. *J. Appl. Polym. Sci.* **2002**, *84*, 2660–2669. [CrossRef]
11. Zhan, G.-D.; Kuntz, J.D.; Wan, J.; Mukherjee, A.K. Single-wall carbon nanotubes as attractive toughening agents in alumina-based nanocomposites. *Nat. Mater.* **2003**, *2*, 38–42. [CrossRef] [PubMed]
12. Hsiao, K.-T.; Alms, J.; Advani, S.G. Use of epoxy/multiwalled carbon nanotubes as adhesives to join graphite fibre reinforced polymer composites. *Nanotechnology* **2003**, *14*, 791. [CrossRef]
13. Yang, J.; Hu, J.; Wang, C.; Qin, Y.; Guo, Z. Fabrication and characterization of soluble multi-walled carbon nanotubes reinforced P (MMA-co-EMA) composites. *Macromol. Mater. Eng.* **2004**, *289*, 828–832. [CrossRef]
14. De Zhang, W.; Shen, L.; Phang, I.Y.; Liu, T. Carbon nanotubes reinforced nylon-6 composite prepared by simple melt-compounding. *Macromolecules* **2004**, *37*, 256–259. [CrossRef]
15. Zhao, C.; Hu, G.; Justice, R.; Schaefer, D.W.; Zhang, S.; Yang, M.; Han, C.C. Synthesis and characterization of multi-walled carbon nanotubes reinforced polyamide 6 via in situ polymerization. *Polymer* **2005**, *46*, 5125–5132. [CrossRef]
16. Chen, G.-X.; Kim, H.-S.; Park, B.H.; Yoon, J.-S. Multi-walled carbon nanotubes reinforced nylon 6 composites. *Polymer* **2006**, *47*, 4760–4767. [CrossRef]
17. Deng, C.F.; Wang, D.Z.; Zhang, X.X.; Li, A.B. Processing and properties of carbon nanotubes reinforced aluminum composites. *Mater. Sci. Eng. A* **2007**, *444*, 138–145. [CrossRef]
18. Broza, G.; Piszczek, K.; Schulte, K.; Sterzynski, T. Nanocomposites of poly (vinyl chloride) with carbon nanotubes (CNT). *Compos. Sci. Technol.* **2007**, *67*, 890–894. [CrossRef]
19. Mathur, R.B.; Pande, S.; Singh, B.P.; Dhami, T.L. Electrical and mechanical properties of multi-walled carbon nanotubes reinforced PMMA and PS composites. *Polym. Compos.* **2008**, *29*, 717–727. [CrossRef]
20. Ramana, G.V.; Padya, B.; Kumar, R.N.; Prabhakar, K.V.P.; Jain, P.K. Mechanical properties of multi-walled carbon nanotubes reinforced polymer nanocomposites. *Indian J. Eng. Mater. Sci.* **2010**, *17*, 331–337.
21. Peigney, A.; Garcia, F.L.; Estournes, C.; Weibel, A.; Laurent, C. Toughening and hardening in double-walled carbon nanotube/nanostructured magnesia composites. *Carbon N. Y.* **2010**, *48*, 1952–1960. [CrossRef]
22. Bikiaris, D. Microstructure and properties of polypropylene/carbon nanotube nanocomposites. *Materials* **2010**, *3*, 2884–2946. [CrossRef]
23. Rahmanian, S.; Thean, K.S.; Suraya, A.R.; Shazed, M.A.; Salleh, M.A.M.; Yusoff, H.M. Carbon and glass hierarchical fibers: Influence of carbon nanotubes on tensile, flexural and impact properties of short fiber reinforced composites. *Mater. Des.* **2013**, *43*, 10–16. [CrossRef]
24. Hasan, M.; Lee, M. Enhancement of the thermo-mechanical properties and efficacy of mixing technique in the preparation of graphene/PVC nanocomposites compared to carbon nanotubes/PVC. *Prog. Nat. Sci. Mater. Int.* **2014**, *24*, 579–587. [CrossRef]
25. Garg, M.; Sharma, S.; Mehta, R. Pristine and amino functionalized carbon nanotubes reinforced glass fiber epoxy composites. *Compos. Part A Appl. Sci. Manuf.* **2015**, *76*, 92–101. [CrossRef]
26. Rajeshwari, P. Microstructure and mechanical properties of multiwall carbon nanotubes reinforced polymer composites. *Mater. Today Proc.* **2015**, *2*, 3598–3604. [CrossRef]
27. Venugopal, G.; Veetil, J.C.; Raghavan, N.; Singh, V.; Kumar, A.; Mukkannan, A. Nano-dynamic mechanical and thermal responses of single-walled carbon nanotubes reinforced polymer nanocomposite thinfilms. *J. Alloys Compd.* **2016**, *688*, 454–459. [CrossRef]
28. Tarfaoui, M.; Lafdi, K.; El Moumen, A. Mechanical properties of carbon nanotubes based polymer composites. *Compos. Part B Eng.* **2016**, *103*, 113–121. [CrossRef]
29. Zhao, Z.; Teng, K.; Li, N.; Li, X.; Xu, Z.; Chen, L.; Niu, J.; Fu, H.; Zhao, L.; Liu, Y. Mechanical, thermal and interfacial performances of carbon fiber reinforced composites flavored by carbon nanotube in matrix/interface. *Compos. Struct.* **2017**, *159*, 761–772. [CrossRef]
30. Deep, N.; Mishra, P. Evaluation of mechanical properties of functionalized carbon nanotube reinforced PMMA polymer nanocomposite. *Karbala Int. J. Mod. Sci.* **2018**, *4*, 207–215. [CrossRef]
31. Boroujeni, A.Y.; Al-Haik, M. Carbon nanotube–Carbon fiber reinforced polymer composites with extended fatigue life. *Compos. Part B Eng.* **2019**, *164*, 537–545. [CrossRef]
32. Han, K.; Zhou, W.; Qin, R.; Wang, G.; Ma, L.-H. Effects of carbon nanotubes on open-hole carbon fiber reinforced polymer composites. *Mater. Today Commun.* **2020**, *24*, 101106. [CrossRef]
33. Vanyorek, L.; Sikora, E.; Balogh, T.; Román, K.; Marossy, K.; Pekker, P.; Szabó, T.J.; Viskolcz, B.; Fiser, B. Nanotubes as polymer composite reinforcing additive materials–A comparative study. *Arab. J. Chem.* **2020**, *13*, 3775–3782. [CrossRef]
34. Su, C.; Wang, X.; Ding, L.; Yu, P. Enhancement of mechanical behavior of resin matrices and fiber reinforced polymer composites by incorporation of multi-wall carbon nanotubes. *Polym. Test.* **2021**, *96*, 107077. [CrossRef]
35. Vidu, R.; Rahman, M.; Mahmoudi, M.; Enachescu, M.; Poteca, T.D.; Opris, I. Nanostructures: A platform for brain repair and augmentation. *Front. Syst. Neurosci.* **2014**, *8*. [CrossRef]
36. Norizan, M.N.; Abdullah, N.; Demon, S.Z.N.; Halim, N.A.; Azmi, A.F.M.; Knight, V.F.; Mohamad, I.S. The frontiers of functionalized graphene—Based nanocomposites as chemical sensors. *Nanotechnol. Rev.* **2021**, *10*, 330–369. [CrossRef]

37. Nurazzi, N.M.; Abdullah, N.; Demon, S.Z.N.; Halim, N.A.; Mohamad, I.S. The Influence of Reaction Time on Non-Covalent Functionalisation of P3HT/MWCNT Nanocomposites. *Polymers* **2021**, *13*, 1916. [CrossRef] [PubMed]
38. Ma, P.C.; Siddiqui, N.A.; Marom, G.; Kim, J.K. Dispersion and functionalization of carbon nanotubes for polymer-based nanocomposites: A review. *Compos. Part A Appl. Sci. Manuf.* **2010**, *41*, 1345–1367. [CrossRef]
39. Overney, G.; Zhong, W.; Tomanek, D. Structural rigidity and low frequency vibrational modes of long carbon tubules. *Z. Für Phys. D Atoms, Mol. Clust.* **1993**, *27*, 93–96. [CrossRef]
40. Wang, C.Y.; Zhang, L.C. A critical assessment of the elastic properties and effective wall thickness of single-walled carbon nanotubes. *Nanotechnology* **2008**, *19*. [CrossRef]
41. Poncharal, P.; Wang, Z.L.; Ugarte, D.; De Heer, W.A. Electrostatic deflections and electromechanical resonances of carbon nanotubes. *Science* **1999**, *283*, 1513–1516. [CrossRef]
42. Wong, E.W.; Sheehan, P.E.; Lieber, C.M. Nanobeam mechanics: Elasticity, strength, and toughness of nanorods and nanotubes. *Science* **1997**, *277*, 1971–1975. [CrossRef]
43. Daniel, I.M.; Ishai, O.; Daniel, I.M.; Daniel, I. *Engineering Mechanics of Composite Materials*; Oxford University Press: New York, NY, USA, 2006.
44. Fredriksson, T. Carbon Nanotubes: A Theoretical Study of Young's Modulus. Ph.D. Thesis, Karlstad University, Karlstad, Sweden, 2014.
45. Nurazzi, N.M.; Khalina, A.; Sapuan, S.M.; Rahmah, M. Development of sugar palm yarn/glass fibre reinforced unsaturated polyester hybrid composites. *Mater. Res. Express* **2018**, *5*, 045308. [CrossRef]
46. Nurazzi, N.; Khalina, A.; Sapuan, S.; Laila, A.H.D.; Mohamed, R. Curing behaviour of unsaturated polyester resin and interfacial shear stress of sugar palm fibre. *J. Mech. Eng. Sci.* **2017**, *11*, 2650–2664. [CrossRef]
47. Kumar, S.; Sharma, K.; Dixit, A.R. A review on the mechanical properties of polymer composites reinforced by carbon nanotubes and graphene. *Carbon Lett.* **2021**, *31*, 149–165. [CrossRef]
48. Hassan, M.A. Physicaland Thermal Properties of Fiber (S-Type)-Reinforced Compositearaldite Resin (GY 260). *Al-Qadisiyah J. Eng. Sci.* **2012**, *5*, 341–346.
49. Marulanda, J.M. *Carbon Nanotubes Applications on Electron Devices*; InTech Open: London, UK, 2012.
50. Mazumdar, S. *Composites Manufacturing: Materials, Product, and Process Engineering*; CRC Press: Boca Raton, FL, USA, 2001.
51. Arash, B.; Wang, Q.; Varadan, V.K. Mechanical properties of carbon nanotube/polymer composites. *Sci. Rep.* **2014**, *4*, 1–8. [CrossRef] [PubMed]
52. Coleman, J.N.; Khan, U.; Gun'ko, Y.K. Mechanical reinforcement of polymers using carbon nanotubes. *Adv. Mater.* **2006**, *18*, 689–706. [CrossRef]
53. Jian, W.; Lau, D. Understanding the effect of functionalization in CNT-epoxy nanocomposite from molecular level. *Compos. Sci. Technol.* **2020**, *191*, 108076. [CrossRef]
54. Sánchez-Romate, X.F.; Martín, J.; Jiménez-Suárez, A.; Prolongo, S.G.; Ureña, A. Mechanical and strain sensing properties of carbon nanotube reinforced epoxy/poly (caprolactone) blends. *Polymer* **2020**, *190*, 122236. [CrossRef]
55. Sheth, D.; Maiti, S.; Patel, S.; Kandasamy, J.; Chandan, M.R.; Rahaman, A. Enhancement of mechanical properties of carbon fiber reinforced epoxy matrix laminated composites with multiwalled carbon nanotubes. *Fuller. Nanotub. Carbon Nanostruc.* **2020**, *29*, 1–7.
56. Kim, J.K.; Mai, Y.W. *Engineered Interfaces in Fiber Reinforced Composites*; Elsevier: Amsterdam, The Netherlands, 1998.
57. Thostenson, T.E.; Ren, Z.; Chou, T.W. Advances in the science and technology of carbon nanotubes and their composites: A review. *Compos. Sci. Technol.* **2001**, *61*, 1899–1912. [CrossRef]
58. Paramsothy, M. Dispersion, interface, and alignment of carbon nanotubes in thermomechanically stretched polystyrene matrix. *JOM* **2014**, *66*. [CrossRef]
59. Ruoff, R.S.; Tersoff, J.; Lorents, D.C.; Subramoney, S.; Chan, B. Radial deformation of carbon nanotubes by Van Der Waals forces. *Nature* **1993**, *364*. [CrossRef]
60. Venkataraman, A.; Amadi, E.V.; Chen, Y.; Papadopoulos, C. Carbon Nanotube Assembly and Integration for Applications. *Nanoscale Res. Lett.* **2019**, *14*, 220. [CrossRef]
61. Zhu, H.W.; Xu, C.L.; Wu, D.H.; Wei, B.Q.; Vajtai, R.; Ajayan, P.M. Direct synthesis of long single-walled carbon nanotube strands. *Science* **2002**, *296*. [CrossRef]
62. Lau, K.T.; Gu, C.; Hui, D. A critical review on nanotube and nanotube/nanoclay related polymer composite materials. *Compos. Part B Eng.* **2006**, *37*, 425–436. [CrossRef]
63. Wernik, J.M.; Meguid, S.A. On the mechanical characterization of carbon nanotube reinforced epoxy adhesives. *Mater. Des.* **2014**, *59*. [CrossRef]
64. Coleman, J.N.; Khan, U.; Blau, W.J.; Gun'ko, Y.K. Small but strong: A review of the mechanical properties of carbon nanotube-polymer composites. *Carbon N. Y.* **2006**, *44*, 1624–1652. [CrossRef]
65. Sattar, R.; Kausar, A.; Siddiq, M. Advances in thermoplastic polyurethane composites reinforced with carbon nanotubes and carbon nanofibers: A review. *J. Plast. Film Sheeting* **2015**, *31*, 186–224. [CrossRef]
66. Imtiaz, S.; Siddiq, M.; Kausar, A.; Muntha, S.T.; Ambreen, J.; Bibi, I. A Review Featuring Fabrication, Properties and Applications of Carbon Nanotubes (CNTs) Reinforced Polymer and Epoxy Nanocomposites. *Chin. J. Polym. Sci.* **2018**, *36*, 445–461. [CrossRef]

67. Bahun, G.J.; Wang, C.; Adronov, A. Solubilizing single-walled carbon nanotubes with pyrene-functionalized block copolymers. *J. Polym. Sci. Part A Polym. Chem.* **2006**, *44*. [CrossRef]
68. Singh, B.; Lohan, S.; Sandhu, P.S.; Jain, A.; Mehta, S.K. Functionalized carbon nanotubes and their promising applications in therapeutics and diagnostics. In *Nanobiomaterials in Medical Imaging: Applications of Nanobiomaterials*; William Andrew Publishing: Norwich, NY, USA, 2016.
69. Ajori, S.; Ansari, R.; Darvizeh, M. Vibration characteristics of single- and double-walled carbon nanotubes functionalized with amide and amine groups. *Phys. B Condens. Matter* **2015**, *462*. [CrossRef]
70. Afrin, R.; Shah, N.A. Room temperature gas sensors based on carboxyl and thiol functionalized carbon nanotubes buckypapers. *Diam. Relat. Mater.* **2015**, *60*. [CrossRef]
71. Janudin, N.; Abdullah, L.C.; Abdullah, N.; Yasin, F.M.; Saidi, N.M.; Kasim, N.A.M. Characterization of amide and ester functionalized multiwalled Carbon Nanotubes. *Asian J. Chem.* **2018**, *30*. [CrossRef]
72. Chen, J.; Yan, L.; Song, W.; Xu, D. Interfacial characteristics of carbon nanotube-polymer composites: A review. *Compos. Part A Appl. Sci. Manuf.* **2018**, *114*, 149–169. [CrossRef]
73. Liu, W.; Li, L.; Zhang, S.; Yang, F.; Wang, R. Mechanical properties of carbon nanotube/carbon fiber reinforced thermoplastic polymer composite. *Polym. Compos.* **2017**, *38*, 2001–2008. [CrossRef]
74. Harussani, M.M.; Sapuan, S.M.; Rashid, U.; Khalina, A. Development and Characterization of Polypropylene Waste from Personal Protective Equipment (PPE)-Derived Char-Filled Sugar Palm Starch Biocomposite Briquettes. *Polymers* **2021**, *13*, 1707. [CrossRef] [PubMed]
75. Thomason, J.L.; Yang, L. Temperature dependence of the interfacial shear strength in glass-fibre polypropylene composites. *Compos. Sci. Technol.* **2011**, *71*, 1600–1605. [CrossRef]
76. Sen, R.; Zhao, B.; Perea, D.; Itkis, M.E.; Hu, H.; Love, J.; Bekyarova, E.; Haddon, R.C. Preparation of single-walled carbon nanotube reinforced polystyrene and polyurethane nanofibers and membranes by electrospinning. *Nano Lett.* **2004**, *4*. [CrossRef]
77. Chen, W.; Tao, X. Self-organizing alignment of carbon nanotubes in thermoplastic polyurethane. *Macromol. Rapid Commun.* **2005**, *26*. [CrossRef]
78. Lu, X.D.; Huang, Y.D.; Zhang, C.H. Curing behaviour of epoxy resin with a diamine containing heterocyclic rings. *Polym. Polym. Compos.* **2007**, *15*. [CrossRef]
79. Uthaman, A.; Xian, G.; Thomas, S.; Wang, Y.; Zheng, Q.; Liu, X. Durability of an epoxy resin and its carbon fiber-reinforced polymer composite upon immersion in water, acidic, and alkaline solutions. *Polymers* **2020**, *12*, 614. [CrossRef] [PubMed]
80. Uthaman, A.; Lal, H.M.; Li, C.; Xian, G.; Thomas, S. Mechanical and water uptake properties of epoxy nanocomposites with surfactant-modified functionalized multiwalled carbon nanotubes. *Nanomaterials* **2021**, *11*, 1234. [CrossRef] [PubMed]
81. Herceg, T.M.; Yoon, S.H.; Abidin, M.S.Z.; Greenhalgh, E.S.; Bismarck, A.; Shaffer, M.S.P. Thermosetting nanocomposites with high carbon nanotube loadings processed by a scalable powder based method. *Compos. Sci. Technol.* **2016**, *127*, 62–70. [CrossRef]
82. Lopes, M.C.; de Castro, V.G.; Seara, L.M.; Diniz, V.P.A.; Lavall, R.L.; Silva, G.G. Thermosetting polyurethane-multiwalled carbon nanotube composites: Thermomechanical properties and nanoindentation. *J. Appl. Polym. Sci.* **2014**, *131*. [CrossRef]
83. Zahid, S.; Nasir, M.A.; Nauman, S.; Karahan, M.; Nawab, Y.; Ali, H.M.; Khalid, Y.; Nabeel, M.; Ullah, M. Experimental analysis of ILSS of glass fibre reinforced thermoplastic and thermoset textile composites enhanced with multiwalled carbon nanotubes. *J. Mech. Sci. Technol.* **2019**, *33*, 197–204. [CrossRef]
84. Yazik, M.H.M.; Sultan, M.T.H.; Mazlan, N.; Talib, A.R.A.; Naveen, J.; Shah, A.U.M.; Safri, S.N.A. Effect of hybrid multi-walled carbon nanotube and montmorillonite nanoclay content on mechanical properties of shape memory epoxy nanocomposite. *J. Mater. Res. Technol.* **2020**, *9*, 6085–6100. [CrossRef]
85. Islam, I.; Sultana, S.; Kumer Ray, S.; Parvin Nur, H.; Hossain, M. Electrical and tensile properties of carbon black reinforced polyvinyl chloride conductive composites. *C J. Carbon Res.* **2018**, *4*, 15. [CrossRef]
86. Gong, T.; Peng, S.-P.; Bao, R.-Y.; Yang, W.; Xie, B.-H.; Yang, M.-B. Low percolation threshold and balanced electrical and mechanical performances in polypropylene/carbon black composites with a continuous segregated structure. *Compos. Part B Eng.* **2016**, *99*, 348–357. [CrossRef]
87. Naik, P.; Pradhan, S.; Sahoo, P.; Acharya, S.K. Effect of filler loading on mechanical properties of natural carbon black reinforced polymer composites. *Mater. Today Proc.* **2020**, *26*, 1892–1896. [CrossRef]
88. Ojha, S.; Acharya, S.K.; Raghavendra, G. Mechanical properties of natural carbon black reinforced polymer composites. *J. Appl. Polym. Sci.* **2015**, *132*. [CrossRef]
89. Alam, M.K.; Islam, M.T.; Mina, M.F.; Gafur, M.A. Structural, mechanical, thermal, and electrical properties of carbon black reinforced polyester resin composites. *J. Appl. Polym. Sci.* **2014**, *131*, 13. [CrossRef]
90. Jovanović, V.; Samaržija-Jovanović, S.; Budinski-Simendić, J.; Marković, G.; Marinović-Cincović, M. Composites based on carbon black reinforced NBR/EPDM rubber blends. *Compos. Part B Eng.* **2013**, *45*, 333–340. [CrossRef]
91. Li, Y.; Li, R.; Lu, L.; Huang, X. Experimental study of damage characteristics of carbon woven fabric/epoxy laminates subjected to lightning strike. *Compos. Part A Appl. Sci. Manuf.* **2015**, *79*, 164–175. [CrossRef]
92. Wang, H.; Xie, G.; Fang, M.; Ying, Z.; Tong, Y.; Zeng, Y. Electrical and mechanical properties of antistatic PVC films containing multi-layer graphene. *Compos. Part B Eng.* **2015**, *79*, 444–450. [CrossRef]
93. Vadukumpully, S.; Paul, J.; Mahanta, N.; Valiyaveettil, S. Flexible conductive graphene/poly (vinyl chloride) composite thin films with high mechanical strength and thermal stability. *Carbon N. Y.* **2011**, *49*, 198–205. [CrossRef]

94. Chen, G.; Wu, D.; Weng, W.; Yan, W. Preparation of polymer/graphite conducting nanocomposite by intercalation polymerization. *J. Appl. Polym. Sci.* **2001**, *82*, 2506–2513. [CrossRef]
95. Du, X.S.; Xiao, M.; Meng, Y.Z.; Hay, A.S. Synthesis and properties of poly (4,4'-oxybis (benzene) disulfide)/graphite nanocomposites via in situ ring-opening polymerization of macrocyclic oligomers. *Polymer* **2004**, *45*, 6713–6718. [CrossRef]
96. Wen-Ping, W.; Cai-Yuan, P. Preparation and characterization of poly (methyl methacrylate)-intercalated graphite oxide/poly (methyl methacrylate) nanocomposite. *Polym. Eng. Sci.* **2004**, *44*, 2335–2339. [CrossRef]
97. Fang, M.; Wang, K.; Lu, H.; Yang, Y.; Nutt, S. Covalent polymer functionalization of graphene nanosheets and mechanical properties of composites. *J. Mater. Chem.* **2009**, *19*, 7098–7105. [CrossRef]
98. Yasmin, A.; Daniel, I.M. Mechanical and thermal properties of graphite platelet/epoxy composites. *Polymer* **2004**, *45*, 8211–8219. [CrossRef]
99. Kamarudin, S.H.; Abdullah, L.C.; Aung, M.M.; Ratnam, C.T. Mechanical and physical properties of Kenaf-reinforced Poly(lactic acid) plasticized with epoxidized Jatropha Oil. *BioResources* **2019**, *14*, 9001–9020.
100. Chandrasekar, M.; Kumar, T.S.M.; Senthilkumar, K.; Nurazzi, N.M.; Sanjay, M.R.; Rajini, N.; Siengchin, S. Inorganic Nanofillers-Based Thermoplastic and Thermosetting Composites. In *Lightweight Polymer Composite Structures*; Taylor & Francis: Oxfordshire, UK, 2020.
101. Zhou, Y.; Pervin, F.; Lewis, L.; Jeelani, S. Experimental study on the thermal and mechanical properties of multi-walled carbon nanotube-reinforced epoxy. *Mater. Sci. Eng. A* **2007**, *452–453*. [CrossRef]
102. Zakaria, M.R.; Abdul Kudus, M.H.; Md Akil, H.; Thirmizir, M.Z.M.; Abdul Malik, M.F.I.; Othman, M.B.H.; Ullah, F.; Javed, F. Comparative study of single-layer graphene and single-walled carbon nanotube-filled epoxy nanocomposites based on mechanical and thermal properties. *Polym. Compos.* **2019**, *40*. [CrossRef]
103. Caseri, W.R. Nanocomposites of polymers and inorganic particles: Preparation, structure and properties. *Mater. Sci. Technol.* **2006**, *22*, 807–817. [CrossRef]
104. Sapiai, N.; Jumahat, A.; Mahmud, J. Mechanical properties of functionalised CNT filled kenaf reinforced epoxy composites. *Mater. Res. Express* **2018**, *5*, 045034. [CrossRef]
105. Kushwaha, P.K.; Pandey, C.N.; Kumar, R. Study on the effect of carbon nanotubes on plastic composite reinforced with natural fiber. *J. Indian Acad. Wood Sci.* **2014**, *11*, 82–86. [CrossRef]
106. Aryasomayajula, L.; Wolter, K.J. Carbon nanotube composites for electronic packaging applications: A review. *J. Nanotechnol.* **2013**. [CrossRef]
107. Randjbaran, E.; Majid, D.L.; Zahari, R.; Sultan, M.T.H.; Mazlan, N. Effects of volume of carbon nanotubes on the angled ballistic impact for carbon kevlar hybrid fabrics. *Facta Univ. Ser. Mech. Eng.* **2020**, *18*, 229–244. [CrossRef]
108. Randjbaran, E.; Majid, D.L.; Zahari, R.; Sultan, M.T.H.; Mazlan, N. Impacts of Volume of Carbon Nanotubes on Bending for Carbon-Kevlar Hybrid Fabrics. *J. Appl. Comput. Mech.* **2021**, *7*, 839–848. [CrossRef]
109. Nurazzi, N.M.; Asyraf, M.R.M.; Fatimah Athiyah, S.; Shazleen, S.S.; Rafiqah, S.A.; Harussani, M.M.; Kamarudin, S.H.; Razman, M.R.; Rahmah, M.; Zainudin, E.S.; et al. A Review on Mechanical Performance of Hybrid Natural Fiber Polymer Composites for Structural Applications. *Polymers* **2021**, *13*, 2170. [CrossRef]
110. Ilyas, R.A.; Sapuan, M.S.; Norizan, M.N.; Norrrahim, M.N.F.; Ibrahim, R.; Atikah, M.S.N.; Huzaifah, M.R.M.; Radzi, A.M.; Izwan, S.; Azammi, A.M.N.; et al. Macro to nanoscale natural fiber composites for automotive components: Research, development, and application. In *Biocomposite and Synthetic Composites for Automotive Applications*; Sapuan, M.S., Ilyas, R.A., Eds.; Woodhead Publishing Series: Amsterdam, Netherland, 2020.
111. Benzait, Z.; Trabzon, L. A review of recent research on materials used in polymer–matrix composites for body armor application. *J. Compos. Mater.* **2018**, *52*, 3241–3263. [CrossRef]
112. Hanif, W.Y.W.; Risby, M.S.; Noor, M.M. Influence of Carbon Nanotube Inclusion on the Fracture Toughness and Ballistic Resistance of Twaron/Epoxy Composite Panels. *Procedia Eng.* **2015**, *114*, 118–123. [CrossRef]
113. Mylvaganam, K.; Zhang, L.C. Ballistic resistance capacity of carbon nanotubes. *Nanotechnology* **2007**, *18*, 4–7. [CrossRef]
114. Han, Y.; Elliott, J. Molecular dynamics simulations of the elastic properties of polymer/carbon nanotube composites. *Comput. Mater. Sci.* **2007**, *39*, 315–323. [CrossRef]
115. Zhang, C.L.; Shen, H.S. Temperature-dependent elastic properties of single-walled carbon nanotubes: Prediction from molecular dynamics simulation. *Appl. Phys. Lett.* **2006**, *89*. [CrossRef]
116. Chang, T. A molecular based anisotropic shell model for single-walled carbon nanotubes. *J. Mech. Phys. Solids* **2010**, *58*, 1422–1433. [CrossRef]
117. Ni, Z.; Bu, H.; Zou, M.; Yi, H.; Bi, K.; Chen, Y. Anisotropic mechanical properties of graphene sheets from molecular dynamics. *Phys. B Condens. Matter* **2010**, *405*, 1301–1306. [CrossRef]
118. Shen, L.; Shen, H.S.; Zhang, C.L. Temperature-dependent elastic properties of single layer graphene sheets. *Mater. Des.* **2010**, *31*, 4445–4449. [CrossRef]
119. Fan, Y.; Xiang, Y.; Shen, H.S. Temperature-dependent negative Poisson's ratio of monolayer graphene: Prediction from molecular dynamics simulations. *Nanotechnol. Rev.* **2019**, *8*, 415–421. [CrossRef]
120. Lin, F.; Xiang, Y.; Shen, H.S. Temperature dependent mechanical properties of graphene reinforced polymer nanocomposites—A molecular dynamics simulation. *Compos. Part B Eng.* **2017**, *111*, 261–269. [CrossRef]

121. Huang, B. Carbon nanotubes and their polymeric composites: The applications in tissue engineering. *Biomanufacturing Rev.* **2020**, *5*. [CrossRef]
122. Tanaka, M.; Sato, Y.; Haniu, H.; Nomura, H.; Kobayashi, S.; Takanashi, S.; Okamoto, M.; Takizawa, T.; Aoki, K.; Usui, Y.; et al. A three-dimensional block structure consisting exclusively of carbon nanotubes serving as bone regeneration scaffold and as bone defect filler. *PLoS ONE* **2017**, *12*. [CrossRef] [PubMed]
123. Zare, H.; Ahmadi, S.; Ghasemi, A.; Ghanbari, M.; Rabiee, N.; Bagherzadeh, M.; Karimi, M.; Webster, T.J.; Hamblin, M.R.; Mostafavi, E. Carbon nanotubes: Smart drug/gene delivery carriers. *Int. J. Nanomed.* **2021**, *16*, 1681–1706. [CrossRef]
124. Mittal, M.; Kumar, A. Carbon nanotube (CNT) gas sensors for emissions from fossil fuel burning. *Sens. Actuators B Chem.* **2014**, *203*, 349–362. [CrossRef]
125. Penza, M.; Cassano, G.; Rossi, R.; Alvisi, M.; Rizzo, A.; Signore, M.A.; Dikonimos, T.; Serra, E.; Giorgi, R. Enhancement of sensitivity in gas chemiresistors based on carbon nanotube surface functionalized with noble metal (Au, Pt) nanoclusters. *Appl. Phys. Lett.* **2007**, *90*, 173123. [CrossRef]
126. Li, Y.; Wang, H.; Chen, Y.; Yang, M. A multi-walled carbon nanotube/palladium nanocomposite prepared by a facile method for the detection of methane at room temperature. *Sens. Actuators B Chem.* **2008**, *132*, 155–158. [CrossRef]
127. Penza, M.; Rossi, R.; Alvisi, M.; Cassano, G.; Signore, M.A.; Serra, E.; Giorgi, R. Pt- and Pd-nanoclusters functionalized carbon nanotubes networked films for sub-ppm gas sensors. *Sens. Actuators B Chem.* **2008**, *135*, 289–297. [CrossRef]
128. Zanolli, Z.; Leghrib, R.; Felten, A.; Pireaux, J.-J.; Llobet, E.; Charlier, J.-C. Gas Sensing with Au-Decorated Carbon Nanotubes. *ACS Nano* **2011**, *5*, 4592–4599. [CrossRef]
129. Sinha, M.; Neogi, S.; Mahapatra, R.; Krishnamurthy, S.; Ghosh, R. Material dependent and temperature driven adsorption switching (p- to n- type) using CNT/ZnO composite-based chemiresistive methanol gas sensor. *Sens. Actuators B Chem.* **2021**, *336*, 129729. [CrossRef]
130. Lim, H.-R.; Lee, Y.; Jones, K.A.; Kwon, Y.-T.; Kwon, S.; Mahmood, M.; Lee, S.M.; Yeo, W.-H. All-in-one, wireless, fully flexible sodium sensor system with integrated Au/CNT/Au nanocomposites. *Sens. Actuators B Chem.* **2021**, *331*, 129416. [CrossRef]
131. Li, J.; Lu, Y.; Ye, Q.; Cinke, M.; Han, J.; Meyyappan, M. Carbon Nanotube Sensors for Gas and Organic Vapor Detection. *Nano Lett.* **2003**, *3*, 929–933. [CrossRef]
132. Zhang, X.; Wang, Y.; Gu, M.; Wang, M.; Zhang, Z.; Pan, W.; Jiang, Z.; Zheng, H.; Lucero, M.; Wang, H.; et al. Molecular engineering of dispersed nickel phthalocyanines on carbon nanotubes for selective CO_2 reduction. *Nat. Energy* **2020**, *5*, 684–692. [CrossRef]
133. Ahmad, Z.; Manzoor, S.; Talib, M.; Islam, S.S.; Mishra, P. Self-standing MWCNTs based gas sensor for detection of environmental limit of CO_2. *Mater. Sci. Eng. B* **2020**, *255*, 114528. [CrossRef]
134. Peng, S.; Cho, K. Ab Initio Study of Doped Carbon Nanotube Sensors. *Nano Lett.* **2003**, *3*, 513–517. [CrossRef]
135. Peng, S.; Cho, K. Chemical control of nanotube electronics. *Nanotechnology* **2000**, *11*, 57–60. [CrossRef]
136. Santucci, S.; Picozzi, S.; Di Gregorio, F.; Lozzi, L.; Cantalini, C.; Valentini, L.; Kenny, J.M.; Delley, B. NO2 and CO gas adsorption on carbon nanotubes: Experiment and theory. *J. Chem. Phys.* **2003**, *119*, 10904–10910. [CrossRef]
137. Matranga, C.; Bockrath, B. Hydrogen-Bonded and Physisorbed CO in Single-Walled Carbon Nanotube Bundles. *J. Phys. Chem. B* **2005**, *109*, 4853–4864. [CrossRef]
138. Yao, F.; Duong, D.L.; Lim, S.C.; Yang, S.B.; Hwang, H.R.; Yu, W.J.; Lee, I.H.; Güneş, F.; Lee, Y.H. Humidity-assisted selective reactivity between NO2 and SO2 gas on carbon nanotubes. *J. Mater. Chem.* **2011**, *21*, 4502. [CrossRef]
139. Ingle, N.; Mane, S.; Sayyad, P.; Bodkhe, G.; AL-Gahouari, T.; Mahadik, M.; Shirsat, S.; Shirsat, M.D. Sulfur Dioxide (SO_2) Detection Using Composite of Nickel Benzene Carboxylic (Ni3BTC2) and OH-Functionalized Single Walled Carbon Nanotubes (OH-SWNTs). *Front. Mater.* **2020**, *7*. [CrossRef]
140. Ingle, N.; Sayyad, P.; Deshmukh, M.; Bodkhe, G.; Mahadik, M.; Al-Gahouari, T.; Shirsat, S.; Shirsat, M.D. A chemiresistive gas sensor for sensitive detection of SO_2 employing Ni-MOF modified –OH-SWNTs and –OH-MWNTs. *Appl. Phys. A* **2021**, *127*, 157. [CrossRef]
141. Kuganathan, N.; Chroneos, A. Ru-Doped Single Walled Carbon Nanotubes as Sensors for SO_2 and H_2S Detection. *Chemosensors* **2021**, *9*, 120. [CrossRef]
142. Song, H.; Li, Q.; Zhang, Y. CNT-based sensor array for selective and steady detection of SO_2 and NO. *Mater. Res. Bull.* **2020**, *124*, 110772. [CrossRef]
143. Su, P.-G.; Zheng, Y.-L. Room-temperature ppb-level SO_2 gas sensors based on RGO/WO 3 and MWCNTs/WO 3 nanocomposites. *Anal. Methods* **2021**, *13*, 782–788. [CrossRef] [PubMed]
144. Lin, W.; Li, F.; Chen, G.; Xiao, S.; Wang, L.; Wang, Q. A study on the adsorptions of SO2 on pristine and phosphorus-doped silicon carbide nanotubes as potential gas sensors. *Ceram. Int.* **2020**, *46*, 25171–25188. [CrossRef]
145. Septiani, N.L.W.; Saputro, A.G.; Kaneti, Y.V.; Maulana, A.L.; Fathurrahman, F.; Lim, H.; Yuliarto, B.; Nugraha; Dipojono, H.K.; Golberg, D.; et al. Hollow Zinc Oxide Microsphere–Multiwalled Carbon Nanotube Composites for Selective Detection of Sulfur Dioxide. *ACS Appl. Nano Mater.* **2020**, *3*, 8982–8996. [CrossRef]
146. Zakaria, M.R.; Md Akil, H.; Abdul Kudus, M.H.; Ullah, F.; Javed, F.; Nosbi, N. Hybrid carbon fiber-carbon nanotubes reinforced polymer composites: A review. *Compos. Part B Eng.* **2019**, *176*, 107313. [CrossRef]
147. Jin, F.-L.; Park, S.-J. A review of the preparation and properties of carbon nanotubes-reinforced polymer compositess. *Carbon Lett.* **2011**, *12*, 57–69. [CrossRef]

148. Goldoni, A.; Larciprete, R.; Petaccia, L.; Lizzit, S. Single-Wall Carbon Nanotube Interaction with Gases: Sample Contaminants and Environmental Monitoring. *J. Am. Chem. Soc.* **2003**, *125*, 11329–11333. [CrossRef]
149. Chang, H.; Do Lee, J.; Lee, S.M.; Lee, Y.H. Adsorption of NH_3 and NO_2 molecules on carbon nanotubes. *Appl. Phys. Lett.* **2001**, *79*, 3863–3865. [CrossRef]
150. Yim, W.-L.; Gong, X.G.; Liu, Z.-F. Chemisorption of NO_2 on Carbon Nanotubes. *J. Phys. Chem. B* **2003**, *107*, 9363–9369. [CrossRef]
151. Zhang, Y.; Suc, C.; Liu, Z.; Li, J. Carbon Nanotubes Functionalized by NO_2: Coexistence of Charge Transfer and Radical Transfer. *J. Phys. Chem. B* **2006**, *110*, 22462–22470. [CrossRef]
152. Ricca, A.; Bauschlicher, C.W. The adsorption of NO_2 on (9,0) and (10,0) carbon nanotubes. *Chem. Phys.* **2006**, *323*, 511–518. [CrossRef]
153. Peng, S.; Cho, K.; Qi, P.; Dai, H. Ab initio study of CNT NO_2 gas sensor. *Chem. Phys. Lett.* **2004**, *387*, 271–276. [CrossRef]
154. Mercuri, F.; Sgamellotti, A.; Valentini, L.; Armentano, I.; Kenny, J.M. Vacancy-Induced Chemisorption of NO_2 on Carbon Nanotubes: A Combined Theoretical and Experimental Study. *J. Phys. Chem. B* **2005**, *109*, 13175–13179. [CrossRef] [PubMed]
155. Goldoni, A.; Petaccia, L.; Gregoratti, L.; Kaulich, B.; Barinov, A.; Lizzit, S.; Laurita, A.; Sangaletti, L.; Larciprete, R. Spectroscopic characterization of contaminants and interaction with gases in single-walled carbon nanotubes. *Carbon N. Y.* **2004**, *42*, 2099–2112. [CrossRef]
156. Mäklin, J.; Mustonen, T.; Kordás, K.; Saukko, S.; Tóth, G.; Vähäkangas, J. Nitric oxide gas sensors with functionalized carbon nanotubes. *Phys. Status Solidi* **2007**, *244*, 4298–4302. [CrossRef]
157. Ueda, T.; Norimatsu, H.; Bhuiyan, M.M.H.; Ikegami, T.; Ebihara, K. NO Sensing Property of Carbon Nanotube Based Thin Film Gas Sensors Prepared by Chemical Vapor Deposition Techniques. *Jpn. J. Appl. Phys.* **2006**, *45*, 8393–8397. [CrossRef]
158. Kazemi, N.; Hashemi, B.; Mirzaei, A. Promotional effect of nitric acid treatment on CO sensing properties of SnO_2/MWCNT nanocomposites. *Processing and Application of. Ceramics* **2016**, *10*, 97–105.
159. Mohd Nurazzi, N.; Khalina, A.; Sapuan, S.M.; Dayang Laila, A.H.A.M.; Rahmah, M.; Hanafee, Z. A review: Fibres, polymer matrices and composites. *Pertanika J. Sci. Technol.* **2017**, *25*, 1085–1102.
160. Nurazzi, N.M.; Asyraf, M.R.M.; Khalina, A.; Abdullah, N.; Aisyah, H.A.; Rafiqah, S.A.; Sabaruddin, F.A.; Kamarudin, S.H.; Norrrahim, M.N.F.; Ilyas, R.A.; et al. A Review on Natural Fiber Reinforced Polymer Composite for Bullet Proof and Ballistic Applications. *Polymers* **2021**, *13*, 646. [CrossRef]
161. Aisyah, H.A.; Paridah, M.T.; Sapuan, S.M.; Ilyas, R.A.; Khalina, A.; Nurazzi, N.M.; Lee, S.H.; Lee, C.H. A comprehensive review on advanced sustainable woven natural fibre polymer composites. *Polymers* **2021**, *13*, 471. [CrossRef]
162. Norrrahim, M.N.F.; Yasim-Anuar, T.A.T.; Jenol, M.A.; Mohd Nurazzi, N.; Ilyas, R.A.; Sapuan, S. Performance Evaluation of Cellulose Nanofiber Reinforced Polypropylene Biocomposites for Automotive Applications. In *Biocomposite and Synthetic Composites for Automotive Applications*; Woodhead Publishing Series: Amsterdam, The Netherlands, 2020; pp. 119–215.
163. Lee, C.H.; Khalina, A.; Nurazzi, N.M.; Norli, A.; Harussani, M.M.; Rafiqah, S.; Aisyah, H.A.; Ramli, N. The Challenges and Future Perspective of Woven Kenaf Reinforcement in Thermoset Polymer Composites in Malaysia: A Review. *Polymers* **2021**, *13*, 1390. [CrossRef] [PubMed]
164. Yang, Y.; Boom, R.; Irion, B.; van Heerden, D.J.; Kuiper, P.; de Wit, H. Recycling of composite materials. *Chem. Eng. Process. Process Intensif.* **2012**, *51*. [CrossRef]
165. Subadra, S.P.; Yousef, S.; Griskevicius, P.; Makarevicius, V. High-performance fiberglass/epoxy reinforced by functionalized CNTs for vehicle applications with less fuel consumption and greenhouse gas emissions. *Polym. Test.* **2020**, *86*, 106480. [CrossRef]
166. Zhu, S.; Sheng, J.; Chen, Y.; Ni, J.; Li, Y. Carbon nanotubes for flexible batteries: Recent progress and future perspective. *Natl. Sci. Rev.* **2021**, *8*. [CrossRef]
167. Andrews, J.B.; Cardenas, J.A.; Lim, C.J.; Noyce, S.G.; Mullett, J.; Franklin, A.D. Fully printed and flexible carbon nanotube transistors for pressure sensing in automobile tires. *IEEE Sens. J.* **2018**, *18*, 7875–7880. [CrossRef]
168. Shao, H.Q.; Wei, H.; He, J.H. Dynamic properties and tire performances of composites filled with carbon nanotubes. *Rubber Chem. Technol.* **2018**, *91*, 609–620. [CrossRef]
169. Bhat, A.; Budholiya, S.; Raj, S.A.; Sultan, M.T.H.; Hui, D.; Shah, A.U.M.; Safri, S.N.A. Review on nanocomposites based on aerospace applications. *Nanotechnol. Rev.* **2021**, *10*, 237–253. [CrossRef]
170. Liang, F.; Tang, Y.; Gou, J.; Gu, H.C.; Song, G. Multifunctional nanocomposites with high damping performance for aerospace structures. In Proceedings of the ASME International Mechanical Engineering Congress and Exposition, Lake Buena Vista, FL, USA, 13–19 November 2009; Volume 43840, pp. 267–273.
171. Venkatesan, M.; Palanikumar, K.; Boopathy, S.R. Experimental investigation and analysis on the wear properties of glass fiber and CNT reinforced hybrid polymer composites. *Sci. Eng. Compos. Mater.* **2018**, *25*, 963–974. [CrossRef]
172. Kwon, H.; Bradbury, C.R.; Leparoux, M. Fabrication of functionally graded carbon nanotube-reinforced aluminum matrix composite. *Adv. Eng. Mater.* **2011**, *13*, 325–329. [CrossRef]
173. Laurenzi, S.; de Zanet, G.; Santonicola, M.G. Numerical investigation of radiation shielding properties of polyethylene-based nanocomposite materials in different space environments. *Acta Astronaut.* **2020**, *170*, 530–538. [CrossRef]
174. Li, Z.; Chen, S.; Nambiar, S.; Sun, Y.; Zhang, M.; Zheng, W.; Yeow, J.T. PMMA/MWCNT nanocomposite for proton radiation shielding applications. *Nanotechnology* **2016**, *27*, 234001. [CrossRef]
175. Al-Saleh, M.H.; Sundararaj, U. Electromagnetic interference shielding mechanisms of CNT/polymer composites. *Carbon N. Y.* **2009**, *47*, 1738–1746. [CrossRef]

176. Francis, A.P.; Devasena, T. Toxicity of carbon nanotubes: A review. *Toxicol. Ind. Health* **2018**, *34*, 200–210. [CrossRef]
177. Donaldson, K.; Poland, C.A.; Murphy, F.A.; MacFarlane, M.; Chernova, T.; Schinwald, A. Pulmonary toxicity of carbon nanotubes and asbestos—similarities and differences. *Adv. Drug Deliv. Rev.* **2013**, *65*, 2078–2086. [CrossRef] [PubMed]
178. Mercer, R.R.; Hubbs, A.F.; Scabilloni, J.F.; Wang, L.; Battelli, L.A.; Friend, S.; Castranova, V.; Porter, D.W. Pulmonary fibrotic response to aspiration of multi-walled carbon nanotubes. *Part. Fibre Toxicol.* **2011**, *8*, 1–12. [CrossRef] [PubMed]
179. Dong, J.; Ma, Q. Advances in mechanisms and signaling pathways of carbon nanotube toxicity. *Nanotoxicology* **2015**, *9*, 658–676. [CrossRef] [PubMed]
180. Sousa, S.P.; Baptista, J.S.; Ribeiro, M. Polymer nano and submicro composites risk assessment. *Int. J. Work. Cond.* **2014**, *7*, 103–119.
181. National Institute for Occupational Safety and Health DHHS (NIOSH). *Current Intelligence Bulletin 65: Occupational Exposure to Carbon Nanotubes and Nanofibers*; National Institute for Occupational Safety and Health DHHS: Cincinnati, OH, USA, 2013.
182. Poland, C.A.; Duffin, R.; Kinloch, I.; Maynard, A.; Wallace, W.A.; Seaton, A.; Stone, V.; Brown, S.; MacNee, W.; Donaldson, K. Carbon nanotubes introduced into the abdominal cavity of mice show asbestos-like pathogenicity in a pilot study. *Nat. Nanotechnol.* **2008**, *3*, 423. [CrossRef] [PubMed]
183. Kim, J.E.; Lim, H.T.; Minai-Tehrani, A.; Kwon, J.T.; Shin, J.Y.; Woo, C.G.; Choi, M.; Baek, J.; Jeong, D.H.; Ha, Y.C.; et al. Toxicity and clearance of intratracheally administered multiwalled carbon nanotubes from murine lung. *J. Toxicol. Environ. Heal. Part A* **2010**, *73*, 1530–1543. [CrossRef]
184. Heister, E.; Brunner, E.W.; Dieckmann, G.R.; Jurewicz, I.; Dalton, A.B. Are carbon nanotubes a natural solution? Applications in biology and medicine. *ACS Appl. Mater. Interfaces* **2013**, *5*, 1870–1891. [CrossRef] [PubMed]
185. Donaldson, K.; Poland, C.; Bonner, J.; Duffin, R. *The Toxicology of Carbon Nanotubes*; Cambridge University Press: Cambridge, UK, 2012.
186. Dazon, C.; Witschger, O.; Bau, S.; Fierro, V.; Llewellyn, P.L. Toward an operational methodology to identify industrial-scaled nanomaterial powders with the volume specific surface area criterion. *Nanoscale Adv.* **2019**, *1*, 3232–3242. [CrossRef]
187. Carrero-Sanchez, J.C.; Elias, A.L.; Mancilla, R.; Arrellin, G.; Terrones, H.; Laclette, J.P.; Terrones, M.J.N.L. Biocompatibility and toxicological studies of carbon nanotubes doped with nitrogen. *Nano Lett.* **2006**, *6*, 1609–1616. [CrossRef]
188. Taylor, A.J.; McClure, C.D.; Shipkowski, K.A.; Thompson, E.A.; Hussain, S.; Garantziotis, S.; Parsons, G.N.; Bonner, J.C. Atomic layer deposition coating of carbon nanotubes with aluminum oxide alters pro-fibrogenic cytokine expression by human mononuclear phagocytes in vitro and reduces lung fibrosis in mice in vivo. *PLoS ONE* **2014**, *9*, e106870. [CrossRef]
189. Bottini, M.; Bruckner, S.; Nika, K.; Bottini, N.; Bellucci, S.; Magrini, A.; Bergamaschi, A.; Mustelin, T. Multi-walled carbon nanotubes induce T lymphocyte apoptosis. *Toxicol. Lett.* **2006**, *160*, 121–126. [CrossRef]
190. Bianco, A.; Kostarelos, K.; Partidos, C.D.; Prato, M. Biomedical applications of functionalised carbon nanotubes. *Chem. Commun.* **2005**, *5*, 571–577. [CrossRef] [PubMed]
191. Kam, N.W.S.; O'Connell, M.; Wisdom, J.A.; Dai, H. Carbon nanotubes as multifunctional biological transporters and near-infrared agents for selective cancer cell destruction. *Proc. Natl. Acad. Sci. USA* **2005**, *102*, 11600–11605. [CrossRef]
192. Yang, S.T.; Wang, X.; Jia, G.; Gu, Y.; Wang, T.; Nie, H.; Ge, C.; Wang, H.; Liu, Y. Long-term accumulation and low toxicity of single-walled carbon nanotubes in intravenously exposed mice. *Toxicol. Lett.* **2008**, *181*, 182–189. [CrossRef]
193. Pantarotto, D.; Singh, R.; McCarthy, D.; Erhardt, M.; Briand, J.P.; Prato, M.; Kostarelos, K.; Bianco, A. Functionalized carbon nanotubes for plasmid DNA gene delivery. *Angew. Chem. Int. Ed.* **2004**, *116*, 5354–5358. [CrossRef]
194. Wick, P.; Manser, P.; Limbach, L.K.; Dettlaff-Weglikowska, U.; Krumeich, F.; Roth, S.; Stark, W.J.; Bruinink, A. The degree and kind of agglomeration affect carbon nanotube cytotoxicity. *Toxicol. Lett.* **2007**, *168*, 121–131. [CrossRef] [PubMed]
195. Lu, Q.; Moore, J.M.; Huang, G.; Mount, A.S.; Rao, A.M.; Larcom, L.L.; Ke, P.C. RNA polymer translocation with single-walled carbon nanotubes. *Nano Lett.* **2004**, *4*, 2473–2477. [CrossRef]
196. Dumortier, H.; Lacotte, S.; Pastorin, G.; Marega, R.; Wu, W.; Bonifazi, D.; Briand, J.P.; Prato, M.; Muller, S.; Bianco, A. Functionalized carbon nanotubes are non-cytotoxic and preserve the functionality of primary immune cells. *Nano Lett.* **2006**, *6*, 1522–1528. [CrossRef]
197. Wang, H.; Wang, J.; Deng, X.; Sun, H.; Shi, Z.; Gu, Z.; Liu, Y.; Zhaoc, Y. Biodistribution of carbon single-wall carbon nanotubes in mice. *J. Nanosci. Nanotechnol.* **2004**, *4*, 1019–1024. [CrossRef]
198. Kam, N.W.S.; Liu, Z.; Dai, H. Carbon nanotubes as intracellular transporters for proteins and DNA: An investigation of the uptake mechanism and pathway. *Angew. Chem. Int. Ed.* **2006**, *45*, 577–581. [CrossRef] [PubMed]
199. Yehia, H.N.; Draper, R.K.; Mikoryak, C.; Walker, E.K.; Bajaj, P.; Musselman, I.H.; Daigrepont, M.C.; Dieckmann, G.R.; Pantano, P. Single-walled carbon nanotube interactions with HeLa cells. *J. Nanobiotechnol.* **2007**, *5*, 1–17. [CrossRef]
200. Cui, D.; Tian, F.; Ozkan, C.S.; Wang, M.; Gao, H. Effect of single wall carbon nanotubes on human HEK293 cells. *Toxicol. Lett.* **2005**, *155*, 73–85. [CrossRef]
201. Patlolla, A.K.; Berry, A.; Tchounwou, P.B. Study of hepatotoxicity and oxidative stress in male Swiss-Webster mice exposed to functionalized multi-walled carbon nanotubes. *Mol. Cell. Biochem.* **2011**, *358*. [CrossRef]

MDPI
St. Alban-Anlage 66
4052 Basel
Switzerland
www.mdpi.com

Nanomaterials Editorial Office
E-mail: nanomaterials@mdpi.com
www.mdpi.com/journal/nanomaterials

Disclaimer/Publisher's Note: The statements, opinions and data contained in all publications are solely those of the individual author(s) and contributor(s) and not of MDPI and/or the editor(s). MDPI and/or the editor(s) disclaim responsibility for any injury to people or property resulting from any ideas, methods, instructions or products referred to in the content.

www.ingramcontent.com/pod-product-compliance
Lightning Source LLC
LaVergne TN
LVHW070432100526
838202LV00014B/1582